Artificial Intelligence and Computer Vision for Ecological Informatics

Editors:

Siddharth Singh Chouhan
Assistant Professor, VIT Bhopal University
Kothri Kalan, Sehore, Madhya Pradesh, India

Uday Pratap Singh
Professor, Department of Mathematics
Central University of Jammu, Jammu, India

Akash Saxena
Professor, Central University of Haryana
Mahendergarh, India

Akhilesh Tiwari
Professor & Head at Department of Information Technology
Madhav Institute of Technology & Science (MITS), Gwalior

Sanjeev Jain
Vice Chancellor, Central University of Jammu
Jammu, India

CRC Press
Taylor & Francis Group
Boca Raton London New York

CRC Press is an imprint of the
Taylor & Francis Group, an **informa** business
A SCIENCE PUBLISHERS BOOK

First edition published 2026
by CRC Press
2385 NW Executive Center Drive, Suite 320, Boca Raton FL 33431

and by CRC Press
4 Park Square, Milton Park, Abingdon, Oxon, OX14 4RN

CRC Press is an imprint of Taylor & Francis Group, LLC

Library of Congress Cataloging-in-Publication Data (applied for)

ISBN: 978-1-032-83643-0 (hbk)
ISBN: 978-1-032-85574-5 (pbk)
ISBN: 978-1-003-51877-8 (ebk)

DOI: 10.1201/9781003518778

Typeset in Times New Roman
by Prime Publishing Services

Preface

The rapid advancement of Artificial Intelligence (AI) and Computer Vision has revolutionized the way we perceive and analyze the natural world. In the field of Ecological Informatics, these technologies offer groundbreaking opportunities to monitor, model, and understand ecosystems with unprecedented accuracy and efficiency. This book, Artificial Intelligence and Computer Vision for Ecological Informatics, explores the synergy between AI-driven computational techniques and ecological data analysis, bridging the gap between technology and environmental science.

In recent years, the increasing availability of remote sensing data, camera trap images, and automated monitoring systems has created vast amounts of ecological information. Traditional methods of analysis, while effective, often struggle to keep pace with the scale and complexity of modern ecological datasets. AI and Computer Vision provide powerful solutions, enabling automated species identification, habitat classification, environmental monitoring, and predictive modeling. These techniques not only enhance research capabilities but also support conservation efforts by providing real-time insights into ecosystem dynamics.

This book is structured to provide both theoretical foundations and practical applications of AI and Computer Vision in ecological research. It is intended for researchers, students, and professionals in fields such as ecology, environmental science, computer science, and data analytics who seek to harness the power of AI in ecological studies. Readers will find discussions on key AI methodologies, case studies demonstrating real-world applications, and insights into the future of ecological informatics in an increasingly data-driven world.

By integrating AI and Computer Vision into ecological research, we take a crucial step toward a deeper understanding of our environment, empowering scientists and conservationists to make informed decisions for sustainable management and biodiversity conservation. We hope this book serves as a valuable resource and inspires further innovation in this exciting interdisciplinary domain.

Contents

Preface iii

1. Drone Importance and their Necessity in Future Generation Agriculture 1

2. Advances in Wildfire Spread Detection and Prediction: Techniques, Challenges, and Applications 9

3. Leveraging Remote Sensing for Agriculture Mapping: Techniques, Applications, and Future Directions 36

4. Basil Crop Detection Using Computer Vision and Deep Learning Approach 67

5. Remote Sensing for Agriculture Mapping 83

6. Advances of Remote Sensing Technologies in Agriculture: Current Progress and Future Perspectives 91

7. Remote Sensing for Sustainable Agriculture: A Machine Learning Approach to Optimizing Farm Yield and Economic Returns 116

8. Leveraging AI and CV Technologies to Advance Water Quality Assessment in Ecological Informatics 151

9. Tracing Plant Growth Patterns: Employing Artificial Intelligence and Computer Vision for Explicit Mapping 172

10. AI-Driven Circular Economy: Innovations in Agro and Food Waste Management 185

11. Advancements in Machine Learning for Water Quality Assessment 202

12. Enhancing Agricultural Support with AI in the Farmer ChatBot Framework 219

13. Federated Learning: A Game-Changer in Agricultural Decision-Making and Precision Farming 231

14. Clean Streams, Clear Futures: AI Innovations in Water Quality Monitoring 246

15. Transforming Waste into Resources: AI's Impact on Wastewater Treatment **259**

16. Soil Moisture Evaluation by Artificial Intelligence and Computer Vision **274**

17. Crop Yield Estimation **294**

18. Soil Fertility Evaluation **305**

19. Advancements in Soil Moisture Evaluation: Sensors, Remote Sensing and Artificial Intelligence **316**

20. Sustainable Agriculture: Economic Perspectives on AI and ML in Crop Yield Estimation **348**

Index **393**

About the Authors **395**

Chapter 1

Drone Importance and their Necessity in Future Generation Agriculture

Siddharth Singh Chouhan, [1] ***Rajneesh Kumar Patel,*** [1]
Anju Shukla [1,*] **and** ***Uday Pratap Singh*** [2]

Introduction

Various factors are responsible for distorting agricultural production. Factors such as climatic shifts, limited or nonexistent arable land, water scarcity, plant diseases, population growth, industrialization, and many more have occurred in the last few years. These biotic or abiotic components not only disrupt the quality and quantity of an agricultural product but also deteriorate the economic growth of farmers as well as the nation (Radoglou-Grammatikis et al., 2020; Tahir et al., 2016; Li et al., 2021). The field location depicted in Fig. 1(a), and 1(b) shows a cluster of mango trees in healthy and diseased conditions. This is an impeccable example that concludes that, if an appropriate diagnosis process and preventive countermeasures had been adopted at the initial stage of the pathogen attack, several trees could have been saved (Li et al., 2021; Sinha and Dhanalakshmi, 2022; Heidari et al., 2023; Kangunde et al., 2021).

Plant monitoring is essential to sustain ecological requirements. However, there are several other factors responsible for balancing the overall agricultural industry. This includes nutrient deficiency estimation, vegetation indexes analysis, crop surveillance, pre and post-farming analysis, water quality assessment, weed and

[1] School of Computing Science and Engineering, VIT Bhopal University, Bhopal-Indore Highway, Kothrikalan, Sehore, Madhya Pradesh, 466114, India.

[2] Department of Mathematics, Central University of Jammu, Jammu, Jammu and Kashmir, 180001, India.

* Corresponding author: anjushukla.iiitb@gmail.com

(a) Healthy trees

(b) Unhealthy trees

Fig. 1. Environmental scene image.

pesticide management, waste management, soil quality assessment, plant density estimation, warehouse management, production or supply chain management, crop disease diagnosis, counting of fruits and flowers, quality and quantity estimation, assistance in demographic variations, gathering spatial and temporal information and many more (Luo et al., 2023; Olson and Anderson, 2021; Khan et al., 2022; Chouhan et al., 2024; Kumar et al., 2024; Solanki et al., 2023).

Need for Drones

Drones have been widely espoused in Precision Agriculture (PA). They are capable of offering solutions to many problems that a farmer experiences on and off the field. These applications and their valuable impact have been extensively observed in the past few years. Because of this productivity, efficiency, and profitability in the agriculture sector have been boosted. Drones in agribusiness save time, help troubleshoot potential problems and offer better data and cost-saving solutions. Farm ranchers and other agriculture business owners have already started to adopt technological aspects for their assistance (van Wynsberghe and Comes, 2020; Boukoberine et al., 2019; Hassanalian and Abdelkefi, 2017; Jerhamre et al., 2022). The information collected using drones is analyzed with the help of experts and effective solutions are provided to the farmers. This solution is provided in less time and at a lower cost, therefore enhancing the quality and quantity of yields. The data collected using drones is in the form of high-resolution images. Varying on the necessity of information needed the images collected can be RGB, multispectral, and hyperspectral.

From the perspective of ecological conditions drones are very important. They can fly and gather data even from terrestrial regions which are far beyond the reach of human beings. Some of these applications include the identification of historical wetlands regions (Stahl and Weimann, 2022), mapping of dryland savannahs (Amputu et al., 2023), detection of a mphibian species (Senturk, 2022) (Zhang, 2022) using the hyperspectral images captured with the help of drones for the classification of desert grassland species, identification of maize leaf diseases with the help of aerial images (Ishengoma et al., 2022) (Yang et al., 2022) have used the drone images to extract the soybean planting area, plant species monitoring, mapping, and quantifying with the help of drone images (Baena, 2018), columnar cactus recognition (López-Jiménez et al., 2019) among all. Further, in the subsequent section, some of the impacts of drone applications in agriculture are discussed.

Applications of Drones

Crop Mapping and Surveying

An effective place to start is with drones, they are excellent for monitoring and sensing, and they can swiftly cover a region to examine crop development and soil health. Drones are typically utilized for this task because their sensors can identify the absorbance of a particular wavelength, producing a color contrast image that highlights potentially problematic areas (Prado Osco et al., 2021; Khan et al., 2020; Corcoran et al., 2021).

Seeding and Planting

Due to the high level of skill required to carry out its operations, agriculture is by definition a labor-intensive and time-consuming sector. Particularly seeding, which is a time-consuming operation, requires human labor. To make this strenuous task easier, seeds of several crop varieties are sown using drone technology. Drones can plant seeds quickly and efficiently thanks to their built-in lasers, sensors, tanks, and other characteristics. Seed planting is one of the more recent and less popular uses of drones in agriculture. Currently, automated drone seeders are largely utilized in the forestry sector, but a broader application may be applied in the future. Drone planting makes it possible to replant inaccessible locations without putting workers in danger. With a crew of two operators and 10 drones that can plant 400,000 trees every day, they are also able to plant considerably more effectively (Yang et al., 2022).

Crop Spraying

One of the most popular uses of drones in agriculture is their swift movement to their desired locations. Drones with this skill can apply fertilizer and insecticides to crops to hydrate and provide them with the nutrients they need. Such additives can help crops thrive and stay healthy. To prevent worms, pests, and insects and prolong crop life, drone commanders have complete control over the drone spraying nutrients. In south-east Asia, the use of drones to apply spray treatments is already pervasive; South Korea uses drones for over 30% of its agricultural spraying. Drone sprayers can maneuver across highly challenging terrain, such as steep tea estates at great altitudes. Drone sprayers spare workers from carrying backpack sprayers through fields, which can be harmful to their health. To maximize productivity and reduce chemical costs, drone sprayers offer very tiny spray treatments that can be targeted to specific locations. Currently, laws governing drone sprayers vary greatly from nation to nation (Radoglou-Grammatikis et al., 2020).

Soil and Field Analysis

The condition of the field and the health of the soil can both be monitored via drone field monitoring. Drones can accurately map the field, providing elevation data that enables producers to identify any field abnormalities. The ability to determine drainage patterns and wet/dry patches, which enable more effective watering practices, is facilitated by knowledge of field elevation. Utilizing improved sensors, some sellers and service providers of agricultural drones also provide soil nitrogen level monitoring services. This makes it possible to apply fertilizers precisely, removing troublesome growth areas and enhancing soil health for years to come. Powerful drone capabilities help with the time-consuming procedure of evaluating the soil's health. The character and health of the soil may be monitored, modified, and maintained using drones, which collect and analyze data from monitoring systems. The soil can benefit from drone technology by receiving the nutrients it needs to become healthier. Drones use their data processing and 3D mapping capabilities to carry out this mission of monitoring the condition of the soil. Farmers can learn more about the soil characteristics of their land thanks to the drone survey. In-depth field soil analysis, irrigation, nitrogen-level management, and seed planting patterns can all benefit from data captured by multispectral sensors. Farmers can completely assess their soil conditions thanks to precise photogrammetry and 3D mapping (Olson and Anderson, 2021; Cucho-Padin, 2020).

Irrigation Monitoring

Plant health monitoring is one application for drone imaging that has already been widely used with remarkable effectiveness. The Normalized Difference Vegetation Index (NDVI), a specialized imaging tool used by drones, uses precise color data to assess the health of plants. This enables farmers to keep an eye on crops as they develop so that any issues can be resolved quickly enough to save the plants. Simply said, this graphic demonstrates how NDVI functions. Drones with sensors equipped with hyperspectral, thermal, or multispectral technology can identify regions that are too dry or require farmer improvement. Drone survey provides calculations of the vegetation index for irrigation monitoring, which aids in realizing the health of crops and releasing heat/energy. This helps to increase water efficiency and reveal potential pooling/leaks in irrigation (Khan, 2020).

Crop health is also observed by drones equipped with 'regular' cameras. To monitor crop growth, density, and coloration, many farmers currently utilize satellite images. However, accessing satellite data is expensive and frequently less efficient than closer drone surveillance. Cloud cover and bad lighting conditions are less important with drones than with satellite photography because they fly near fields. While satellite imaging may provide accuracy down to the meter, drone imaging can produce millimeter-level accuracy in image placement. This implies that regions with stand gaps can be identified and replanted as necessary after planting, and disease or pest issues can be quickly identified and dealt with. Monitoring the condition of the plants and identifying bacterial/fungal problems early on are critical. Drones used in agriculture can identify which plants reflect various quantities of NIRS and green light. To track crop health, multispectral images are created with the aid of this data. Crops can be saved by quick monitoring and flaw detection. In the event of crop failure, the farmer can also provide accurate damage documentation for insurance claims (Khan, 2020; Brook et al., 2020).

Crop Damage Assessment

Agricultural drones equipped with multispectral and RGB sensors can also spot the regions which are damaged due to natural calamities or pathogens. This information makes it possible to combat these infestations and helps the government and other regulatory bodies in assessing the loss.

Growth Monitoring

This plantation serves as the ecological system's spine. There have been numerous attempts to cultivate plants in the past. Usually, the first plantation or seeding procedure is carried out. However, those plants are not being cared for in their isolated locations and a lack of information is available. In this task, drones are essential. They can keep a constant eye on the plants. This process is cost-effective and requires less human effort (Khan, 2020; Brook et al., 2020).

Geofencing/Security

Drone security is a rapidly expanding industry apart from agriculture, but it is also very helpful for managing farms. Drones can monitor remote sections of a farm without having to travel there, saving time and enabling regular monitoring of difficult-to-reach areas. Drone cameras can provide a day-long overview of agricultural activities to make sure everything is going properly and to identify where equipment is being used. Instead of adding more security staff, security drones can be used to monitor the fencing and boundaries around more important crops, including cannabis. Using drone cameras to find lost or hurt herd animals in distant grazing regions is another fascinating way that drones are being utilized to protect agricultural animals. It now just takes a few minutes to monitor remote places, as opposed to the hours-long walks that were once required (Khan, 2020; Brook et al., 2020).

Potential Pollinators

Drones are capable of acting as artificial pollinators. Drones may eventually overtake bees as the most crucial pollinators, even though insect vectors remain the most significant pollinators. Researchers are optimistic that drones will be able to transport and disperse pollen seeds in orchards and fields, despite the need for more research in this area ((Khan, 2020; Brook et al., 2020).

Agricultural Research

Intelligent approaches and high-end computing devices have offered academicians and researchers the opportunity to direct their study towards the beneficial aspects of the agricultural sector. Nowadays, it has been observed that, especially due to the low labor cost and efficient results, drone technology has been espoused in almost all applications of precision agriculture. Because the underlying technology is currently in its initial phase, it has numerous aspects to be improved and developed. Among them designing and offering a cost drone is the essential one. The sensors installed on drones also play an important role, they must be selected specific to the application. Overall, the hardware cost and easy installation process should be grounded to make drones affordable. A unique platform to counter various problems associated with the agriculture domain is a necessity, so research has to be directed toward the development of software for a place offering solutions for all the applications (Khan, 2020; Brook et al., 2020; Osco et al., 2021; Gupta et al., 2016).

Conclusion

Unmanned aerial vehicles have made a significant impact on the expansion of the agriculture industry. Their primary contribution consists of reducing labor engrossment, expenditure, and resource consumption efficiently. However, the adaptability of the underlying technology is limited within the range of a few farmers and domain knowledge experts. Several factors including security, privacy, economic, climatic, lack of understanding, trajectory planning, choice of sensors, etc., influence their adaptability. However, to sustain, enhance, and shelter the quality and productivity of the yield outcomes, the focus has to be directed toward the augmented practice of drone technology.

Several factors are responsible for influencing the health and productivity of crop yields. They can be categorized as biotic and abiotic. Timely, precise, and cost-effective mechanisms are a feasible solution to overcome these challenges. The advent of recent technologies more specifically, the accelerated development of drone technology combined with computer vision, deep learning, and high-end computing devices can offer promising solutions to handle the aforementioned threats.

References

Amputu, V., Knox, N., Braun, A., Heshmati, S., Retzlaff, R., Röder, A. and Tielbörger, K. 2023. Unmanned aerial systems accurately map rangeland condition indicators in a dryland savannah. *Ecological Informatics*, 75: 102007. doi: 10.1016/j.ecoinf.2023.102007.

Baena, S., Boyd, D.S. and Moat, J. 2018. UAVs in pursuit of plant conservation—Real world experiences. *Ecological Informatics*, 47: 2–9. doi: 10.1016/j.ecoinf.2017.11.001.

Boukoberine, M.N., Zhou, Z. and Benbouzid, M. 2019. A critical review on unmanned aerial vehicles power supply and energy management: Solutions, strategies, and prospects. *Applied Energy*, 255: 113823. doi: 10.1016/j.apenergy.2019.113823.

Brook, A., De Micco, V., Battipaglia, G., Erbaggio, A., Ludeno, G., Catapano, I. and Bonfante, A. 2020. A smart multiple spatial and temporal resolution system to support precision agriculture from satellite images: Proof of concept on Aglianico vineyard. *Remote Sensing of Environment*, 240: 111679. doi: 10.1016/j.rse.2020. 111679.

Chouhan, S.S., Singh, U.P., Sharma, U. et al. 2024. Classification of different plant species using Deep Learning and Machine Learning algorithms. *Wireless Pers. Commun.*, 136: 2275–2298. https://doi.org/10.1007/s11277-024-11374-y.

Corcoran, E., Winsen, M., Sudholz, A. and Hamilton, G. 2021. Automated detection of wildlife using drones: Synthesis, opportunities and constraints. *Methods in Ecology and Evolution*. doi: 10.1111/2041-210X.13581.

Cucho-Padin, G., Loayza, H., Palacios, S. et al. 2020. Development of low-cost remote sensing tools and methods for supporting smallholder agriculture. *Appl. Geomat.*, 12: 247–263. doi: 10.1007/s12518-019-00292-5.

Gupta, A.K. et al. 2016. Low Contrast image enhancement technique by using fuzzy method. I*nternational Journal of Engineering Research and General Science*, 4(2).

Hassanalian, M. and Abdelkefi, A. 2017. Classifications, applications, and design challenges of drones: A review. *Progress in Aerospace Sciences*, 91: 99–131. doi: 10.1016/j.paerosci.2017.04.003.

Heidari, A., Navimipour, N.J., Unal, M. and Zhang, G. 2023. Machine Learning applications in Internet-of-Drones: Systematic review, recent deployments, and open issues. *ACM Comput. Surv.*, 55: 12, Article 247 (December 2023), 45 pages. https://doi.org/10.1145/3571728.

Ishengoma, F.S., Rai, I.A. and Ngoga, S.R. 2022. Hybrid convolution neural network model for a quicker detection of infested maize plants with fall armyworms using UAV-based images. *Ecological Informatics*, 67: 101502. doi: 10.1016/j.ecoinf.2021.101502.

Jerhamre, E., Carlberg, C.J.C. and van Zoest, V. 2022. Exploring the susceptibility of smart farming: Identified opportunities and challenges. *Smart Agricultural Technology*, 2: 100026. doi: 10.1016/j. atech.2021.100026.

Kangunde, V., Jamisola, R.S. and Theophilus, E.K. 2021. A review on drones controlled in real-time. *Int. J. Dynam. Control*, 9: 1832–1846. doi: 10.1007/s40435-020-00737-5.

Khan, A., Vibhute, A.D., Mali, S. and Patil, C.H. 2022. A systematic review on hyperspectral imaging technology with a machine and deep learning methodology for agricultural applications. *Ecological Informatics*, 69: 101678. doi: 10.1016/j.ecoinf. 2022.101678.

Khan, P.W., Byun, Y.-C. and Latif, M.A. 2020. Clifford geometric algebra-based approach for 3D modeling of agricultural images acquired by UAVs. *IEEE Access*, 8: 226297–226308. doi: 10.1109/ACCESS.2020.3045443.

Kumar, R. et al. 2024. Hybrid approach of cotton disease detection for enhanced crop health and yield. In *IEEE Access*. doi: 10.1109/ACCESS.2024.3419906.

Li, W., Zheng, T., Yang, Z., Li, M., Sun, C. and Yang, X. 2021. Classification and detection of insects from field images using deep learning for smart pest management: A systematic review. *Ecological Informatics*, 66: 101460. doi: 10.1016/j.ecoinf.2021.101460.

López-Jiménez, E., Vasquez-Gomez, J.I., Sanchez-Acevedo, M.A., Herrera-Lozada, J.C. and Uriarte-Arcia, A.V. 2019. Columnar cactus recognition in aerial images using a deep learning approach. *Ecological Informatics*, 52: 131–138. doi: 10.1016/j.ecoinf.2019.05.005.

Luo, J., Li, B. and Leung, C. 2023. A survey of computer vision technologies in urban and controlled-environment agriculture. *ACM Comput. Surv.* Just Accepted (October 2023). https://doi.org/10.1145/3626186.

Olson, D. and Anderson, J. 2021. Review on unmanned aerial vehicles, remote sensors, imagery processing, and their applications in agriculture. *Agronomy Journal.* doi: 10.1002/agj2.20595.

Osco, L.P. et al. 2021. A review on deep learning in UAV remote sensing. *International Journal of Applied Earth Observation and Geoinformation*, 102: 102456. doi: 10.1016/j.jag.2021.102456.

Radoglou-Grammatikis, P., Sarigiannidis, P., Lagkas, T. and Moscholios, I. 2020. A compilation of UAV applications for precision agriculture. *Computer Networks*, 172: 107148. doi: 10.1016/j.comnet.2020.107148.

Senturk, Z.K. 2022. Amphibian species detection in water reservoirs using artificial neural networks for ecology-friendly city planning. *Ecological Informatics*, 69: 101640. doi: 10.1016/j.ecoinf.2022. 101640.

Sinha, B.B. and Dhanalakshmi, R. 2022. Recent advancements and challenges of Internet of Things in smart agriculture: A survey. *Future Generation Computer Systems*, 126: 169–184. doi: 10.1016/j.future. 2021.08.006.

Solanki, S., Singh, U.P. and Chouhan, S.S. 2023. Brain tumor classification using ML and DL approaches. *2023 IEEE 5th International Conference on Cybernetics, Cognition and Machine Learning Applications (ICCCMLA)*, Hamburg, Germany, pp. 204–208. Doi: 10.1109/ICCCMLA58983.2023.10346854.

Stahl, N. and Weimann, L. 2022. Identifying wetland areas in historical maps using deep convolutional neural networks. *Ecological Informatics*, 68: 101557. doi: 10.1016/j.ecoinf.2022.101557.

Tahir, A., Boling, J., Haghbayan, M.-H., Toivonen, H.T. and Plosila, J. 2016. Swarms of unmanned aerial vehicles—A survey. *Journal of Industrial Information Integration*, 16: 100106. doi: 10.1016/j.jii.2019.100106.

van Wynsberghe, A. and Comes, T. 2020. Drones in humanitarian contexts, robot ethics, and the human–robot interaction. *Ethics and Information Technology*, 22: 43–53. doi: 10.1007/s10676-019-09514-1.

Yang, Q., She, B., Huang, L., Yang, Y., Zhang, G., Zhang, M., Hong, Q. and Zhang, D. 2022. Extraction of soybean planting area based on feature fusion technology of multi-source low altitude unmanned aerial vehicle images. *Ecological Informatics*, 70: 101715. doi: 10.1016/j.ecoinf.2022.101715.

Zhang, T., Bi, Y., Du, J., Zhu, X. and Gao, X. 2022. Classification of desert grassland species based on a local-global feature enhancement network and UAV hyperspectral remote sensing. *Ecological Informatics*, 72: 101852. doi: 10.1016/j.ecoinf.2022.101852.

Chapter 2

Advances in Wildfire Spread Detection and Prediction

Techniques, Challenges, and Applications

Rufai Yusuf Zakari,[1,*] ***Kassim Kalinaki***,[2,3]
Zaharaddeen Karami Lawal[4] and ***Najib Abdulrazak***[5]

Introduction

Forests, which are crucial in neutralizing carbon dioxide emissions, face increasing threats due to climate change (Anderegg et al., 2022; Kalinaki et al., 2023a). Research has shown that rising temperatures and elevated atmospheric CO_2 levels extend growing seasons, leading to more significant biomass accumulation and combustible fuel loads. Simultaneously, global warming intensifies lightning activity in temperate regions, posing a considerable wildfire hazard (Succarie et al., 2022). Elevated temperatures enhance atmospheric evaporation, intensifying drought conditions and exacerbating wildfire vulnerability, particularly in dense forests having ample ignitable materials. Notably, the seasonal distribution of wildfires is shifting, with recent Australian fire seasons experiencing a marked increase in late-season wildfires

[1] Department of Computer Science, School of Science, and Information Technology (SSIT), Skyline University Nigeria, 1080, Nigeria. https://orcid.org/0000-0002-4645-6412
[2] Department of Computer Science, Islamic University in Uganda (IUIU) P. O. Box 2555, Mbale, Uganda.
[3] Borderline Research Laboratory, Kampala, Uganda. https://orcid.org/0000-0001-8630-9110
[4] Department of Computer Science, Federal University Dutse, Jigawa State, Nigeria. https://orcid.org/0000-0003-3011-5581
[5] Department of Computer Science, Bayero University, Kano.
* Corresponding author: rufaig6@gmail.com

fueled by heightened burn intensity and consequent CO_2 emissions. Rising global temperatures have increased the likelihood of severe wildfires, particularly in areas experiencing prolonged dry spells (Pérez-Sánchez et al., 2019). With ongoing global warming, we anticipate more frequent severe weather occurrences, including intense heat and prolonged hot spells. These conditions will likely create environments increasingly prone to widespread and recurrent wildfires. Intense heat episodes further dry out the landscape, creating an ideal environment for these destructive blazes (Gaudreau et al., 2016). This, in turn, leads to amplified carbon dioxide emissions from wildfires, further triggering climate change and perpetuating a cyclical pattern of severe wildfires (McColl-Gausden et al., 2022). This feedback loop, illustrated in Fig. 1, highlights the interconnectedness of wildfires and climate change.

Cutting-edge technologies, sophisticated modeling, enhanced sensing systems, and improved observation methods have significantly enhanced our ability to detect and forecast wildfire progression (Ladi et al., 2022). The concurrent evolution of computational theory, mathematical innovations, and increased processing capabilities have armed us with powerful tools for data collection and analysis (Ding et al., 2022). Implementing these innovative technologies has markedly improved our capacity to tackle the complex issues inherent in wildfire control and prevention (Coen et al., 2020). Applications of AI-driven approaches, such as ML, have gained popularity in recent years and have become versatile tools for technological advancement (Rolnick et al., 2019). ML, a subfield of AI, employs algorithms to use big datasets to identify patterns (Jafar Ali Ibrahim et al., 2022; Kalinaki et al., 2024a; Kalinaki et al., 2023b). It is rapidly evolving and valuable for understanding complex networks, discovering knowledge patterns in big data, and future predictions

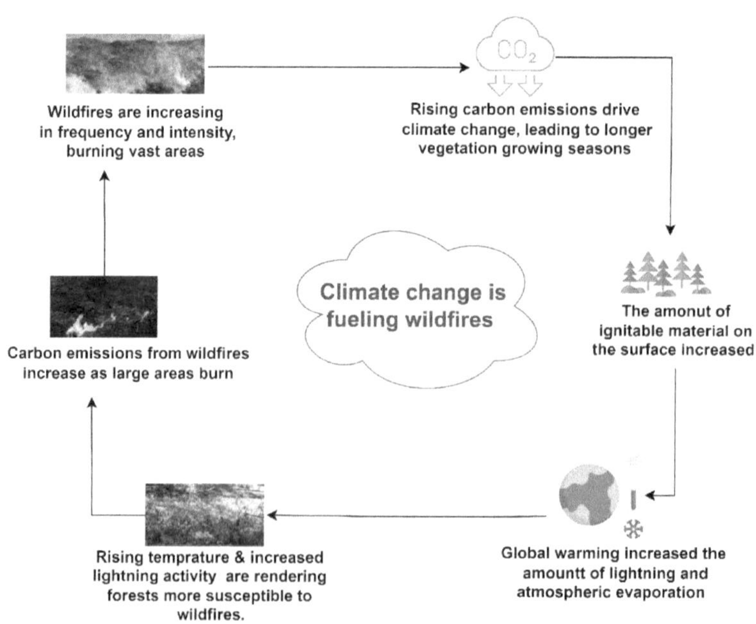

Fig. 1. The loop between forest wildfires and global climate change.

(Ding et al., 2022). ML techniques are indispensable in various application fields, including wildfire spread detection, urban planning, resource distribution, energy requirement forecasting, food security forecasting, and air quality monitoring and forecasting (Sreenath et al., 2022). Given that wildfire spread detection and prediction are highly complex and require research from multiple disciplines (Yandouzi et al., 2022), the use of ML techniques has proven to be a valuable tool for understanding and forecasting issues related to wildfires (Massetti et al., 2019). Generally, ML approaches typically fall into two main categories: supervised and unsupervised. Supervised methods map input data to known outcomes, using labeled datasets to discern patterns that can predict results for new inputs. Unsupervised techniques, on the other hand, explore data without predetermined outcomes, potentially uncovering novel relationships or distant correlations in wildfire propagation models (Kalinaki et al., 2023c; Umunnakwe et al., 2022). ML methods can capture strong nonlinearities, influencing many aspects of wildfire behavior (Ghali and Akhloufi, 2023), challenging to elaborate using conventional regression approaches (Ghali et al., 2022).

Wildfires represent a significant environmental and socio-economic challenge worldwide. Their occurrence has increased due to climate change, anthropogenic activities, and natural weather patterns (Küçükarslan, 2024). The ramifications of wildfires are multifaceted and severe, affecting ecosystems, human health, infrastructure, and economies. Ecologically, wildfires can result in loss of biodiversity, degradation of habitats, and alteration of natural processes. From a human perspective, wildfires pose direct threats to life and property, cause displacement of communities, and result in significant economic losses.

Furthermore, the release of smoke and particulate matter during wildfires has substantial health implications, contributing to respiratory and cardiovascular diseases (Dillis et al., 2022; Wang et al., 2023a). Early detection and accurate prediction of wildfires are crucial for effective mitigation and management. Prompt detection allows for rapid response, reducing the potential for large-scale spread and minimizing damage. Accurate prediction models enable resource allocation, strategic planning, and proactive measures, such as controlled burns and evacuation plans. The ability to anticipate wildfire behavior and spread can significantly enhance emergency response efforts, safeguarding human lives and minimizing ecological and economic impacts (Wang et al., 2023b).

RS plays a vital role in identifying and monitoring wildfires. It employs various data collection methods, including satellite imagery, aerial photography, and diverse sensor systems, to obtain information about active fires, smoke plumes, and areas affected by burns (Charizanos and Demirhan, 2023). This technology enables broad-scale wildfire detection and tracking, offering critical insights for strategic planning and resource deployment. Analyzing data from satellites, meteorological stations, and ground-based sensors is fundamental to effective wildfire surveillance, requiring sophisticated processing and interpretation techniques (Kalinaki et al., 2024b). Advanced analytical methods, including image analysis, statistical modeling, and machine learning algorithms, are employed to recognize fire indicators, spot unusual patterns, and deliver prompt, actionable intelligence for wildfire response teams (Casas et al., 2023; Ding et al., 2023).

This chapter aims to provide a comprehensive overview of the latest advancements in wildfire spread detection and prediction techniques. The chapter's contributions are:

- To explore the fundamental principles of wildfire behavior and spread dynamics, laying the groundwork for understanding the complexities involved.
- To explore remote sensing methodologies, including satellite imagery and unmanned aerial vehicles (UAVs), and examine how these technologies contribute to wildfire detection.
- To investigate the role of ML and DL algorithms in revolutionizing wildfire detection and prediction, enabling rapid analysis of vast datasets.
- To discuss predictive modeling approaches that integrate various environmental factors, topographical features, and fuel characteristics to forecast wildfire behavior and spread patterns.
- To address the challenges and limitations encountered in current wildfire detection and prediction systems, such as data accuracy, computational resources, and operational constraints.
- To illustrate the practical implications of these technologies in wildfire management and emergency response efforts through real-world case studies and applications.
- To highlight future research directions and opportunities for innovation in wildfire management strategies.

Fundamental Principles of Wildfire Behavior and Spread Dynamics

Understanding the factors influencing wildfire ignition and spread is crucial for developing effective detection and prediction systems. These factors include fuel types and characteristics, topographical features, and weather conditions.

Fuel Types and Characteristics

The types of fuels available in an area significantly affect wildfire behavior. Wildfires can ignite and spread through various fuels, including grasses, shrubs, trees, and accumulated dead vegetation. The moisture content of these fuels plays a crucial role in their combustibility (Iban and Sekertekin, 2022). Drier fuels ignite more easily and burn more intensely compared to those with higher moisture content. Additionally, the fuel load, which refers to the amount of available fuel, determines the potential intensity of a wildfire. Areas with higher fuel loads can sustain larger, more intense fires. The spatial distribution of fuels, including their vertical and horizontal continuity, also influences how quickly a fire spreads. For instance, continuous fuels can facilitate rapid fire spread, while discontinuities can slow down or halt the fire (Kadir et al., 2023).

Topographical Features

Topographical features such as slope, aspect, elevation, and natural barriers significantly impact wildfire behavior. Fires tend to move faster uphill due to the preheating of fuels above the fire, making steeper slopes conducive to more rapid fire spread (Piralilou et al., 2022). The direction a slope faces, known as its aspect, affects the amount of sunlight and wind exposure, influencing fuel drying and fire behavior. Elevation impacts weather patterns, temperature, and fuel types, all critical to fire dynamics. Natural barriers like rivers, cliffs, and previously burned areas can act as obstacles to fire spread, providing potential control points for firefighting efforts (Kadir et al., 2023; Tonini et al., 2020).

Weather Conditions

Weather conditions are among the most critical factors influencing wildfire ignition and spread (Bhowmik et al., 2023). Higher temperatures preheat fuels and reduce their moisture content, making them more susceptible to ignition. Conversely, lower humidity levels dry out fuels, increasing fire risk, while higher humidity can suppress fire activity (Wang et al., 2023c). Wind plays a crucial role in fire spread by directing the flames, carrying embers, and providing a continuous oxygen supply, with strong winds leading to rapid and often unpredictable fire behavior (Bouguettaya et al., 2022a). Precipitation can dampen fuels and reduce fire risk; however, prolonged dry periods can create conditions conducive to wildfires. Lastly, lightning strikes are a natural ignition source for wildfires, particularly in dry conditions, adding another layer of complexity to wildfire prediction and management (Bouguettaya et al., 2022b).

Datasets

Overview of Datasets

The foundation of any ML or DL model is the dataset used for training and evaluation. In wildfire detection and prediction, datasets typically comprise various data types, including satellite imagery, weather records, topographical information, and historical fire occurrences (Rezaie et al., 2023). These datasets are sourced from numerous agencies and platforms, such as NASA's Moderate Resolution Imaging Spectroradiometer (MODIS) and Visible Infrared Imaging Radiometer Suite (VIIRS), the European Space Agency's Sentinel satellites, and local meteorological stations. Comprehensive and high-quality datasets are crucial for building robust models that accurately detect and predict wildfires (Rodrigues and De la Riva, 2014).

Data Preprocessing: Refining raw data for analysis is a crucial preliminary stage known as data preprocessing. This multi-step process enhances data quality and readies the information for effective use in ML algorithms.

Data Cleaning: Data cleaning involves identifying and correcting errors or inconsistencies within the dataset. This includes handling missing values, correcting erroneous data points, and removing duplicates. For instance, satellite images may contain gaps or corrupted pixels, which must be addressed to ensure accurate analysis (van Klompenburg et al., 2020).

Data Normalization and Standardization: Normalization adjusts values to fall within a 0 to 1 range, whereas standardization alters the data distribution to achieve a mean of zero and a standard deviation of one (Bouguettaya et al., 2022b). These scaling approaches are crucial for algorithms affected by input feature magnitudes, as they ensure each feature has a comparable influence on the model's training process (Vyas et al., 2021).

Temporal and Spatial Alignment: Wildfire data often includes time-series and spatial information that must be aligned accurately (Zhang et al., 2023). Temporal alignment ensures that data points from different sources are synchronized in time, while spatial alignment ensures that data from various geographic locations are correctly mapped. For example, aligning weather data with corresponding satellite images based on time and location is crucial for accurate model training (Rjoub et al., 2023).

Data Augmentation: Data Augmentation expands the training dataset by applying diverse modifications to existing samples, and generating new, synthetic examples. This can include rotating, flipping, or scaling images and adding noise to simulate different conditions. Data augmentation helps improve model robustness by increasing the diversity of the training dataset, thereby reducing overfitting (Zakari et al., 2024).

Feature Extraction: Feature extraction converts raw data into refined characteristics that machine learning models can efficiently utilize. This step is crucial for highlighting the relevant information and reducing the complexity of the data.

Remote Sensing Features: Remote sensing provides a wealth of data for wildfire analysis, mainly through satellite imagery and UAV data. Various features can be extracted through advanced image processing techniques to aid in wildfire detection and prediction (Zhong et al., 2023). Thermal anomalies, indicating hotspots that could signify potential fire locations, are identified from infrared imagery. Vegetation indices such as the Normalized Difference Vegetation Index (NDVI) are computed to assess vegetation health and fuel availability. Additionally, multispectral and hyperspectral data allow for the analysis of spectral bands, enabling differentiation between burned and unburned areas based on their reflectance properties (Bhowmik et al., 2023).

Meteorological Features: Meteorological data is crucial in understanding and predicting wildfire behavior, as weather conditions significantly influence fire spread (Masoudian et al., 2023). Key meteorological features include temperature, humidity, wind speed, and precipitation. Elevated temperatures raise the probability of fire initiation and propagation, and reduced atmospheric moisture can desiccate plant life, enhancing its flammability. Wind speed and direction are critical factors, as wind can rapidly spread fires and change their direction. Conversely, precipitation, such as rainfall, can dampen fires and reduce their intensity (Pereira et al., 2022).

Topographical Features: Topographical data provides insights into the physical characteristics of the terrain, which profoundly impact wildfire spread. Elevation, slope, and aspect are essential topographical features (Abid et al., 2021). Higher elevations may exhibit different vegetation types and fuel availability, influencing fire behavior. Steeper slopes can accelerate fire spread due to the increased fuel consumption rate. Aspect, indicating the direction a slope faces, affects its exposure to sunlight and wind, further influencing fire behavior (Pan and Zhang, 2023).

Historical Fire Data: Historical fire occurrence data analysis offers valuable insights into past wildfire patterns and trends. Features derived from historical data include fire frequency, burn severity, and recovery time. Fire frequency indicates the number of times fires have occurred in a specific area over time, providing insight into its fire-prone nature. Burn severity measures the intensity of past fires, which can affect current vegetation and fuel conditions. Recovery time refers to the duration for vegetation regenerating after a fire, impacting future fire risk and ecosystem resilience.

Remote Sensing Methodologies for Wildfire Detection

Overview of Remote Sensing Technologies

Remote sensing technologies have revolutionized the field of wildfire detection by providing comprehensive and timely data. These technologies allow for monitoring vast and inaccessible areas, offering critical insights into wildfire dynamics.

Satellite Imagery: Satellite imagery is crucial for wildfire detection, utilizing sensors such as MODIS, VIIRS, and Landsat. These satellites provide valuable data by covering large areas and enabling continuous monitoring of remote and expansive regions (Brando et al., 2019). They offer various data types, including optical, thermal, and radar imagery, each contributing uniquely to wildfire detection (Wang et al., 2021). However, the temporal resolution (frequency of data capture) can be limited, and cloud cover can obstruct optical and thermal sensors, posing challenges for uninterrupted monitoring (Fedirka, 2020).

UAVs: UAVs (drones) have emerged as vital tools in wildfire detection. These devices are outfitted with advanced imaging equipment and sensors, enabling them to collect high-quality visual and thermal information (Bustillo Sánchez et al., 2021). UAVs offer flexibility, high spatial resolution, and the capability to fly under cloud cover and closer to the ground, providing detailed information that satellites may miss. Despite their advantages, UAVs face limitations in flight duration and range, and their deployment can be restricted by adverse weather conditions and regulatory constraints (Linn et al., 2002).

Techniques for Detecting Forest Fires

Satellite-Based Detection

Thermal Sensors (MODIS, VIIRS): These sensors are crucial for identifying thermal anomalies from space. For instance, NASA's MODIS on the Terra and Aqua satellites and VIIRS on the Suomi NPP satellite provide near-real-time data on wildfire hotspots globally (Rodrigues et al., 2014). During the 2020 Australian bushfires, these sensors detected extensive fire activity, helping emergency services monitor and respond to the fires effectively (Huot et al., 2022).

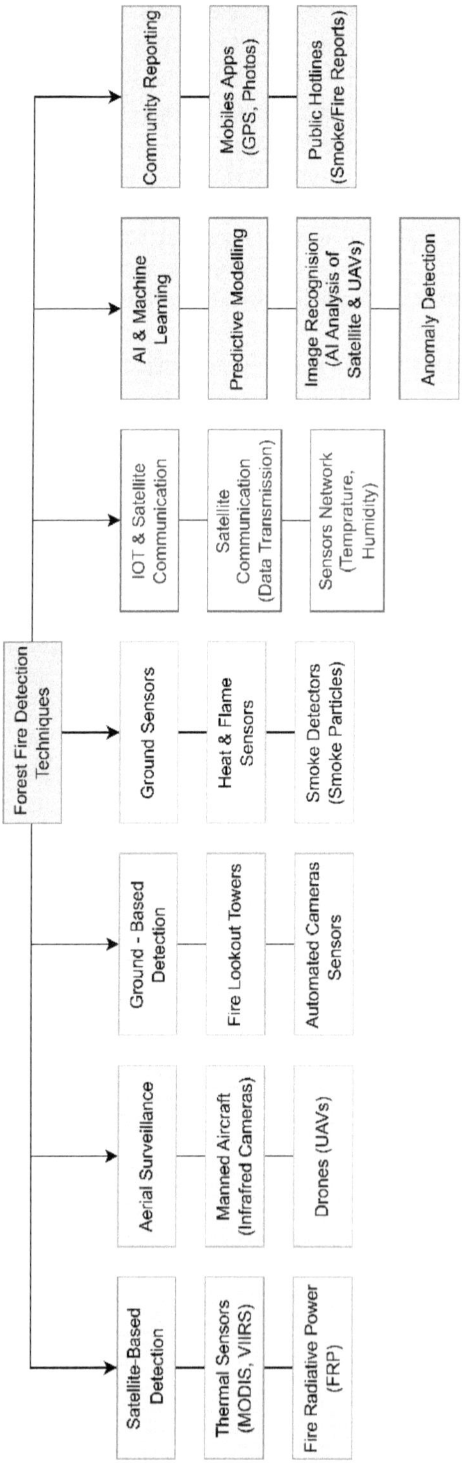

Fig. 2. Forest fire detection techniques.

Fire Radiative Power (FRP): FRP measures the heat energy of fires, indicating their intensity. This technique is used by the European Space Agency's Sentinel-3 satellites to estimate the power of fires in regions like the Amazon rainforest, where large-scale fires can emit significant radiative power, impacting both the environment and human health (Melchiorre and Boschetti, 2018).

Aerial Surveillance

Manned Aircraft (fitted with Infrared Cameras): Infrared-equipped Aircraft are deployed for detailed fire assessments. For example, in California, the U.S. Forest Service uses crewed Aircraft to fly over wildfires, capturing infrared images to map fire perimeters and hotspots, which aids in planning firefighting strategies (Wang et al., 2023a).

Drones (UAVs): Drones offer a flexible and rapid means of wildfire detection. During the 2019 Amazon fires, drones equipped with thermal cameras were used to survey affected areas, providing critical data on fire spread and intensity that was otherwise inaccessible due to dense smoke and challenging terrain (Phan et al., 2022; Chouhan et al., 2024).

Ground-Based Detection

Fire Lookout Towers: These towers are manned by personnel visually monitoring for smoke. In the forests of Montana, USA, lookout towers remain an essential part of wildfire detection, with rangers stationed to observe and report any signs of fire, ensuring early intervention (Ding et al., 2023; Park et al., 2020).

Automated Cameras Sensors: Automated systems, such as those used in Portugal's forested regions, utilize cameras with advanced smoke detection algorithms to identify and alert authorities to potential fires, significantly reducing response times (Ding et al., 2023).

Heat and Flame Sensors: Heat sensors, often placed in fire-prone areas like California's chaparral regions, detect abnormal temperature increases, triggering alarms that notify fire management teams (Wang et al., 2023c).

Smoke Detectors (Smoke Particles): These sensors are deployed in areas like the boreal forests of Canada, where they detect smoke particles in the air, indicating early fire activity and allowing for swift firefighting action (Park et al., 2020).

Internet of Things (IoT) and Satellite Communication

Satellite Communication (Data Transmission): Satellites transmit data from remote sensors to central hubs. For instance, the FireWatch network in Germany uses satellite communications to relay data from forest sensors, enabling real-time monitoring and rapid response to fires (Casas et al., 2023).

Sensors Network (Temperature, Humidity): Networks like those in Spain's Andalusian forests monitor environmental conditions, using data from temperature and humidity sensors to predict fire risk and inform preventive measures, such as controlled burns or public warnings (Mahdi and Mahmood, 2022).

AI and ML

Predictive Modelling: AI algorithms, like those developed by IBM's Watson, analyze weather patterns, vegetation types, and historical fire data to forecast where wildfires are likely to occur (Bouguettaya et al., 2022a). In the western United States, such models allocate firefighting resources more effectively during the fire season (He et al., 2023).

Image Recognition (AI Analysis of Satellite and UAVs): AI tools analyze satellite and drone imagery to identify fire outbreaks. For example, during the 2018 Camp Fire in California, AI systems processed images to locate and track fire spread, aiding emergency response quickly (Chouhan et al., 2024).

Anomaly Detection: ML models detect unusual data patterns indicating fires. For instance, the ALERTWildfire system in Nevada uses ML to monitor camera feeds and identify smoke or fire, providing early warnings to fire departments (Blake et al., 2010).

Community Reporting

Mobile Apps (GPS, Photos): Apps like the one used in British Columbia, Canada, allow residents to report wildfires, providing GPS coordinates and photos. This citizen-sourced information helps fire services promptly verify and respond to new fire incidents (Khaki et al., 2020).

Public Hotlines (Smoke/Fire Reports): Public hotlines in countries like Australia enable residents to report smoke sightings (Leuenberger et al., 2018). These reports are crucial for early detection, allowing authorities to dispatch firefighting resources quickly, as seen during the devastating 2019–2020 bushfire season (Rodrigues et al., 2014).

Techniques for Predicting Forest Fires

Meteorological Modeling

Weather Forecasting Models: These models predict weather conditions that influence wildfire behavior. For instance, the European Centre for Medium-Range Weather Forecasts (ECMWF) provides detailed weather forecasts used by fire management agencies to anticipate wind patterns and humidity levels that could exacerbate fire spread (Yandouzi et al., 2022).

Fire Weather Index (FWI): The FWI is a meteorological tool used globally, such as Canada, to assess fire danger based on weather data such as temperature, wind speed, humidity, and precipitation. This index helps in daily fire risk assessment, guiding fire prevention measures and resource allocation (Bouguettaya et al., 2022a).

Remote Sensing

Satellite Imagery: Satellites like Landsat and Sentinel provide high-resolution images that monitor vegetation health and detect fire scars. For example, during the

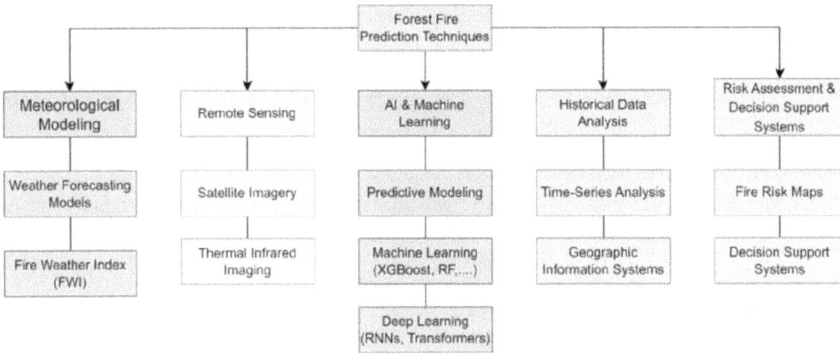

Fig. 3. Forest fire prediction techniques.

2019–2020 Australian bushfires, satellite imagery was crucial for mapping the extent of burned areas and assessing environmental impacts (Jiang et al., 2022).

Thermal Infrared Imaging: This technique detects heat from active fires. The GOES-R series of satellites used by the NOAA in the United States provides thermal infrared data, enabling near real-time monitoring of wildfire hotspots and aiding in quick response efforts (Yan et al., 2023).

AI and Machine Learning

Predictive Modeling: ML models analyze vast datasets to predict fire spread. For instance, California's fire management agencies use predictive models to forecast fire behavior, integrating factors like vegetation type, topography, and weather conditions (Cao, 2003).

ML (XGBoost, RF, etc.): Algorithms like XGBoost and Random Forest are employed to predict fire occurrence and spread. Researchers at Stanford University have developed models using these techniques to improve the accuracy of wildfire risk assessments (Cao, 2003; Thissen et al., 2003).

DL (RNNs, Transformers): DL models, such as recurrent neural networks (RNNs) and transformers, are used for complex pattern recognition in fire spread prediction. These models can process sequential data, effectively analyzing time-series fire behavior data (Zakari et al., 2024).

Historical Data Analysis

Time-Series Analysis: This method involves studying historical fire data to identify trends and patterns (Wu et al., 2022). In Australia, time-series analysis of past wildfire incidents helps understand seasonal variations and improves preparedness strategies (Jimeng, 2022).

Geographic Information Systems: GIS technology maps and analyzes spatial data related to wildfires. In regions like California, GIS integrates various data layers, such as vegetation, weather, and topography, to assess fire risks and plan mitigation measures (Bhowmik et al., 2023).

Risk Assessment and Decision Support Systems

Fire Risk Maps: These maps visualize areas at high risk of wildfires. For example, the US Forest Service produces fire risk maps highlighting vulnerable regions, aiding in targeted fire prevention efforts and resource allocation (He et al., 2023).

Decision Support Systems: These systems help fire managers make informed decisions (Lin et al., 2023). The Fire Information for Resource Management System (FIRMS) developed by NASA provides real-time fire data and decision support tools, enhancing the ability of authorities to respond effectively to wildfire threats (Koh, 2023).

ML and DL in Wildfire Detection and Prediction

Introduction to ML and DL

ML and DL have transformed various fields, including wildfire detection and prediction, by enabling the analysis of large, complex datasets (Pourghasemi et al., 2020). ML involves training algorithms to recognize patterns and make predictions based on data, while DL, a subset of ML, uses neural networks with multiple layers to model intricate relationships. These technologies are crucial for processing and interpreting vast amounts of data from diverse sources, including satellite imagery, weather data, and sensor networks, to enhance wildfire management strategies (Ding et al., 2023).

Algorithms and Models Used in Wildfire Detection

In wildfire detection, several ML and DL algorithms have proven effective in identifying the presence and characteristics of wildfires from remote sensing data.

Convolutional Neural Networks (CNNs): CNNs are widely used in image recognition tasks because they automatically learn spatial hierarchies of features. In wildfire detection, CNNs analyze satellite and UAV imagery to identify fire hotspots and smoke plumes. Their ability to process large images efficiently makes them ideal for real-time monitoring systems, providing rapid and accurate detection (Chouhan et al.).

Support Vector Machines (SVMs): SVMs are used for classification tasks and applied to wildfire detection by classifying different land cover types and identifying areas with active fires (Iban and Sekertekin, 2022). SVMs can handle high-dimensional data and are effective in scenarios where the number of features is large relative to the number of observations (Gao et al., 2023).

Random Forests: These employ an ensemble approach, generating numerous decision trees during the learning phase. They produce outputs based on the most frequent class prediction across all trees for classification tasks. In wildfire detection, Random Forests are used to classify remote sensing data and predict fire occurrences by learning from historical fire data and environmental variables (Gao et al., 2023; Iban and Sekertekin, 2022).

Algorithms and Models Used in Wildfire Forecasting

Predicting wildfire spread and behavior involves forecasting based on various environmental and meteorological factors. Several ML and DL models have been developed for this purpose.

Recurrent Neural Networks (RNNs): These networks, particularly Long Short-Term Memory (LSTM), are used to model temporal dependencies in sequential data. In wildfire prediction, RNNs and LSTMs analyze time-series data from weather stations, historical fire records, and fuel conditions to forecast fire spread and intensity (Ghorbanzadeh et al., 2019).

Gradient Boosting Machines (GBMs): These powerful ensemble learning methods are used for regression and classification tasks. In wildfire prediction, GBMs combine predictions from multiple weak learners to improve accuracy, making them suitable for modeling the complex interactions between various factors influencing wildfire behavior (Tehrany et al., 2019).

Transformer Models: Transformer models, known for their effectiveness in natural language processing, are increasingly applied in wildfire prediction. They excel in capturing long-range dependencies and can process vast datasets efficiently. By integrating meteorological data, fuel characteristics, and topographical features. Transformer models can accurately predict wildfire spread (Tehrany et al., 2019; Tien Bui et al., 2017).

Transfer Learning and Domain Adaptation

These are advanced techniques that enhance ML and DL models' performance by leveraging knowledge from related tasks or domains.

Transfer Learning: This leverages a model initially trained on an extensive dataset and then adapted the model to a specific task using a more limited dataset through fine-tuning. In wildfire detection and prediction, transfer learning allows models trained on extensive global datasets to be adapted to particular regions, improving their accuracy and reducing the need for large, labeled datasets (Ding et al., 2023).

Domain Adaptation: This addresses the problem of applying models trained in one domain to another with different characteristics. By aligning the distributions of source and target domains, domain adaptation techniques improve model performance on new datasets (Singh et al., 2022). In wildfire prediction, this approach is crucial for adapting models trained in one geographic region to work effectively in another with various ecological conditions (Rahimi et al., 2021).

Challenges and Limitations in Wildfire Detection and Prediction

Data Accuracy and Availability

Incomplete or Inconsistent Data: Wildfire prediction models rely on various data sources, including weather data, satellite imagery, and historical fire records (Bruzzone, 2019). Inconsistent data quality or data gaps can significantly reduce predictions' accuracy. Remote areas might lack adequate sensor coverage, leading to incomplete datasets that fail to capture critical early signs of fire risk (Pan and Zhang, 2023).

Resolution of Data: High-resolution data is essential for accurate fire detection and prediction. However, such data can be expensive and logistically challenging, especially for large areas. Satellite imagery may have limitations in temporal resolution, missing rapid changes in fire conditions or small-scale fires (He et al., 2023).

Real-Time Data Integration: Integrating real-time data from diverse sources such as ground sensors, weather stations, and satellites is complex and requires robust data management systems. Delays in data acquisition and processing can hinder timely detection and response efforts (Wang et al., 2023b).

Computational Resources and Efficiency

High Computational Demands: Advanced predictive models, especially those based on ML and DL, require significant computational power for training and real-time operation (Rim et al., 2018). High-performance computing resources are needed to handle large datasets and complex simulations, which may not be readily available in all regions (Phan et al., 2022).

Scalability: Scaling models to cover extensive areas with diverse environmental conditions can be challenging. Models must be adapted and optimized for different regions, requiring substantial computational resources and expertise (Kale et al., 2022). Ensuring that predictive models can run efficiently on available hardware while maintaining accuracy is a constant challenge (Kumar et al., 2024).

Energy Consumption: The computational power required for running advanced models can lead to high energy consumption, which is both costly and environmentally taxing (Kadir et al., 2023).

Operational Constraints and Real-Time Application

Timeliness and Speed: For wildfire detection and prediction to be compelling, they must operate in real-time or near real-time (Umunnakwe et al., 2022). Any delays in data processing, analysis, or dissemination can reduce the effectiveness of early warning systems. Rapidly changing fire conditions require models to update predictions frequently, which challenges maintaining computational efficiency (Elia et al., 2020).

Integration with Existing Systems: Integrating new technologies and predictive models with existing fire management and emergency response systems can be challenging (Dillis et al., 2022). Compatibility issues and the need for personnel training can hinder smooth implementation (Khajuria et al., 2018b). Operationalizing predictive models within the workflow of fire management agencies requires robust systems for data communication, decision support, and actionable intelligence (Wang et al., 2023a).

Resource Allocation: Effective wildfire response depends on the availability and efficient allocation of resources such as firefighting personnel, equipment, and aerial support. Predictive models must provide actionable insights that can be used to optimize resource deployment (Kale et al., 2022). Limited resources can constrain the ability to act on predictions, particularly in regions with fewer financial and logistical capabilities (Wang et al., 2023c).

Legal, Ethical, and Social Considerations

Privacy Concerns: Using drones, satellite surveillance, and ground sensors raises privacy concerns. Ensuring that these technologies are used to respect individuals' privacy rights is essential. There must be clear guidelines and regulations governing data collection, storage, and use to prevent misuse (Massetti et al., 2019).

Ethical Use of AI: The use of AI in wildfire prediction involves ethical considerations, such as the potential for algorithmic bias (Phelps and Woolford, 2021; Rishickesh et al., 2019). Models trained on historical data might reflect and perpetuate existing biases in fire management practices. Ensuring transparency and accountability in developing and deploying AI-based systems is crucial to maintaining public trust (Mandallaz and Ye, 1997).

Social Impacts: Predictive models might influence land management and zoning policies, affecting local communities, particularly those living in fire-prone areas (Apostolakis et al., 2022). Decisions based on model predictions need to consider the social and economic impacts on these communities. Engaging with communities to ensure they understand and trust technology is essential for successfully implementing predictive systems (McColl-Gausden et al., 2022).

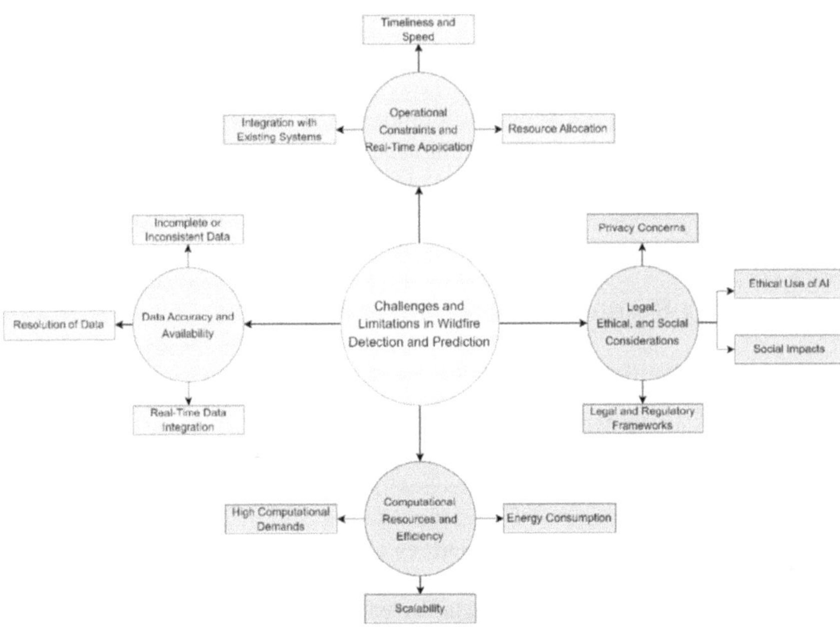

Fig. 4. Challenges in forest fire detection and prediction.

Legal and Regulatory Frameworks: Different legal and regulatory requirements across regions regarding data usage, surveillance, and fire management practices (Tuyen et al., 2021). Navigating these frameworks can be challenging. Different legal standards and political considerations can complicate international and regional cooperation for data sharing and fire management (Tien Bui et al., 2017).

Practical Applications and Real-World Case Studies in Wildfire Detection and Prediction

The practical applications of wildfire detection and prediction technologies span various aspects of wildfire management, from emergency response and resource allocation to community preparedness and risk communication. Below are some key implementations and real-world case studies demonstrating the effectiveness and impact of these technologies.

Implementation in Wildfire Management Strategies

Advanced Predictive Modeling

Case Study: Australia's Bushfire Prediction System: In Australia, advanced predictive models like the Phoenix RapidFire system simulate fire spread and behavior under different scenarios (Cheng et al., 2022). This tool integrates weather forecasts, vegetation maps, and topographic data for accurate fire predictions. It has been instrumental in developing effective fire management strategies and conducting controlled burns to reduce fuel loads (Rubí et al., 2022).

Real-Time Monitoring and Early Warning Systems

Case Study: California's FireMap: California has implemented FireMap, a real-time monitoring system that uses satellite data, ground sensors, and aerial surveillance to detect wildfires early. The system alerts fire management agencies immediately, allowing for rapid response and coordination. This system has significantly improved the state's ability to manage and contain wildfires (Rubí et al., 2022).

Emergency Response and Resource Allocation

Optimized Resource Deployment

Case Study: Fort McMurray Wildfire, Canada: During the Fort McMurray wildfire in 2016, predictive modeling tools helped optimize the deployment of firefighting resources. The models provided forecasts on fire spread, enabling authorities to position firefighting teams and equipment strategically. This approach minimized response times and improved the efficiency of firefighting efforts, ultimately saving lives and property (Rim et al., 2018).

Seventeen wildfires were raging in Fort McMurray, Canada, as of May 12, 2016. One of those fires was still regarded as "out of control" by provincial authorities.

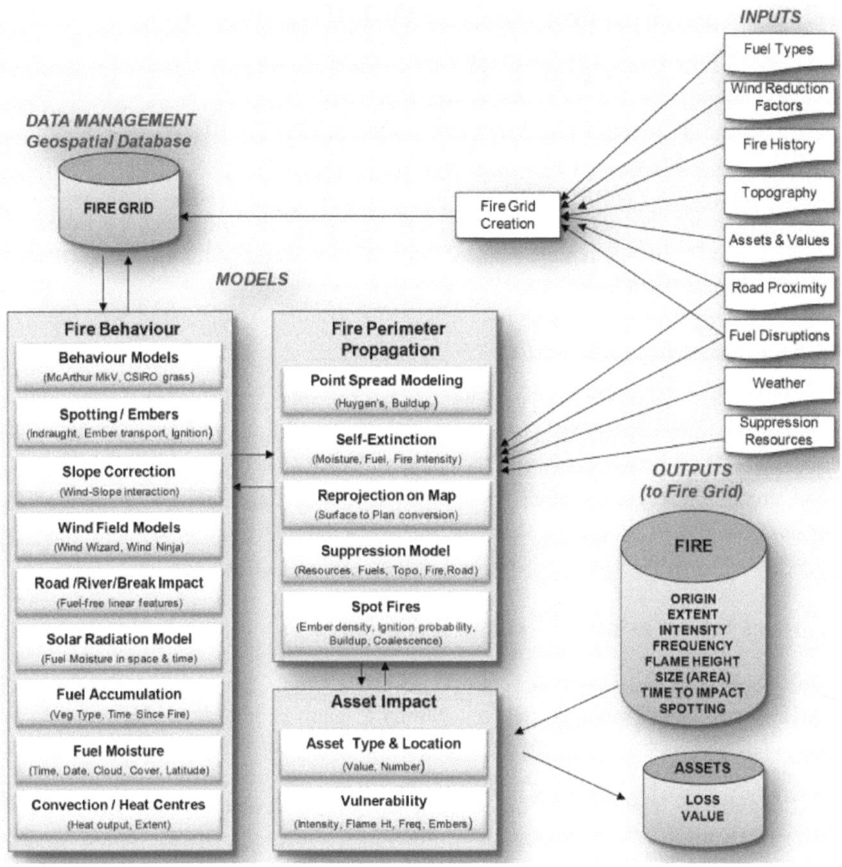

Fig. 5. Australian forest fire prediction system (Auditor-General, 2020).

At least 2,410 square kilometers (930 square miles) of land have burned due to fires in that region of Alberta as of May 12. This photo of the burn scar was taken on May 12 by the Landsat 8 satellite's Operational Land Imager (OLI). Green light, near-infrared, and shortwave infrared (OLI bands 7-5-3) combine to create the false color image. Short- and near-wave infrared helps highlight the hot patches connected to active flames (red) by penetrating clouds and smoke (white). Areas that have been burned become brown when combined.

Aerial Support Coordination

Case Study: Mediterranean Region's FireHub: The European Union's FireHub (2019) platform integrates satellite data and weather information to support aerial firefighting operations across the Mediterranean region. By predicting fire spread and intensity, FireHub assists in coordinating aerial water drops and other aerial firefighting tactics. This has proven effective in managing large-scale fires and protecting critical infrastructure.

Community Preparedness and Risk Communication

Public Awareness Campaigns

Case Study: Ready, Set, Go! Program, United States: The Ready, Set, Go! program in the United States educates communities in fire-prone areas about wildfire risks and preparedness measures. The program uses data from predictive models to inform residents about the potential for wildfires and provides guidelines on evacuation planning and home hardening. This initiative has increased community resilience and preparedness for wildfire events.

Risk Communication Tools

Case Study: Spain's eFIRECOM System: Spain's eFIRECOM system is a community-focused platform that provides real-time updates on fire risk levels, evacuation routes, and safety tips. It integrates data from wildfire prediction models and communicates this information through mobile apps and social media. This system has enhanced public safety and ensured the timely dissemination of critical information during fire emergencies.

Success Stories and Lessons Learned

One notable success story is using AI to predict the Camp Fire in California. In 2018, researchers employed ML algorithms to analyze historical weather data and fire incidents, successfully predicting the areas most at risk for wildfires, like the Camp Fire. AI provided actionable insights crucial in pre-emptive resource allocation and community alerts. From this experience, several lessons were learned. Firstly, data integration is key; combining various data sources such as satellite imagery, weather forecasts, and historical fire data significantly enhances prediction accuracy. Secondly, community involvement is critical; engaging communities in understanding and using predictive tools increases preparedness and compliance with evacuation orders, improving overall safety.

Another success story comes from Australia, where predictive models have been used to guide prescribed burns in fire-prone areas (Qiao et al., 2023). These controlled burns help reduce the fuel load and prevent more enormous, uncontrollable wildfires. The experience in Australia highlighted the importance of proactive measures; predictive models are invaluable in planning and executing preventive measures such as controlled burns (Linn et al., 2002). These models help mitigate the risk of large-scale wildfires and support the development of more effective fire management strategies. These success stories underline the transformative potential of predictive technologies in wildfire management and the importance of integrating advanced data analysis and community engagement into these efforts. Adaptive Management: Continually refining models based on real-world outcomes improves their reliability and effectiveness (Eskandari et al., 2021).

Future Directions and Research Opportunities in Wildfire Detection and Prediction Emerging Technologies and Innovations

Emerging technologies and innovations are set to revolutionize wildfire detection and prediction. AI and ML are at the forefront, with DL models like CNNs and RNNs enhancing our ability to predict complex wildfire behaviors and interactions with environmental variables (He et al., 2023). Generative Adversarial Networks (GANs) offer the potential to simulate various wildfire scenarios, providing valuable data for training predictive models and preparing for diverse outcomes. IoT and sensor networks are also crucial, with smart sensors deployed in high-risk areas to monitor real-time environmental conditions such as temperature, humidity, and wind speed. UAVs equipped with thermal imaging cameras provide rapid, high-resolution data collection over vast areas, enabling precise monitoring and early detection of wildfires. Satellite and aerial technologies continue to advance, with hyperspectral imaging improving early detection and monitoring capabilities and CubeSats offering frequent, high-resolution imagery to complement data from larger satellites (Wang et al., 2023a).

Integration of Multisource Data and Interdisciplinary Approaches

Integrating multisource data and interdisciplinary approaches is vital for enhancing wildfire prediction accuracy. Data fusion techniques combine information from satellites, ground sensors, weather stations, social media, and historical fire records to create comprehensive datasets for more accurate predictions (Deng et al., 2023). Cross-disciplinary collaboration brings together meteorology, ecology, and computer science expertise to develop more robust models. Big data analytics play a significant role, with real-time data processing systems capable of handling vast amounts of information and cloud computing platforms providing scalable solutions for storing and analyzing large datasets (Pan and Zhang, 2023). Human and social factors must also be considered, with data on human activities such as land use, infrastructure, and recreational activities integrated to account for anthropogenic influences on wildfire risk. Community-based monitoring engages local communities in data collection and monitoring efforts, leveraging citizen science to enhance data quality and coverage (Pereira et al., 2022).

Enhancing Model Accuracy and Predictive Capabilities

Enhancing model accuracy and predictive capabilities is crucial for effective wildfire management. Improved environmental models, such as high-resolution climate models, provide detailed weather forecasts and ecological conditions essential for accurate wildfire predictions (Mahdi and Mahmood, 2022). Dynamic vegetation models that simulate vegetation growth and fuel accumulation help predict fire behavior and spread (Wang et al., 2023b). Advanced computational techniques, including quantum computing and parallel processing, offer the potential to solve complex problems in wildfire prediction that are currently infeasible with classical computers and to handle large-scale simulations more efficiently. Validation and calibration through field experiments and controlled burns ensure that predictive models accurately reflect real-world conditions. Additionally, methods to quantify and communicate uncertainties in model predictions are essential for making informed decisions under uncertainty (Park et al., 2020; Phan et al., 2022).

Policy Implications and Stakeholder Engagement

Effective policy development and stakeholder engagement are critical for translating technological advances into practical wildfire management strategies (Kumalawati et al., 2021). Evidence-based policies should be formulated based on scientific evidence and predictive model outputs to improve wildfire management and mitigation strategies. Establishing regulatory frameworks that support the use of advanced technologies and data sharing across agencies and organizations is essential. Interagency cooperation fosters collaboration among government agencies, fire management organizations, researchers, and private sector partners to share data, resources, and expertise (Singh et al., 2022). Engaging local communities in wildfire preparedness and response planning ensures that policies and strategies are aligned with local needs and contexts. Public awareness and education are also vital, with effective communication strategies developed to inform the public about wildfire risks and preparedness measures using data-driven insights from predictive models. Training and capacity-building initiatives for stakeholders, including fire managers, policymakers, and community leaders, enhance their understanding and use of predictive technologies (Sachithra and Subhashini, 2023).

Conclusion

This comprehensive examination of wildfire detection and prediction underscores the transformative potential of advanced technologies and interdisciplinary approaches in mitigating wildfire risks and enhancing management strategies. Emerging technologies such as AI, machine learning, IoT, drones, and advanced satellite imaging have shown remarkable promise in improving the accuracy and timeliness of wildfire detection and prediction. These innovations enable more precise monitoring, early detection, and the ability to simulate complex wildfire behaviors, providing critical insights for proactive management and response. Integrating multisource data from satellites, ground sensors, weather stations, and human activities, combined with cross-disciplinary collaboration, enhances the robustness of predictive models. This holistic approach improves the accuracy of predictions and allows for a more comprehensive understanding of wildfire dynamics, incorporating environmental, meteorological, and anthropogenic factors. Advanced computational techniques, including DL, quantum computing, and big data analytics, are pivotal in refining predictive models. Continuous validation and calibration through field experiments and the development of methods to quantify uncertainties are essential to ensure that these models accurately reflect real-world conditions and can be relied upon for effective decision-making. Effective policy development and stakeholder engagement are crucial for translating technological advances into practical wildfire management strategies. Evidence-based policies, regulatory frameworks that support technology use, interagency cooperation, and community involvement are all vital components of a comprehensive approach to wildfire management.

Declaration

Ethical Responsibilities of Authors

"All authors have read, understood, and have complied as applicable with the statement on "Ethical responsibilities of Authors" as found in the Instructions for Authors".

Ethical Approval

Not applicable.

Use of AI Tool

The corresponding author declares that no AI tool has been used to prepare this manuscript.

Conflict of Interest

On behalf of all the authors the corresponding author states that there is no conflict of interest to disclose.

Competing Interests

Not applicable.

Authors' Contributions

Author[1]: Conceptualization, Methodology, Writing - Original Draft, Writing - Review & Editing, **Author**[2]: Investigation, Writing - Review & Editing, **Author**[3]: Writing - Original Draft, Writing - Review & Editing, **Author**[4]: Writing - Review & Editing.

Funding

No funding has been received.

References

Abid, S.K., Sulaiman, N., Chan, S.W., Nazir, U., Abid, M., Han, H., Ariza-Montes, A. and Vega-Muñoz, A. 2021. Toward an integrated disaster management approach: How artificial intelligence can boost disaster management. In *Sustainability (Switzerland)*, 13(22). https://doi.org/10.3390/su132212560.

Anderegg, W.R.L., Wu, C., Acil, N., Carvalhais, N., Pugh, T.A.M., Sadler, J.P. and Seidl, R. 2022. A climate risk analysis of Earth's forests in the 21st century. *Science*, 377(6610). https://doi.org/10.1126/science.abp9723.

Apostolakis, A., Girtsou, S., Giannopoulos, G., Bartsotas, N.S. and Kontoes, C. 2022. Estimating next day's forest fire risk via a complete machine learning methodology. *Remote Sensing*, 14(5). https://doi.org/10.3390/rs14051222.

Auditor-General, V. 2020. Reducing Bushfire Risks. www.audit.vic.gov.au.

Bhowmik, R.T., Jung, Y.S., Aguilera, J.A., Prunicki, M. and Nadeau, K. 2023. A multi-modal wildfire prediction and early-warning system based on a novel machine learning framework. *Journal of Environmental Management*, 341. https://doi.org/10.1016/j.jenvman.2023.117908.

Blake, W.H., Theocharopoulos, S.P., Skoulikidis, N., Clark, P., Tountas, P., Hartley, R. and Amaxidis, Y. 2010. Wildfire impacts on hillslope sediment and phosphorus yields. *Journal of Soils and Sediments*, 10(4). https://doi.org/10.1007/s11368-010-0201-y.

Bouguettaya, A., Zarzour, H., Kechida, A. and Taberkit, A.M. 2022a. Deep learning techniques to classify agricultural crops through UAV imagery: A review. In *Neural Computing and Applications*, 34(12). https://doi.org/10.1007/s00521-022-07104-9.

Bouguettaya, A., Zarzour, H., Taberkit, A.M. and Kechida, A. 2022b. A review on early wildfire detection from unmanned aerial vehicles using deep learning-based computer vision algorithms. In *Signal Processing* (Vol. 190). https://doi.org/10.1016/j.sigpro.2021.108309.

Brando, P.M., Paolucci, L., Ummenhofer, C.C., Ordway, E.M., Hartmann, H., Cattau, M.E., Rattis, L., Medjibe, V., Coe, M.T. and Balch, J. 2019. Droughts, wildfires, and forest carbon cycling: A pantropical synthesis. In *Annual Review of Earth and Planetary Sciences* (Vol. 47). https://doi.org/10.1146/annurev-earth-082517-010235.

Bruzzone, S. 2019. Wildfire forecasting: Between criminal act and unintentional events. *Culture and Organization*, 25(1). https://doi.org/10.1080/14759551.2016.1213730.

Bustillo Sánchez, M., Tonini, M., Mapelli, A. and Fiorucci, P. 2021. Spatial assessment of wildfires susceptibility in santa cruz (Bolivia) using random forest. *Geosciences (Switzerland)*, 11(5). https://doi.org/10.3390/geosciences11050224.

Cao, L. 2003. Support vector machines experts for time series forecasting. *Neurocomputing*, 51. https://doi.org/10.1016/S0925-2312(02)00577-5.

Casas, E., Ramos, L., Bendek, E. and Rivas-Echeverria, F. 2023. Assessing the effectiveness of YOLO architectures for smoke and wildfire detection. *IEEE Access*, 11. https://doi.org/10.1109/ACCESS.2023.3312217.

Charizanos, G. and Demirhan, H. 2023. Bayesian prediction of wildfire event probability using normalized difference vegetation index data from an Australian forest. *Ecological Informatics*, 73. https://doi.org/10.1016/j.ecoinf.2022.101899.

Cheng, S., Jin, Y., Harrison, S.P., Quilodrán-Casas, C., Prentice, I.C., Guo, Y.K. and Arcucci, R. 2022. Parameter flexible wildfire prediction using machine learning techniques: Forward and inverse modelling. *Remote Sensing*, 14(13). https://doi.org/10.3390/rs14133228.

Chouhan, S.S., Singh, U.P., Sharma, U. and Jain, S. 2021. Leaf disease segmentation and classification of *Jatropha Curcas* L. and *Pongamia Pinnata* L. biofuel plants using computer vision based approaches. *Measurement*, 171: 108796. https://doi.org/10.1016/J.MEASUREMENT.2020.108796.

Chouhan, S.S., Singh, U.P., Sharma, U. and Jain, S. 2024. Classification of different plant species using deep learning and machine learning algorithms. *Wireless Personal Communications*, 136(4): 2275–2298. https://doi.org/10.1007/S11277-024-11374-Y/FIGURES/9.

Coen, J.L., Schroeder, W., Conway, S. and Tarnay, L. 2020. Computational modeling of extreme wildland fire events: A synthesis of scientific understanding with applications to forecasting, land management, and firefighter safety. *Journal of Computational Science*, 45. https://doi.org/10.1016/j.jocs.2020.101152.

Deng, J., Wang, W., Gu, G., Chen, Z., Liu, J., Xie, G., Weng, S., Ding, L. and Li, C. 2023. Wildfire susceptibility prediction using a multisource and spatiotemporal cooperative approach. *Earth Science Informatics*, 16(4). https://doi.org/10.1007/s12145-023-01104-6.

Dillis, C., Butsic, V., Moanga, D., Parker-Shames, P., Wartenberg, A. and Grantham, T.E. 2022. The threat of wildfire is unique to cannabis among agricultural sectors in California. *Ecosphere*, 13(9). https://doi.org/10.1002/ecs2.4205.

Ding, C., Zhang, X., Chen, J., Ma, S., Lu, Y. and Han, W. 2022. Wildfire detection through deep learning based on Himawari-8 satellites platform. *International Journal of Remote Sensing*, 43(13). https://doi.org/10.1080/01431161.2022.2119110.

Ding, Y., Wang, M., Fu, Y., Zhang, L. and Wang, X. 2023. A wildfire detection algorithm based on the dynamic brightness temperature threshold. *Forests*, 14(3). https://doi.org/10.3390/f14030477.

Elia, M., D'Este, M., Ascoli, D., Giannico, V., Spano, G., Ganga, A., Colangelo, G., Lafortezza, R. and Sanesi, G. 2020. Estimating the probability of wildfire occurrence in Mediterranean landscapes using Artificial Neural Networks. *Environmental Impact Assessment Review*, 85. https://doi.org/10.1016/j.eiar.2020.106474.

Eskandari, S., Pourghasemi, H.R. and Tiefenbacher, J.P. 2021. Fire-susceptibility mapping in the natural areas of Iran using new and ensemble data-mining models. *Environmental Science and Pollution Research*, 28(34). https://doi.org/10.1007/s11356-021-13881-y.

Fedirka, A. 2020. *The Geopolitical Consequences of Australia's Wildfires*. https://geopoliticalfutures.com.

FireHub. 2019. *BEYOND Center: First Assessment of Burned Areas in Evia using the Sentinel 2 Satellite*. http://beyond-eocenter.eu/index.php/web-services/firehub.

Gao, C., Lin, H. and Hu, H. 2023. Forest-fire-risk prediction based on random forest and backpropagation neural network of heihe area in Heilongjiang Province, China. *Forests*, 14(2). https://doi.org/10.3390/f14020170.

Gaudreau, J., Perez, L. and Drapeau, P. 2016. BorealFireSim: A GIS-based cellular automata model of wildfires for the boreal forest of Quebec in a climate change paradigm. *Ecological Informatics*, 32. https://doi.org/10.1016/j.ecoinf.2015.12.006.

Ghali, R. and Akhloufi, M.A. 2023. Deep Learning approaches for wildland fires using satellite remote sensing data: Detection, mapping, and prediction. In *Fire* (Vol. 6, Issue 5). https://doi.org/10.3390/fire6050192.

Ghali, R., Akhloufi, M.A. and Mseddi, W.S. 2022. Deep Learning and transformer approaches for UAV-based wildfire detection and segmentation. *Sensors*, 22(5). https://doi.org/10.3390/s22051977.

Ghorbanzadeh, O., Kamran, K.V., Blaschke, T., Aryal, J., Naboureh, A., Einali, J. and Bian, J. 2019. Spatial prediction of wildfire susceptibility using field survey gps data and machine learning approaches. *Fire*, 2(3). https://doi.org/10.3390/fire2030043.

He, H., Zhang, Z., Jia, Q., Huang, L., Cheng, Y. and Chen, B. 2023. Wildfire detection for transmission line based on improved lightweight YOLO. *Energy Reports*, 9. https://doi.org/10.1016/j.egyr.2022.10.435.

Huot, F., Hu, R.L., Goyal, N., Sankar, T., Ihme, M. and Chen, Y.F. 2022. Next day wildfire spread: A machine learning dataset to predict wildfire spreading from remote-sensing data. *IEEE Transactions on Geoscience and Remote Sensing*, 60. https://doi.org/10.1109/TGRS.2022.3192974.

Iban, M.C. and Sekertekin, A. 2022. Machine learning based wildfire susceptibility mapping using remotely sensed fire data and GIS: A case study of Adana and Mersin provinces, Turkey. *Ecological Informatics*, 69. https://doi.org/10.1016/j.ecoinf.2022.101647.

Jafar Ali Ibrahim, S., Prathyusha, Y., Akhila, B., Rajini, M. and Naveen Sai, U. 2022. A machine learning-based prediction model for preterm birth in rural areas. *South Asian Journal of Engineering and Technology*, 12(3). https://doi.org/10.26524/sajet.2022.12.46.

Jiang, N., Li, P. and Feng, Z. 2022. Remote sensing of swidden agriculture in the tropics: A review. In *International Journal of Applied Earth Observation and Geoinformation* (Vol. 112). https://doi.org/10.1016/j.jag.2022.102876.

Jimeng, S. 2022. Time-series-forecasting-using-various-deep-learning-models. *International Journal of Computer and Systems Engineering*, 16(6).

Kadir, E.A., Kung, H.T., AlMansour, A.A., Irie, H., Rosa, S.L. and Fauzi, S.S.M. 2023. Wildfire hotspots forecasting and mapping for environmental monitoring based on the long short-term memory networks deep learning algorithm. *Environments - MDPI*, 10(7). https://doi.org/10.3390/environments10070124.

Kale, M.P., Mishra, A., Pardeshi, S., Ghosh, S., Pai, D.S. and Roy, P.S. 2022. Forecasting wildfires in major forest types of India. *Frontiers in Forests and Global Change*, 5. https://doi.org/10.3389/ffgc.2022.882685.

Kalinaki, K., Acaru, S.F., Kugonza, J. and Nsubuga, R. 2024a. Towards an intelligent tomorrow: Machine Learning enabling sustainable development. In *Methodologies, Frameworks, and Applications of Machine Learning* (pp. 66–89). https://doi.org/10.4018/979-8-3693-1062-5.CH004.

Kalinaki, K., Malik, O.A. and Ching Lai, D.T. 2023a. FCD-AttResU-Net: An improved forest change detection in Sentinel-2 satellite images using attention residual U-Net. *International Journal of Applied Earth Observation and Geoinformation*, 122: 103453. https://doi.org/10.1016/J.JAG.2023.103453.

Kalinaki, K., Malik, O.A., Lai, D.T.C., Sukri, R.S. and Wahab, R.B.H.A. 2023b. Spatial-temporal mapping of forest vegetation cover changes along highways in Brunei using deep learning techniques and Sentinel-2 images. *Ecological Informatics*, 77: 102193. https://doi.org/10.1016/J.ECOINF.2023.102193.

Kalinaki, K., Shafik, W., Gutu, T.J.L. and Malik, O.A. 2023c. Computer vision and machine learning for smart farming and agriculture practices. In *Artificial Intelligence Tools and Technologies for Smart Farming and Agriculture Practices* (pp. 79–100). IGI Global. https://doi.org/10.4018/978-1-6684-8516-3.ch005.

Kalinaki, K., Yahya, U., Malik, O.A. and Lai, D.T.C. 2024b. A review of big data analytics and Artificial Intelligence in Industry 5.0 for smart decision-making. In *Human-Centered Approaches in Industry 5.0: Human-Machine Interaction, Virtual Reality Training, and Customer Sentiment Analysis* (pp. 24–47). https://doi.org/10.4018/979-8-3693-2647-3.ch002.

Khajuria, R. Sunanda and Singh, S. 2018a. A survey of existing leaf disease techniques using artificial neural network. *International Journal of Computer Trends and Technology*, 59(1): 52–62. https://doi.org/10.14445/22312803/IJCTT-V59P109.

Khajuria, R. Sunanda and Singh, S. 2018b. A survey of existing leaf disease techniques using artificial neural network. *International Journal of Computer Trends and Technology*, 59(1): 52–62. https://doi.org/10.14445/22312803/IJCTT-V59P109.

Khaki, S., Wang, L. and Archontoulis, S.V. 2020. A CNN-RNN framework for crop yield prediction. *Frontiers in Plant Science*, 10. https://doi.org/10.3389/fpls.2019.01750.

Koh, J. 2023. Gradient boosting with extreme-value theory for wildfire prediction. *Extremes*, 26(2). https://doi.org/10.1007/s10687-022-00454-6.

Küçükarslan, A.B. 2024. An overview of machine learning (ML) techniques applied to forest fire studies. *Eurasian Journal of Forest Science*. https://doi.org/10.31195/ejejfs.1386306.

Kumalawati, R., Salamiah, Yuliarti, A. and Murliawan, K.H. 2021. Potential mapping agricultural commodities to mitigation of food problem in the future. In *Geojournal of Tourism and Geosites* (Vol. 33, Issue 4). https://doi.org/10.30892/gtg.334spl05-596.

Kumar, R., Kumar, A., Bhatia, K., Nisar, K.S., Chouhan, S.S., Maratha, P. and Tiwari, A.K. 2024. Hybrid approach of cotton disease detection for enhanced crop health and yield. *IEEE Access*. https://doi.org/10.1109/ACCESS.2024.3419906.

Ladi, T., Jabalameli, S. and Sharifi, A. 2022. Applications of machine learning and deep learning methods for climate change mitigation and adaptation. *Environment and Planning B: Urban Analytics and City Science*, 49(4). https://doi.org/10.1177/23998083221085281.

Leuenberger, M., Parente, J., Tonini, M., Pereira, M.G. and Kanevski, M. 2018. Wildfire susceptibility mapping: Deterministic vs. stochastic approaches. *Environmental Modelling and Software*, 101. https://doi.org/10.1016/j.envsoft.2017.12.019.

Lin, X., Li, Z., Chen, W., Sun, X. and Gao, D. 2023. Forest fire prediction based on long- and short-term time-series network. *Forests*, 14(4). https://doi.org/10.3390/f14040778.

Linn, R., Reisner, J., Colman, J.J. and Winterkamp, J. 2002. Studying wildfire behavior using FIRETEC. *International Journal of Wildland Fire*, 11(3-4). https://doi.org/10.1071/wf02007.

Mahdi, A.S. and Mahmood, S.A. 2022. An edge computing environment for early wildfire detection. *Annals of Emerging Technologies in Computing*, 6(3). https://doi.org/10.33166/AETiC.2022.03.005.

Mandallaz, D. and Ye, R. 1997. Prediction of forest fires with Poisson models. *Canadian Journal of Forest Research*, 27(10). https://doi.org/10.1139/x97-103.

Masoudian, S., Sharples, J., Jovanoski, Z., Towers, I. and Watt, S. 2023. Incorporating stochastic wind vectors in wildfire spread prediction. *Atmosphere*, 14(11). https://doi.org/10.3390/atmos14111609.

Massetti, A., Rüdiger, C., Yebra, M. and Hilton, J. 2019. The Vegetation Structure Perpendicular Index (VSPI): A forest condition index for wildfire predictions. *Remote Sensing of Environment*, 224. https://doi.org/10.1016/j.rse.2019.02.004.

McColl-Gausden, S.C., Bennett, L.T., Clarke, H.G., Ababei, D.A. and Penman, T.D. 2022. The fuel–climate–fire conundrum: How will fire regimes change in temperate eucalypt forests under climate change? *Global Change Biology*, 28(17). https://doi.org/10.1111/gcb.16283.

Melchiorre, A. and Boschetti, L. 2018. Global analysis of burned area persistence time with MODIS data. *Remote Sensing*, 10(5). https://doi.org/10.3390/rs10050750.

Pan, M. and Zhang, S. 2023. Visualization of prediction methods for wildfire modeling using CiteSpace: A bibliometric analysis. *Atmosphere*, 14(6). https://doi.org/10.3390/atmos14061009.

Park, M., Tran, D.Q., Jung, D. and Park, S. 2020. Wildfire-detection method using densenet and cyclegan data augmentation-based remote camera imagery. *Remote Sensing*, 12(22). https://doi.org/10.3390/rs12223715.

Pereira, J., Mendes, J., Júnior, J.S.S., Viegas, C. and Paulo, J.R. 2022. A review of genetic algorithm approaches for wildfire spread prediction calibration. In *Mathematics* (Vol. 10, Issue 3). https://doi.org/10.3390/math10030300.

Pérez-Sánchez, J., Jimeno-Sáez, P., Senent-Aparicio, J., Díaz-Palmero, J.M. and Cabezas-Cerezo, J. de D. 2019. Evolution of burned area in forest fires under climate change conditions in southern Spain using ANN. *Applied Sciences (Switzerland)*, 9(19). https://doi.org/10.3390/app9194155.

Phan, T.C., Quach, N.D.K., Nguyen, T.T., Nguyen, T.T., Jo, J. and Nguyen, Q.V.H. 2022. Real-time wildfire detection with semantic explanations. *Expert Systems with Applications*, 201. https://doi.org/10.1016/j.eswa.2022.117007.

Phelps, N. and Woolford, D.G. 2021. Comparing calibrated statistical and machine learning methods for wildland fire occurrence prediction: A case study of human-caused fires in Lac La Biche, Alberta, Canada. *International Journal of Wildland Fire*, 30(11). https://doi.org/10.1071/WF20139.

Piralilou, S.T., Einali, G., Ghorbanzadeh, O., Nachappa, T.G., Gholamnia, K., Blaschke, T. and Ghamisi, P. 2022. A google earth engine approach for wildfire susceptibility prediction fusion with remote sensing data of different spatial resolutions. *Remote Sensing*, 14(3). https://doi.org/10.3390/rs14030672.

Pourghasemi, H.R., Gayen, A., Lasaponara, R. and Tiefenbacher, J.P. 2020. Application of learning vector quantization and different machine learning techniques to assessing forest fire influence factors and spatial modelling. *Environmental Research*, 184. https://doi.org/10.1016/j.envres.2020.109321.

Qiao, Y., Jiang, W., Wang, F., Su, G., Li, X. and Jiang, J. 2023. FireFormer: An efficient transformer to identify forest fire from surveillance cameras. *International Journal of Wildland Fire*. https://doi.org/10.1071/WF22220.

Qiao, Y., Wang, T., Huang, Q., Guo, H., Zhang, H., Xu, Q., Shen, Q. and Ling, N. 2024. Core species impact plant health by enhancing soil microbial cooperation and network complexity during community coalescence. *Soil Biology and Biochemistry*, 188. https://doi.org/10.1016/j.soilbio.2023.109231.

Rahimi, E., Barghjelveh, S., Dong, P., Pirlar, M.A. and Jahanbakhshian, M.M. 2021. PollMap: A software for crop pollination mapping in agricultural landscapes. *Journal of Ecology and Environment*, 45(1). https://doi.org/10.1186/s41610-021-00210-0.

Rezaie, F., Panahi, M., Bateni, S.M., Lee, S., Jun, C., Trauernicht, C. and Neale, C.M.U. 2023. Development of novel optimized deep learning algorithms for wildfire modeling: A case study of Maui, Hawai'i. *Engineering Applications of Artificial Intelligence*, 125. https://doi.org/10.1016/j.engappai.2023.106699.

Rim, C.B., Om, K.C., Ren, G., Kim, S.S., Kim, H.C. and Kang-Chol, O. 2018. Establishment of a wildfire forecasting system based on coupled weather–Wildfire modeling. *Applied Geography*, 90. https://doi.org/10.1016/j.apgeog.2017.12.011.

Rishickesh, R., Shahina, A. and Nayeemulla Khan, A. 2019. Predicting forest fires using supervised and ensemble machine learning algorithms. *International Journal of Recent Technology and Engineering*, 8(2). https://doi.org/10.35940/ijrte.B2878.078219.

Rjoub, D., Alsharoa, A. and Masadeh, A. 2023. Unmanned-aircraft-system-assisted early wildfire detection with air quality sensors. *Electronics (Switzerland)*, 12(5). https://doi.org/10.3390/electronics12051239.

Rodrigues, M. and De la Riva, J. 2014. An insight into machine-learning algorithms to model human-caused wildfire occurrence. *Environmental Modelling and Software*, 57. https://doi.org/10.1016/j.envsoft.2014.03.003.

Rodrigues, M., de la Riva, J. and Fotheringham, S. 2014. Modeling the spatial variation of the explanatory factors of human-caused wildfires in Spain using geographically weighted logistic regression. *Applied Geography*, 48. https://doi.org/10.1016/j.apgeog.2014.01.011.

Rubí, J.N.S., de Carvalho, P.H.P. and Gondim, P.R.L. 2022. Forestry 4.0 and Industry 4.0: Use case on wildfire behavior predictions. *Computers and Electrical Engineering*, 102. https://doi.org/10.1016/j.compeleceng.2022.108200.

Sachithra, V. and Subhashini, L.D.C.S. 2023. How artificial intelligence uses to achieve the agriculture sustainability: Systematic review. *Artificial Intelligence in Agriculture*, 8. https://doi.org/10.1016/j.aiia.2023.04.00

Singh, G., Singh, S., Sethi, G. and Sood, V. 2022. Deep learning in the mapping of agricultural land use using Sentinel-2 satellite data. *Geographies*, 2(4). https://doi.org/10.3390/geographies204004.

Sreenath, K., Hemalatha Reddy, G., Venkata Ranga Reddy, K., Sneha Sandhya, G. and Mahaboob Basha, M.D. 2022. Deep learning approach for uav-based wildfire detection and segmentation. *South Asian Journal of Engineering and Technology*, 12(3). https://doi.org/10.26524/sajet.2022.12.45.

Succarie, A., Xu, Z. and Wang, W. 2022. The variation and trends of nitrogen cycling and nitrogen isotope composition in tree rings: the potential for fingerprinting climate extremes and bushfires. *Journal of Soils and Sediments*, 22(9). https://doi.org/10.1007/s11368-022-03260-6.

Tehrany, M.S., Jones, S., Shabani, F., Martínez-Álvarez, F. and Tien Bui, D. 2019. A novel ensemble modeling approach for the spatial prediction of tropical forest fire susceptibility using LogitBoost machine learning classifier and multi-source geospatial data. *Theoretical and Applied Climatology*, 137(1-2). https://doi.org/10.1007/s00704-018-2628-9.

Thissen, U., Van Brakel, R., De Weijer, A.P., Melssen, W.J. and Buydens, L.M.C. 2003. Using support vector machines for time series prediction. *Chemometrics and Intelligent Laboratory Systems*, 69(1-2). https://doi.org/10.1016/S0169-7439(03)00111-4.

Tien Bui, D., Bui, Q.T., Nguyen, Q.P., Pradhan, B., Nampak, H. and Trinh, P.T. 2017. A hybrid artificial intelligence approach using GIS-based neural-fuzzy inference system and particle swarm optimization for forest fire susceptibility modeling at a tropical area. *Agricultural and Forest Meteorology*, 233. https://doi.org/10.1016/j.agrformet.2016.11.002.

Tonini, M., D'andrea, M., Biondi, G., Esposti, S.D., Trucchia, A. and Fiorucci, P. 2020. A machine learning-based approach for wildfire susceptibility mapping. The case study of the liguria region in italy. *Geosciences (Switzerland)*, 10(3). https://doi.org/10.3390/geosciences10030105.

Tuyen, T.T., Jaafari, A., Yen, H.P.H., Nguyen-Thoi, T., Phong, T. Van, Nguyen, H.D., Van Le, H., Phuong, T.T.M., Nguyen, S.H., Prakash, I. and Pham, B.T. 2021. Mapping forest fire susceptibility using spatially explicit ensemble models based on the locally weighted learning algorithm. *Ecological Informatics*, 63. https://doi.org/10.1016/j.ecoinf.2021.101292.

Umunnakwe, A., Parvania, M., Nguyen, H., Horel, J.D. and Davis, K.R. 2022. Data-driven spatio-temporal analysis of wildfire risk to power systems operation. *IET Generation, Transmission and Distribution*, 16(13). https://doi.org/10.1049/gtd2.12463.

van Klompenburg, T., Kassahun, A. and Catal, C. 2020. Crop yield prediction using machine learning: A systematic literature review. In *Computers and Electronics in Agriculture* (Vol. 177). https://doi.org/10.1016/j.compag.2020.105709.

Vyas, S., Dalhaus, T., Kropff, M., Aggarwal, P. and Meuwissen, M.P.M. 2021. Mapping global research on agricultural insurance. *Environmental Research Letters*, 16(10). https://doi.org/10.1088/1748-9326/ac263d.

Wang, D., Guan, D., Zhu, S., Kinnon, M. Mac, Geng, G., Zhang, Q., Zheng, H., Lei, T., Shao, S., Gong, P. and Davis, S.J. 2021. Economic footprint of California wildfires in 2018. *Nature Sustainability*, 4(3). https://doi.org/10.1038/s41893-020-00646-7.

Wang, L., Zhang, H., Zhang, Y., Hu, K. and An, K. 2023a. A deep learning-based experiment on forest wildfire detection in machine vision course. *IEEE Access*, 11. https://doi.org/10.1109/ACCESS.2023.3262701.

Wang, W., Zhao, F., Wang, Y., Huang, X. and Ye, J. 2023b. Seasonal differences in the spatial patterns of wildfire drivers and susceptibility in the southwest mountains of China. *Science of the Total Environment*, 869. https://doi.org/10.1016/j.scitotenv.2023.161782.

Wang, X., Pan, Z., Gao, H., He, N. and Gao, T. 2023c. An efficient model for real-time wildfire detection in complex scenarios based on multi-head attention mechanism. *Journal of Real-Time Image Processing*, 20(4). https://doi.org/10.1007/s11554-023-01321-8.

Wu, H., Hu, T., Liu, Y., Zhou, H., Wang, J. and Long, M. 2022. *TimesNet: Temporal 2D-Variation Modeling for General Time Series Analysis*. http://arxiv.org/abs/2210.02186.

Yan, C.S., Chen, Y.W., Yang, H.M. and Ahokas, E. 2023. Optical spectrum analyzers and typical applications in astronomy and remote sensing. In *Review of Scientific Instruments* (Vol. 94, Issue 8). https://doi.org/10.1063/5.0138963.

Yandouzi, M., Grari, M., Idrissi, I., Moussaoui, O., Azizi, M., Ghoumid, K. and Elmiad, A.K. 2022. Review on forest fires detection and prediction using deep learning and drones. In *Journal of Theoretical and Applied Information Technology* (Vol. 100, Issue 12).

Zakari, R.Y., Shafik, W., Kalinaki, K. and Iheaturu, C.J. 2024. Internet of Forestry Things (IoFT) technologies and applications in forest management. In *Advanced IoT Technologies and Applications in the Industry 4.0 Digital Economy*. https://doi.org/10.1201/9781003434269-15.

Zhang, X., Guo, Y., Tian, X. and Bai, Y. 2023. Enhancing crop mapping precision through multi-temporal Sentinel-2 image and spatial-temporal neural networks in northern slopes of tianshan mountain. *Agronomy*, 13(11). https://doi.org/10.3390/agronomy13112800.

Zhong, C., Cheng, S., Kasoar, M. and Arcucci, R. 2023. Reduced-order digital twin and latent data assimilation for global wildfire prediction. *Natural Hazards and Earth System Sciences*, 23(5). https://doi.org/10.5194/nhess-23-1755-2023.

Chapter 3

Leveraging Remote Sensing for Agriculture Mapping

Techniques, Applications, and Future Directions

Rufai Yusuf Zakari,[1,*] ***Zaharaddeen Karami Lawal,***[2,3] ***Kassim Kalinaki***[4,5] **and** ***Wasswa Shafik***[6]

Introduction

Remote Sensing (RS) has revolutionized various fields, including agriculture, by providing unprecedented insights and data critical for modern farming practices (Zhang et al., 2023a). Using RS in agriculture involves gathering and processing data from spaceborne platforms (satellites), airborne platforms (aircraft) and ground-based platforms – unmanned aerial vehicles (UAVs), to monitor and manage agricultural activities. This approach allows for the detailed observation of large farmland areas, enabling the assessment of crop health, soil conditions, and environmental factors that affect agricultural productivity (Kluger et al., 2021). The importance and

[1] Department of Computer Science, School of Science, and Information Technology (SSIT), Skyline University Nigeria, 1080, Nigeria. https://orcid.org/0000-0002-4645-6412

[2] Department of Computer Science, Federal University Dutse, Jigawa State, Nigeria.

[3] Faculty of Integrated Technologies, Universiti Brunei Darussalam, Brunei. https://orcid.org/0000-0003-3011-5581

[4] Department of Computer Science, Islamic University in Uganda (IUIU) P. O. Box 2555, Mbale, Uganda.

[5] Borderline Research Laboratory, Kampala, Uganda. https://orcid.org/0000-0001-8630-9110

[6] Dig Connectivity Research Laboratory, Kampala, 600040, Uganda. https://orcid.org/0000-0002-9320-3186

* Corresponding author: rufaig6@gmail.com

benefits of RS for agricultural mapping are manifold. RS offers accurate, up-to-date, and detailed information for tracking plant development, detecting signs of crop distress, and forecasting harvest quantities (Anuradha, 2024; Zhang et al., 2023a). This information is vital for making informed resource allocation, pest control, and irrigation management decisions. Furthermore, RS enhances the efficiency of precision agriculture by enabling site-specific management practices, which results in enhanced crop yields and diminished environmental impact (Hunt et al., 2020). The ability to monitor large and inaccessible areas also supports better planning and disaster management, guaranteeing food security and sustainable agrarian practices (Verma et al., 2023).

Fusing RS techniques with other advanced tools, including Geographic Information Systems (GIS) and ML algorithms, has further amplified its capabilities in agricultural mapping (Karnawat et al., 2020). GIS enables the spatial analysis and visualization of RS data, making it cheaper to detect patterns and trends in crop health and soil conditions (Verma et al., 2023). ML algorithms can swiftly analyze enormous datasets with high precision, generating forecasts and spotting unusual patterns that might elude human observation. Together, these technologies enable a more holistic approach to farm management, where data-driven decisions can be made to optimize productivity and sustainability. For example, ML models can anticipate insect infestations or pathogen proliferation by analyzing records and the present environmental factors, enabling growers to implement preventive strategies (Sachithra and Subhashini, 2023). Additionally, integrating RS data with GIS enables targeted agriculture by generating high-resolution maps. These guide the precise application of nutrients, pest control substances, and irrigation, optimizing resource use and effectiveness.

Moreover, the evolution of RS technology is paving the way for more innovative applications in agriculture. Advances in sensor technology, including the development of hyperspectral and thermal sensors, provide more detailed and specific information about crop conditions and soil health (Shanmugapriya et al., 2019). Hyperspectral sensors can detect subtle differences in plant physiology that indicate stress or nutrient deficiencies, while thermal sensors can identify areas of water stress by measuring surface temperatures. These advancements allow for even more precise monitoring and management of agricultural activities (Chuphal et al., 2024). Furthermore, drones equipped with advanced sensors are becoming increasingly popular, offering high-resolution data and real-time monitoring capabilities (Khaki et al., 2020). Integrating RS data with other agricultural techniques, such as automated machinery and Internet of Things (IoT) sensors, is also on the horizon, promising a future where farming is even more sustainable, efficient, and responsive to environmental changes. This continual innovation in RS technology underscores its critical role in the future of agriculture, driving progress toward more resilient and productive farming systems (Zakari et al., 2024). This chapter aims to provide an exhaustive review of the role of RS in agriculture. We explore the fundamental principles of RS and the different platforms and types of sensors used for agricultural mapping. The chapter will also cover data acquisition techniques, preprocessing steps, and the methodologies employed for image classification. Feature extraction and selection, crucial for accurate analysis, will be discussed, along with the various applications of RS in agriculture, including crop health monitoring, yield prediction, and precision agriculture.

To illustrate the practical implications of RS, a presentation of real-world case studies showcasing its use in different agricultural settings will be provided. Additionally, the chapter will address the challenges faced in RS for agriculture, such as data availability, spatial and temporal resolution limitations, and validation issues. Finally, we will explore emerging trends and future directions in the field, highlighting advances in sensor technology, integration with the IoT, and the potential of big data analytics (BDA) and AI to enhance agricultural practices further (Alabdulatif et al., 2023; Kalinaki et al., 2024a). In summary, RS transforms agriculture by providing valuable data and insights that support efficient and sustainable farming. This chapter will summarize the key points discussed and offer recommendations for researchers, practitioners, and policymakers to harness the full potential of RS in agriculture, paving the way for future developments and innovations in this critical field.

The main objectives of this chapter are to:

- Provide an exhaustive overview of RS principles and technologies used in agriculture.
- Discuss the different RS platforms and sensors employed for agricultural mapping.
- Describe the data acquisition and preprocessing techniques involved in RS for agriculture.
- Explore RS applications in agriculture, such as crop health monitoring, anomaly detection, and yield prediction.
- Highlight real-world case studies showcasing the successful implementation of RS in agriculture.
- Discuss the challenges and limitations of RS for agriculture.
- Explore emerging trends and future directions in RS technology for agriculture.

Chapter Organization

The rest of the chapter is structured as follows: after the introduction in Section 1, Section 2 depicts the fundamental principles of RS. Section 3 provides an overview of RS platforms. Section 4 elaborates on the types of Sensors Used in Agricultural Mapping. Section 5 discusses the data acquisition techniques and challenges. Section 6 details the feature extraction and selection techniques. Section 7 discusses various RS applications in agriculture. Section 8 details the different real-world applications of RS in bettering agriculture. Section 9 discusses the challenges of RS in agriculture. The emerging trends of RS in agriculture are presented in Section 10, and the conclusion is in Section 11.

Fundamental Principles of RS

Definition and Key Concepts

RS is the science of acquiring information about the Earth's surface without direct physical contact. It involves the use of sensors to detect and measure the physical properties of objects or areas from a distance. Key concepts in RS include electromagnetic radiation, sensors, platforms, and data processing.

RS Platforms

RS platforms are vehicles or devices that carry sensors and collect data. The leading platforms used in RS for agriculture are satellites, UAVs (UAVs), and aircraft (Okengwu et al., 2023). Satellites orbit the Earth and provide global coverage, making them ideal for monitoring large agricultural areas (Steven and Clark, 1990; Victor et al., 2024). UAVs are smaller, uncrewed aircraft that can fly at lower altitudes, providing higher-resolution imagery for smaller areas. Aircraft, including manned and unmanned options, are used for more specific or targeted data collection (Victor et al., 2024).

Types of Remote Sensing

RS can be categorized into two main types: active and passive. Active RS involves signal transmission from the sensor to the target and measuring the reflected or backscattered signal (Ullo and Sinha, 2021). Examples include radar and lidar systems, often used for terrain mapping and vegetation analysis. Passive RS relies on detecting natural energy reflected or emitted from the Earth's surface (Khanal et al., 2020). Passive sensors measure radiation intensity in different wavelengths, such as visible, infrared, and microwave. Examples include optical sensors, which capture reflected sunlight to create images, and thermal sensors, which detect emitted infrared radiation to assess temperature (Victor et al., 2024). Figure 1 present the type of remote sensing.

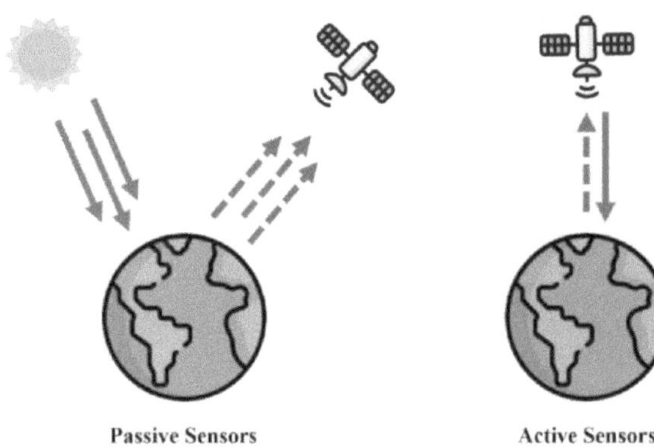

Fig. 1. Types of Remote Sensing.

Key Principles of RS Technology

Electromagnetic Spectrum

RS relies on the interplay between the Earth's surface and electromagnetic (EM) radiation (Okengwu et al., 2023). The EM spectrum encompasses various energy

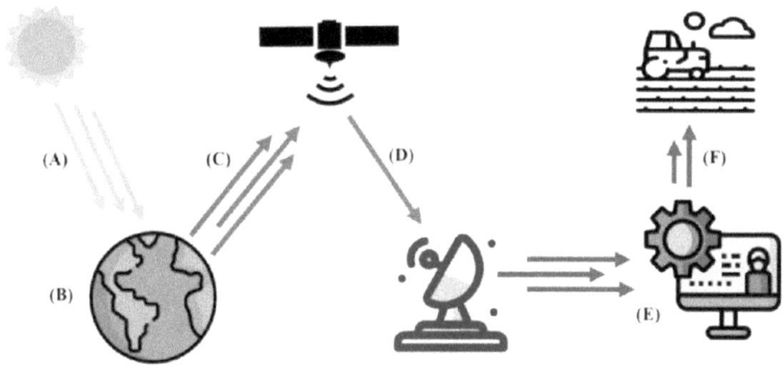

Fig. 2. Satellite RS process applied to agricultural monitoring processes.

waves, spanning from compact gamma radiation to expansive radio signals. Specialized detectors are engineered to capture distinct wavelength ranges within this spectrum (Khanal et al., 2020).

Energy Interactions

As EM waves interact with Earth's terrain, they may be absorbed, pass through, or bounce back. The degree of energy retention or reflection varies based on the characteristics of the surface material encountered (Victor et al., 2024).

Spatial Resolution

This refers to the level of detail in an image. Higher spatial resolution means more detail can be seen in the image. Spatial resolution is determined by the sensor's capabilities and the altitude of the platform (Karnawat et al., 2020).

Temporal Resolution

Temporal resolution refers to how often a sensor can revisit the same area to collect data. Higher temporal resolution allows for more frequent monitoring of changes over time (Okengwu et al., 2023).

Spectral Resolution

Spectral precision denotes a detector's capacity to differentiate various EM wave frequencies (Ullo and Sinha, 2021). More spectrally sensitive instruments can extract finer details about the makeup and state of observed features or regions (Shanmugapriya et al., 2019).

Radiometric Resolution

Radiometric resolution refers to the sensitivity of a sensor to detect variations in radiation intensity. Higher radiometric resolution allows for more subtle differences in the reflectance or emission of energy to be detected (Jiang et al., 2022).

Overview of RS Platforms

RS platforms are the carriers or vehicles that house the sensors used for collecting data. These platforms can be classified based on altitude, coverage area, and operational capabilities.

Satellite Platforms (Space-based)

Satellite platforms orbit the Earth and provide extensive coverage, making them ideal for large-scale agricultural monitoring. Different types of orbits, such as geostationary and polar orbits, offer unique advantages for RS (Khanal et al., 2020). Geostationary satellites remain fixed over a single point on the Earth's surface, providing continuous monitoring of the same area, which is helpful for weather monitoring and tracking large-scale agricultural trends (Oymatov et al., 2023). On the other hand, Polar-orbiting satellites pass over the Earth's poles and cover the entire globe as the Earth rotates. They provide high-resolution data and revisit the same area regularly, which is crucial for monitoring changes in agricultural fields over time. Examples of agrarian satellite platforms include the Landsat series, Sentinel-2, and MODIS, which avail crucia data for crop monitoring, soil assessment, and land use planning (Jiang et al., 2022).

Aerial Platforms

Aerial platforms, including manned aircraft, offer high-resolution data and can be deployed flexibly (Tang and Shao, 2015). Manned aircraft are suitable for covering large areas quickly. They can be equipped with various sensors to capture detailed images and data, making them ideal for regional agricultural surveys and mapping projects.

Ground-Based Platforms

Ground-based platforms include fixed installations and mobile systems that provide highly detailed and accurate data at specific locations. Fixed installations, such as weather stations and ground-based spectroradiometers, continuously monitor specific parameters like soil moisture, temperature, and crop growth (Bouguettaya et al., 2022). These provide valuable ground truth data that can be used to validate and calibrate RS data from other platforms. Mobile systems, mounted on vehicles or handheld devices, can be moved to different locations to collect data on-demand. They are helpful for detailed field surveys and validate RS data with high precision. Ground-based platforms offer the highest level of detail and accuracy, making them indispensable for specific agricultural applications, such as soil analysis, crop scouting, and localized environmental monitoring (Li et al., 2023). Other ground-based platforms include UAVs or drones which are increasingly becoming popular for precision agriculture (Kattenborn et al., 2021). They can fly at low altitudes, capturing high-resolution images and data and providing spatial information about crop health and field conditions. UAVs are particularly useful for small to medium-sized farms where precise and frequent monitoring is necessary. However, their limited flight time and range can be a drawback for larger areas. The flexibility and cost-effectiveness of UAVs make them an essential tool for modern agricultural practices, allowing for timely and targeted interventions (Bouguettaya et al., 2022).

Types of Sensors Used in Agricultural Mapping

Different types of sensors are utilized in RS to capture a wide range of data relevant to agricultural mapping. These sensors can be broadly categorized based on the data they collect from optical, thermal, radar, and LiDAR sensors.

Optical Sensors

Optical sensors detect light reflected from the Earth's surface across numerous wavelengths, such as visible, near-infrared (NIR), and shortwave infrared (SWIR) (Nurdin et al., 2022). These sensors frequently gauge plant vitality, soil characteristics, and crop status. Instruments that record data across several distinct spectral ranges are often employed to compute plant health indicators, like the NDVI (Normalized Difference Vegetation Index) (Nurdin et al., 2022). Hyperspectral sensors, which capture information across numerous fine spectral divisions offer intricate spectral data, enabling crop identification, nutrient shortage detection, and stress monitoring in plants. These instruments find use in scientific studies and precision farming, providing sophisticated tools for in-depth examination and informed choices (Rahimi et al., 2021).

Thermal Sensors

Thermal sensors calculate the thermal infrared radiation emitted by the Earth's surface, providing information about surface temperature. In agriculture, thermal sensors monitor soil moisture levels, detect plant water stress, and manage irrigation systems efficiently (Kluger et al., 2021). For example, sensors on the Landsat 8 satellite provide thermal data applicable in assessing crop water use and manage water resources. Thermal imagery helps farmers optimize irrigation schedules, minimize water consumption, and enhance crop yields by guaranteeing that plants get a sufficient amount of water in a timely manner (Kumalawati et al., 2021).

Radar Sensors

Radar sensors, or Synthetic Aperture Radar (SAR), emit microwave signals and measure the reflected signals to capture Earth's surface information (Javaid et al., 2023). Radar sensors can see through cloud cover and function around the clock, making them crucial for uninterrupted surveillance (Vyas et al., 2021). In agriculture, SAR sensors measure soil moisture, monitor crop growth, and detect changes in field conditions. Examples include the Sentinel-1 satellites, which provide high-resolution radar imagery. The ability to obtain data under all weather conditions and during both day and night makes radar sensors a reliable source of information for agricultural monitoring, particularly in regions with frequent cloud cover (Anuradha, 2024).

LiDAR Sensors

Light Detection and Ranging (LiDAR) sensors emit light pulses to gauge distances to Earth's features, creating accurate three-dimensional representations of landscape and plant formations (Kumar et al., 2024). LiDAR is used in agriculture to create detailed topographic maps, assess crop canopy structure, and monitor changes in land use (Shafik, 2024). Airborne LiDAR systems can capture high-resolution elevation data, which is helpful for smart agriculture and land management. By providing detailed information about the physical structure of crops and the landscape, LiDAR sensors empower farmers to make informed decisions about planting, irrigation, and crop management practices (Verma et al., 2023).

Data Acquisition Techniques and Challenges

Collecting raw data from various sensors mounted on different platforms is a crucial step in RS for agricultural mapping. The quality and relevance of the data acquired directly influence the accuracy and effectiveness of agricultural analysis and decision-making (Benediktsson and Wu, 2021).

Exploring Data Acquisition Techniques

Satellite Imagery

Satellite imagery is a cornerstone of RS for agricultural mapping, offering extensive temporal and spatial coverage. Prominent satellite data sources include MODIS, Landsat, Sentinel-2, and WorldView (Singh et al., 2022). These satellites provide multispectral data for monitoring agricultural parameters such as crop health, soil moisture, and land use changes. One of the significant advantages of satellite imagery is its high temporal resolution, allowing frequent observations of agricultural fields. However, challenges such as cloud cover and atmospheric disturbances can affect data quality. Additionally, the spatial resolution of some satellite images may be insufficient for detailed analysis in smaller fields (Li et al., 2023).

Aerial Photography

Aerial photography, conducted via aircraft or drones, offers high spatial resolution images that can be tailored to specific agricultural needs. Customizing flight paths and collecting data at precise times makes aerial photography valuable for detailed agricultural mapping ("RS in Applications of Geoinformation," 2022). The high-resolution images captured can provide insights into crop conditions, pest infestations, and field variability. However, the coverage area is typically limited compared to satellite imagery. Regulatory restrictions, weather dependency, and the operational costs of flying aircraft or drones can pose significant challenges to aerial data acquisition (Tey et al., 2024).

UAVs

UAVs (drones) are increasingly famous in agricultural RS due to their flexibility and cost-effectiveness (Zhang et al., 2023b). They can capture very high-resolution images, providing spatial information about crop health and field conditions (Chouhan et al., 2024). UAVs are particularly useful for small to medium-sized farms where precise and frequent monitoring is necessary. However, their reduced flight time and range can be a drawback for larger areas. Additionally, processing the large volumes of data collected by UAVs requires significant computational resources. Compliance with flight regulations and obtaining the necessary permissions can also be challenging (Anuradha, 2024).

Ground-Based Sensors

Ground-based sensors, such as spectroradiometers and LIDAR, provide highly accurate and detailed data for agricultural mapping (Ullo and Sinha, 2021). These sensors typically gather information on soil properties, crop growth stages, and plant health. The main advantage of ground-based sensors is their precision, which can complement data from aerial and satellite sources. However, their spatial coverage is limited, making them suitable for smaller areas or validating data obtained from other RS methods. The high equipment cost and the labor-intensive nature of ground-based data collection are notable challenges ("RS in Applications of Geoinformation," 2022).

Challenges in Data Acquisition

Data Quality

Ensuring high-quality data is crucial for accurate agricultural mapping. Common issues include noise, missing values, and inconsistencies across different datasets (Khanal et al., 2020). Noise can result from sensor errors or environmental factors, while missing values may arise due to gaps in data collection. Inconsistent data formats and resolutions from various sources can complicate integration and analysis. Effective data cleaning and standardization techniques are essential to address these quality issues (Kashyap and Kumar, 2021b). Figure 3 illustrate the challenges in data acquisition.

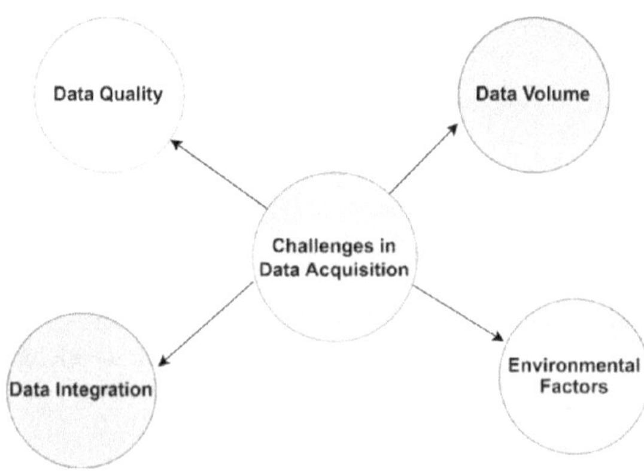

Fig. 3. Datasets challenges.

Data Volume

RS generates large volumes of data, especially when using high-resolution images and frequent temporal observations (Bei et al., 2022). Managing and processing this data requires substantial storage capacity and computational power. Efficient data handling techniques and robust infrastructure are necessary to cope with the sheer volume of information, ensuring timely and accurate analysis for agricultural applications (Khanal et al., 2020).

Data Integration

Combining data from multiple sources, such as satellite images, aerial photographs, and ground-based sensors, can enhance the overall accuracy and detail of agricultural mapping (Kashyap and Kumar, 2021a). However, integrating these diverse datasets poses significant challenges due to differences in spatial resolution, temporal frequency, and data formats. Harmonizing these datasets requires sophisticated algorithms and careful preprocessing to ensure compatibility and reliability of the integrated information (Yan et al., 2023).

Environmental Factors

Environmental conditions such as weather, seasonal variations, and atmospheric disturbances can significantly impact the quality and availability of RS data (Kashyap and Kumar, 2021b). Cloud cover can obstruct satellite and aerial imagery, while atmospheric conditions can introduce distortions. Seasonal changes in vegetation and soil moisture also affect data consistency. Addressing these challenges requires advanced correction techniques and adaptive data acquisition strategies (Steven and Clark, 1990).

Data Preprocessing Steps

Data Cleaning

Data cleaning is the initial step in preprocessing, aimed at removing noise, correcting errors, and handling missing values (Liu, 2020). Noise can be reduced using filtering techniques, while errors are corrected by comparing data against known standards or reference datasets. Missing values can be addressed through interpolation or imputation methods. Effective data cleaning ensures the reliability and accuracy of subsequent analyses (Zakari et al., 2024).

Geometric Correction

Geometric correction adjusts for distortions in RS data caused by sensor movement, Earth's rotation, and terrain variations. Ground Control Points (GCPs) and orthorectification techniques are commonly used to align images accurately with geographic coordinates. This step is crucial for ensuring that spatial relationships in the data are accurate, enabling precise mapping and analysis.

Radiometric Correction

Radiometric correction involves adjusting pixel values for sensor noise, atmospheric conditions, and lighting variations. Techniques like dark object subtraction, atmospheric models, and radiometric calibration standardize the reflectance values (Escandón-Panchana et al., 2024). This correction ensures the data accurately represents the observed features' true spectral characteristics, facilitating reliable interpretation and analysis (Afxentiou et al., 2021).

Image Enhancement

Image enhancement techniques improve the visual appearance of RS data, making it easier to identify and analyze specific features (Zhang et al., 2023a). Histogram equalization, contrast adjustment, and edge enhancement are commonly used. Enhanced images can reveal details that are not readily visible in raw data, aiding in detecting and interpreting regional features (Escandón-Panchana et al., 2024).

Image Classification

Image classification categorizes pixels into different land cover classes based on spectral properties (Sachithra and Subhashini, 2023). Supervised classification techniques, such as Random Forest and Support Vector Machines (SVM), require labeled training data to guide the classification process (Singh et al., 2022). Unsupervised classification methods, like k-means clustering, group pixels into classes based on similarities in their spectral signatures without prior labeling. DL approaches, such as Convolutional Neural Networks (CNNs), offer advanced capabilities for automated classification with high accuracy (Afxentiou et al., 2021).

Data Fusion

Data fusion integrates information from multiple sources to enhance the overall accuracy and completeness of agricultural mapping (Oymatov et al., 2023). Techniques such as multisensory fusion combine data from different sensors, temporal fusion integrates observations over time, and spatial fusion merges data from various spatial resolutions. Effective data fusion provides a comprehensive view of agricultural conditions, supporting informed decision-making (Escandón-Panchana et al., 2024).

Feature Extraction and Selection

Feature extraction techniques are vital in enhancing the performance of prediction models in RS applications (Escandón-Panchana et al., 2024). Although filter methods are simple and computationally efficient, wrapper methods and embedding methods offer more customized approaches by directly evaluating the influence of feature subsets on model performance, as illustrated.

Techniques for Feature Extraction

As presented below, there are four broad techniques for feature extraction in terms of its application to agriculture.

Spectral Indices

These are essential tools for feature extraction in RS in the quest for sustainable agricultural practices. These indices take advantage of the reflectance homes of various wavelengths of light to evaluate the soil qualities and the plants. Soil Adjusted Vegetation Index (SAVI) and Enhanced Vegetation Index (EVI) help with climatic and soil background effects, respectively, giving more exact assessments in challenging settings (hard-to-reach areas) (Shanmugapriya et al., 2019). The Normalized Difference Vegetation Index (NDVI) compares the reflectance of near-infrared and red light to analyze plants' health and thickness. Therefore, by using spectral indices, agricultural specialists can attain beneficial insights concerning plant health, biomass, and anxiety degrees, making it possible for notified decision-making concerning watering, fertilization, and insect monitoring (Chuphal et al., 2024).

Principal Component Analysis (PCA)

A highly successful technique for removing attributes in RS, particularly in hyperspectral data. Principal Component Analysis (PCA) converts the original spectral bands into orthogonal components known as main elements. These primary elements effectively capture the highest amount of variation in the information (Ladi et al., 2022). PCA reduces the dimensionality of information by retaining the most significant components and discarding the less interesting ones while preserving crucial information. The dimensionality reduction simplifies future analysis tasks, such as categorization and clustering, and improves the comprehensibility of the material (Peters et al., 2002). PCA is useful in agricultural applications for identifying patterns and trends in hyperspectral pictures. It helps recognize different types of plants (Chouhan et al., 2024), disease outbreaks, and soil qualities.

Texture Analysis

Since data appears in different formats, text analysis further helps obtain spatial information from RS images using approaches that remove spectral-based attributes. Texture evaluation considers the spatial arrangement of pixels inside an image rather than focusing solely on spectral properties (Grossiord, 2020). The Gray Level Co-occurrence Matrix (GLCM) quantifies textural properties such as contrast, entropy, and homogeneity. These attributes provide valuable information on surface roughness, soil structure, and crop patterns. Structure evaluation is advantageous in differentiating between different land cover types, detecting small variations in crop conditions, and identifying anomalies such as pests or diseases (Chouhan et al., 2021; Qiao et al., 2024; Khajuria et al., 2018). Thus, by integrating texture analysis into feature elimination procedures, agricultural specialists foster a significantly more comprehensive comprehension of agricultural landscapes and make even more educated decisions regarding management.

Object-Based Image Analysis (OBIA)

This approach moves away from traditional pixel-based techniques and focuses on extracting features from RS data. OBIA categorizes pictures into meaningful objects or areas using spectral, spatial, and contextual information. By extracting features such as shape, dimension, and spectral residential qualities from these objects, OBIA allows for a more context-specific and semantically relevant evaluation than pixel-based approaches (van Klompenburg et al., 2020). OBIA is a technique used in farming to accurately identify boundaries between different areas, types of crops, and zones for effective management. This technology plays a crucial role in implementing precision agriculture methods. Moreover, OBIA enhances the precision of categorization and mapping tasks by incorporating spatial relationships between objects, resulting in more robust and interpretable results (Kumar et al., 2024a). Thus, OBIA is widely utilized in monitoring agriculture, planning land utilization, and monitoring natural resources, where the quality of imagery can be determined as per Fig. 4.

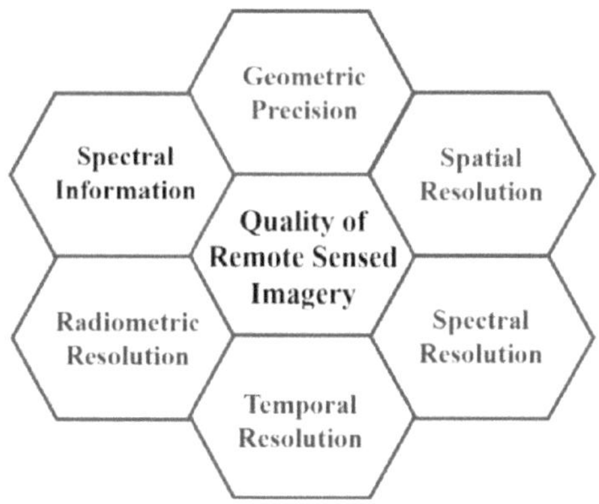

Fig. 4. Quality of remote sensed imagery.

Feature Selection Methods

These features are extracted using the most practical approach presented in this section.

Filter Methods

They are frequently employed for feature selection in RS applications because of their straightforwardness and effectiveness. These methods assess the significance of variables regardless of the selected predictive model. The correlation coefficient is a commonly employed filtering approach that quantifies the linear association between each characteristic and the target variable (Chuphal et al., 2024). Features that have high correlation coefficients are deemed to be more predictive and are kept for subsequent studies. Another illustration is the Chi-Square Test, which evaluates the lack of dependence between category characteristics and the target variable in classification tasks. Features with substantial Chi-Square values are chosen, demonstrating their importance in distinguishing across different classes. Filter methods are beneficial due to their high computational efficiency and capacity to handle big datasets (Ullo and Sinha, 2021). However, they may fail to consider the relationships between features, which could result in less-than-ideal selections of feature subsets.

Wrapper Methods

These are used in assessing feature subsets by measuring their influence on model performance, resulting in higher computational complexity compared to filter methods. These methods evaluate the quality of feature subsets by directly training and testing a predictive model using various combinations of features (Singh et al., 2022). Recursive Feature Elimination (RFE) is a widely used technique that systematically eliminates the least significant features in a model based on performance indicators like accuracy or error rate. Forward and backward selection are alternative wrapping approaches that begin with either an empty or complete set of features. They progressively include or exclude features based on their impact on the model's performance (Nadeem et al., 2022). Wrapper approaches have the advantage of considering feature interactions, but they can be computationally expensive, particularly for datasets with various dimensions.

Embedded Methods

Embedded methods incorporate feature selection directly into training the model, providing a smooth and integrated approach to feature selection. These strategies utilize regularization techniques or inherent feature importance measures in the predictive model to choose relevant features (Tey et al., 2024). Lasso, also known as the Least Absolute Shrinkage and Selection Operator, is a widely used embedded method that uses L1 regularization to penalize the absolute magnitude of feature coefficients. This regularization technique effectively forces some coefficients to become zero, resulting in selecting a subset of features. Tree-based approaches, such as decision trees and Random Forests, naturally conduct feature selection by dividing nodes according to feature relevance metrics such as Gini impurity or information gain (Kumar et al., 2024b). Embedded approaches are beneficial because of their simplicity and efficiency, as they incorporate feature selection into the model training process in forestry and agriculture, as illustrated in Fig. 5.

RS Applications in Agriculture

There are considerable applications of RS in wages for sustainable agriculture, including crop health monitoring and growth, plant growth anomalies and stress conditions, and biophysical parameters estimation, among others, which are elaborated.

Crop Monitoring and Health Assessment

Remote noticing offers a continual, large-scale view of crop fields, crucial for effective crop surveillance and health evaluation. By using multispectral and hyperspectral imaging, farmers can discover refined alterations in plant health that are not noticeable to the naked eye (Singh et al., 2022). Plant indices such as the NDVI

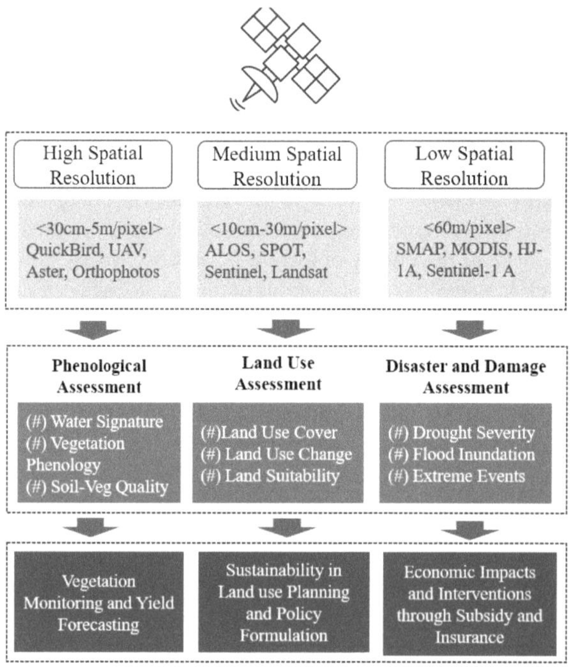

Fig. 5. Application of RS imageries in forestry and agriculture.

can suggest chlorophyll material and total plant growth robustness. Thermal imaging can identify water tension before it becomes aesthetically noticeable, allowing for timely watering. Furthermore, remote noticing helps in the early discovery of bug invasions and conditions by determining abnormalities in plant growth patterns and thermal anomalies (Nadeem et al., 2022). These abilities allow proactive management methods to minimize crop losses and optimize input usage. By incorporating RS information with ground-based observations, farmers can confirm the remote information, resulting in even more precise medical diagnoses and treatments (Tey et al., 2024).

Precision Agriculture

This kind of farming leverages remote noticing to manage field irregularity successfully, maximizing the use of resources such as water, fertilizers, and chemicals. Integrated with global positioning system innovation, high-resolution RS information enables the production of comprehensive area maps that highlight variants in soil buildings, plant health and wellness, and wetness degrees (Shafik, 2024). Variable Price Technology (VRT) can then apply inputs precisely where needed, decreasing waste and enhancing effectiveness. For instance, NDVI maps can guide differential plant food applications, ensuring that the required quantity is used in specific areas, thus boosting nutrient uptake and minimizing overflow. Similarly, dampness maps originated from thermal imaging, or SAR, can notify watering timetables, promoting water conservation (Shafik et al., 2023). Remote noticing additionally sustains real-time surveillance and adaptive monitoring, allowing farmers to respond swiftly to altering problems. This targeted technique enhances efficiency and profitability and lessens environmental effects, supporting sustainable agriculture practices that are critical when faced with environmental change and resource scarcity (Ullo and Sinha, 2021).

Soil Analysis

The SAR and hyperspectral imaging assess dirt dampness content, texture, organic matter, and nutrient degrees. This info is crucial for recognizing soil fertility and making informed decisions about dirt management (Qiao et al., 2024). Geographical maps will determine dry areas requiring watering or zones susceptible to waterlogging that might need drain services. Hyperspectral information can discover variations in soil raw material and nutrients, directing the application of garden composts or fertilizers to enhance soil health and wellness (Jafarbiglu and Pourreza, 2022). RS aids in mapping soil disintegration and compaction, which is essential for preserving dirt structure and fertility. By supplying a spatial introduction of dirt problems, RS supports accurate farming practices that maximize soil monitoring. This brings about boosted plant development, decreased input costs, and much better sustainability (Chuphal et al., 2024). Moreover, continual tracking allows for the analysis of soil wellness in time, assisting in lasting planning and conservation efforts.

Yield Prediction

Regarding yield production and prediction, RS enhances the precision of crop yield forecasts by supplying extensive and prompt information on crop growth stages, health, and environmental problems (Jafarbiglu and Pourreza, 2022; Kumar et al., 2024b). Evaluating historical RS information along with current period monitoring, models' forecasts yield high accuracy. NDVI and Boosted Plants Index (EVI) are the most popular technologies originating from multispectral and hyperspectral imagery that offer an understanding right into plant biomass and health and wellness, which are directly correlated with yield potential (Tey et al., 2024). Thermal imaging can keep track of temperature tension, while SAR information supplies info on dirt dampness levels, both of which impact crop efficiency. Integrating these data points with weather forecasts and ground fact information enables the growth of durable yield prediction versions (Abid et al., 2021). These predictions are indispensable for farmers and agriculture in planning harvests, managing supply chains, and making market choices. Exact return forecasts additionally sustain food security preparation and policymaking by governments and global organizations, guaranteeing that sufficient sources are assigned to meet future food needs (Shanmugapriya et al., 2019).

Disaster Management

RS plays a vital role in catastrophe monitoring within the farming field by supplying fast and precise analyses of damages triggered by natural disasters such as floodings, dry spells, storms, and wildfires. Satellite imagery and airborne data use real-time information on the level and intensity of such events, allowing for prompt feedback and recovery efforts (Singh et al., 2022). SAR penetrates the cloud cover to provide pictures of flooded areas, while thermal imaging can spot fire hotspots. Post-disaster, remote noticing assists in mapping impacted areas, estimating plant losses, and assessing soil and facility damage. This information is essential for collaborating on emergency relief, insurance policy cases, and recovery programs (Qiao et al., 2024). It further uses information to support the advancement of very early warning systems that predict the possibility of extreme weather events, allowing farmers to take safety nets. A thorough risk monitoring structure can be established by incorporating RS with various other information resources, such as atmospheric and hydrological versions, boosting the resilience of farming systems to catastrophes and climate adjustment (Chuphal et al., 2024).

Case Studies

This section presents selected real-world applications of RS in bettering agriculture.

Precision Agriculture in the United States of America

In the Midwest of the United States, a large corn farm executed accurate agriculture methods, using RS and general practitioner modern technology like AI to ensure precision fertilization. Multispectral imaging from drones was used to create comprehensive NDVI maps, highlighting variations in crop vitality throughout the areas (Kumar et al., 2024a; Tey et al., 2024). These maps determined locations with reduced NDVI worth, suggesting nutrient deficiencies. After that, the farm utilized VRT to apply fertilizers differentially based on the NDVI data. This strategy caused an over 15% boost in corn yield, an over 20% reduction in fertilizer usage, and lessened nutrient overflow, thus lowering environmental pollution (Singh et al., 2022). Integrating RS and VRT substantially increased productivity and environmental sustainability.

Drought Monitoring and Management in India

In the drought-prone region of Maharashtra, India, RS up was employed to check and manage water sources better. Satellite thermal imaging offered data on land surface temperatures, plant water stress, and anxiety, while SAR was used to gauge the degree of dirt dampness (Kumar et al., 2024b). This detail was integrated into a central data source available to local farmers and farming agencies to ensure drought monitoring and water management. The combination of thermal and SAR data permitted the advancement of an identical early warning system for dry spell problems, allowing farmers to apply water-saving actions (Chuphal et al., 2024). Optimized irrigation schedules, notified by dirt moisture maps, preserved water sources, and lowered plant losses by 30% throughout dry spell durations, enhancing strength, optimized irrigation, and improved crop resilience and sustainability.

Enhancing Maize Production in Zambia

Within Sub-Saharan Africa, agricultural technologies have been highly used. In Zambia's Central Province, smallholder maize farmers faced obstacles connected to irregular plant health and wellness surveillance and ineffective resource use, affecting returns and profitability (Shafik, 2024). A project was launched to incorporate remote noticing technology into their farming methods, using drones with multispectral cameras to catch high-resolution pictures and generate NDVI maps. These maps highlighted areas of varying plant vigor, allowing targeted treatments for bugs, conditions, and nutrient management (Mwape et al., 2024). Therefore, maize yields increased by an average of 20%, with decreased agrochemical expenses and minimized ecological effects. The task also included training neighborhood farmers, enhancing their capability to utilize remote noticing modern technology individually, therefore boosting food safety and promoting sustainable farming methods through improved crop health monitoring, capacity building, and cost savings (Wang et al., 2021).

Disaster Response in Southeast Asia

The Mekong Delta in Vietnam often experiences seasonal floods that impact rice manufacturing. To mitigate these impacts, RS modern technology was employed for flood influence analysis. High-resolution satellite imagery and Artificial Aperture Radar (SAR) information were used to map swamped locations, assess crop damages, and guide emergency response initiatives (Ullo and Sinha, 2021). RS made it possible to swiftly identify swamped areas and accurately evaluate plant losses, supplying important details for disaster action groups. Accurate flooding maps facilitated the adequate allotment of alleviation sources, such as food, water, and clinical materials, to one of the most affected areas (Jafarbiglu and Pourreza, 2022). Post-flood evaluation supported recuperation planning, consisting of replanting crops and reconstruction of farming infrastructure, therefore minimizing recuperation time for affected farmers and improving the strength of the farming community. Figure 6 illustrates some additional applications of RS within the agricultural sector.

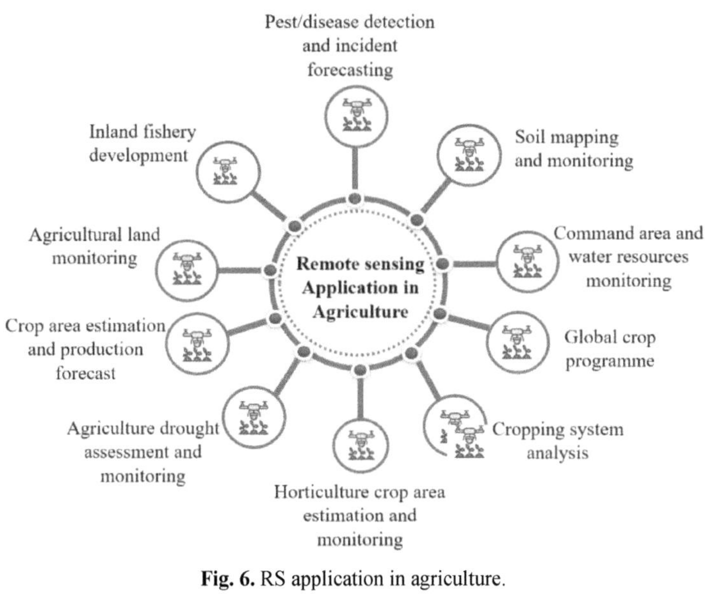

Fig. 6. RS application in agriculture.

Challenges of RS in Agriculture

RS has emerged as a powerful tool for monitoring and managing agricultural practices, providing valuable insights into crop health, soil conditions, and environmental factors that impact crop yield. However, the effective utilization of RS in agriculture faces several challenges that must be addressed. This section highlights the critical challenges in RS for agriculture, including data availability and accessibility, spatial and temporal resolution limitations, validation and accuracy assessment, cost and resource constraints, and technical and operational challenges.

Data Availability and Accessibility

Various factors can hinder the acquisition of high-quality RS data for agricultural applications. Cloud cover and atmospheric conditions can obscure the Earth's surface, leading to data gaps or low-quality imagery, which is particularly problematic for time-sensitive agrarian monitoring (Martos et al., 2021). Additionally, sensor limitations in spectral and radiometric resolution may affect the ability to accurately detect and quantify specific crop characteristics or environmental factors (Jafarbiglu and Pourreza, 2022). Data sharing policies and licensing restrictions imposed by data providers can limit access to particular data sources, creating barriers for small-scale farmers or research institutions with limited budgets (Wood et al., 2024). Furthermore, the complexity of data discovery and retrieval mechanisms may require specialized skills and knowledge to navigate various data repositories and platforms effectively (Li et al., 2021).

Spatial and Temporal Resolution Limitations

Agricultural applications often require high spatial resolution for detailed crop monitoring and frequent temporal resolution for tracking crop growth and conditions throughout the growing season. However, balancing these two aspects can be challenging. High spatial resolution imagery, essential for identifying crop patterns, monitoring plant health, and detecting small-scale features, often comes at the cost of reduced temporal resolution or a limited coverage area (Wu et al., 2023). Conversely, broad-scale imagery with frequent revisit times, such as sentinel-2 satellites, lack the necessary detail for precise analysis of individual fields or crops (Li et al., 2022). Reconciling this trade-off between spatial and temporal resolutions is a significant challenge in RS for agriculture.

Validation and Accuracy Assessment

Validating RS data and assessing its accuracy is crucial for reliable decision-making in agricultural applications. Ground-truthing or collecting in-situ data for comparison and calibration can be labor-intensive, time-consuming, and logistically challenging, especially for large-scale agrarian operations or remote areas (Bennion et al., 2019). Vegetation heterogeneity, including variations in crop types, growth stages, and management practices, can introduce complexities in interpreting and validating RS data (Han et al., 2024). Additionally, sensor calibration and atmospheric effects, such as aerosols and water vapor, can introduce uncertainties and errors in the data, requiring robust calibration and atmospheric correction techniques (Wang et al., 2021). Developing standardized validation protocols and accuracy assessment methods can be challenging due to various agricultural systems, crop types, and environmental conditions across different regions.

Cost and Resource Constraints

The adoption and utilization of RS technologies in agriculture can be hindered by cost and resource constraints. Acquiring high-resolution imagery and advanced RS products can be costly, particularly for small-scale farmers or organizations with limited budgets (Wu et al., 2023). Furthermore, specialized software and hardware for data processing, analysis, and visualization can be expensive and require regular upgrades or maintenance (Kalinaki et al., 2023). Skilled personnel with expertise in RS, image analysis, and agricultural applications can be scarce and costly to hire and retain, especially in resource-constrained regions. Limited access to computing infrastructure, such as high-performance computing clusters or cloud-based platforms, can also hinder the ability to efficiently process and analyze large volumes of RS data.

Technical and Operational Challenges

RS in agriculture involves various technical and operational challenges. Data processing and analysis often require significant computational resources and specialized software, which can be resource-intensive and challenging to maintain, especially in remote or rural areas with limited infrastructure (Kalinaki et al., 2023). Integrating RS data with other sources of information, such as weather data, soil data, and agricultural management practices, is crucial for comprehensive decision-making but can be complex and require sophisticated data fusion techniques (Ahmad et al., 2022). Automating the detection and classification of crop types, growth stages, and stress conditions from RS data is another challenge attributed to the variability in crop characteristics and environmental conditions (Wu et al., 2023). Additionally, operational challenges, such as timely data acquisition, processing, and information dissemination to farmers, are complex and require robust workflows and communication channels (Hariri and Goli, 2024).

Scalability and Generalization

RS techniques developed for specific agricultural regions or crop types may not be easily scalable or generalizable to other areas or crops. The diversity of farming practices, environmental conditions, and crop characteristics across different geographic regions can pose significant challenges in applying RS methods universally (Barbosa Júnior et al., 2024). Factors such as soil types, climate patterns, and farming practices can vary substantially, requiring adaptations or new approaches to utilize RS data effectively. Additionally, the transferability of algorithms and models trained on specific datasets to new regions or crop varieties can be limited, leading to potential inaccuracies or biases in the analysis (Rusňák et al., 2023).

Integration with Precision Agriculture Technologies

The effective integration of RS data with precision agriculture technologies, such as variable rate technology (VRT) and precision planting systems, presents challenges (Marinello et al., 2023). RS data must be processed and analyzed promptly to provide actionable information that can be seamlessly incorporated into precision agriculture workflows. This requires robust data processing pipelines, efficient data management systems, and interoperability between software and hardware components. Furthermore, the spatial resolution of RS data may not always align with the operational scales of precision agriculture equipment, necessitating spatial interpolation for downscaling techniques (Zhu et al., 2024).

Stakeholder Engagement and Adoption

Effective stakeholder engagement and widespread adoption of RS technologies in agriculture can be challenging. Farmers, agricultural extension services, and other stakeholders may lack awareness or understanding of RS's benefits and potential applications (Barbosa Júnior et al., 2024). Overcoming cultural barriers, addressing data privacy and ownership concerns, and providing adequate training and support are crucial for successful adoption. Additionally, the complexity of RS data and analysis methods may pose barriers for non-technical users, highlighting the need for user-friendly interfaces and decision-support tools tailored to the needs of diverse stakeholders (San Emeterio de la Parte et al., 2024). Table 1 summarises the challenges and proposed considerations.

Emerging Trends and Future Directions of RS in Agriculture

RS for agriculture is rapidly evolving and driven by technological advancements, data analytics, and the integration of various disciplines. As we look toward the future, several emerging trends and directions are shaping the landscape of agricultural RS applications. This section explores the following key areas: advances in sensor technology, integration with the Internet of Things (IoT) and smart farming, big data analytics and cloud computing, ML and AI advancements, and policy and regulatory considerations.

Advances in Sensor Technology

Developing new and improved RS sensors is a driving force in enhancing agricultural monitoring and management. High-resolution multispectral and hyperspectral sensors are becoming more accessible, enabling detailed analysis of crop health, nutrient deficiencies, and stress conditions (Ram et al., 2024). Additionally, integrating thermal and radar sensors provides valuable information on soil moisture, crop water status, and biomass estimation (Abdulraheem et al., 2023). Advances in miniaturization and the development of lightweight and low-cost sensors have paved the way for UAVs (UAVs) and drones to play a significant role in precision agriculture, offering high-resolution and timely data acquisition at the field scale (Pungavi and Praveenkumar, 2024).

Table 1. Summary of the challenges and proposed considerations.

Challenge	Summary of Challenge	Summary of Solution	References
Data Availability and Accessibility	Cloud cover, atmospheric conditions, sensor limitations, data sharing policies, and data discovery mechanisms can limit access to high-quality RS data for agricultural applications.	Improve data sharing policies, enhance data discovery tools, and develop advanced atmospheric correction and data fusion techniques.	(Jafarbiglu and Pourreza, 2022; Li et al., 2021; Martos et al., 2021; Wood et al., 2024)
Spatial and Temporal Resolution Limitations	Balancing high spatial resolution for detailed crop monitoring and frequent temporal resolution for tracking crop growth and conditions can be challenging.	Develop multi-sensor fusion techniques, leverage UAVs (UAVs), or explore new satellite constellations to improve resolution.	(Li et al., 2022; Wu et al., 2023)
Validation and Accuracy Assessment	Ground-truthing, vegetation heterogeneity, sensor calibration, and atmospheric effects can introduce uncertainties and challenges in validating agricultural RS data.	Develop standardized validation protocols, robust calibration techniques, and leverage complementary data sources for accuracy assessment.	(Bennion et al., 2019; Han et al., 2024; Wang et al., 2021)
Cost and Resource Constraints	Acquiring high-resolution imagery, advanced products, specialized software, hardware, and skilled personnel can be costly and resource-intensive.	Explore cost-effective open-source solutions, leverage cloud computing, and invest in capacity building and training programs.	(Kalinaki et al., 2023; Wu et al., 2023)
Technical and Operational Challenges	Data processing, analysis, integration with other data sources, automation, and timely dissemination of information pose technical and operational challenges.	Develop efficient data processing pipelines, implement robust data fusion techniques, and establish effective stakeholder communication channels.	(Ahmad et al., 2022; Hariri and Goli, 2024; Kalinaki et al., 2023)
Scalability and Generalization	RS techniques developed for specific regions or crop types may not be easily scalable or generalizable to other areas or crops due to diversity in farming practices, environmental conditions, and crop characteristics.	Develop adaptable and transferable algorithms, leverage ML techniques, and incorporate local knowledge and data sources.	(Barbosa Júnior et al., 2024; Rusňák et al., 2023)

Table 1 contd. ...

...Table 1 contd.

Challenge	Summary of Challenge	Summary of Solution	Reference
Integration with Precision Agriculture Technologies	Integrating RS data with precision agriculture technologies, such as variable rate technology (VRT) and precision planting systems, requires interoperability, spatial alignment, and timely data processing.	Develop robust data processing pipelines, efficient data management systems, and spatial interpolation or downscaling techniques.	(Marinello et al., 2023; Zhu et al., 2024)
Stakeholder Engagement and Adoption	Overcoming cultural barriers, addressing data privacy and ownership concerns, providing training and support, and developing user-friendly interfaces and decision support tools are crucial for widespread adoption of RS technologies in agriculture.	Engage stakeholders early, develop tailored training programs, address data privacy concerns, and design user-friendly interfaces and decision support tools.	(Barbosa Júnior et al., 2024; San Emeterio de la Parte et al., 2024)

Integration with the Internet of Things (IoT) and Smart Farming

Integrating RS with the Internet of Things (IoT) and intelligent farming technologies revolutionizes agricultural practices (Dhal et al., 2024; Kalinaki et al., 2024b). IoT-enabled sensors deployed in fields can collect real-time data on soil moisture, temperature, and other environmental parameters combined with RS data for comprehensive monitoring and decision-making (Rajak et al., 2023). Smart farming systems can leverage this integrated data to optimize irrigation schedules, apply targeted pest and disease management strategies, and precisely allocate resources based on crop requirements. The seamless exchange of information between RS platforms, IoT devices, and farm management systems will enable data-driven precision agriculture practices.

Big Data Analytics and Cloud Computing

The increasing volume and complexity of RS data for agriculture necessitate the adoption of big data analytics and cloud computing solutions (Kalinaki et al., 2024c; Rao et al., 2023). RS platforms generate massive amounts of data, which require robust storage, processing, and analysis capabilities. Cloud computing infrastructures offer scalable and on-demand computing resources, enabling efficient data processing, analysis, and storage. Additionally, cloud-based platforms facilitate data sharing and collaboration among researchers, extension services, and farmers, enabling the development of decision-support tools and knowledge-sharing platforms (Kalinaki et al., 2024d; Rao et al., 2023).

ML and AI Advancements

ML and AI (AI) techniques are increasingly important RS for agriculture (Kalinaki et al., 2024b). These advanced algorithms can be trained on large datasets to automatically identify crop types, detect anomalies, classify land cover, and accurately predict crop yields. DL models, in particular, have demonstrated remarkable performance in image classification, object detection, and pattern recognition tasks, enabling more accurate and efficient analysis of RS data (Kalinaki et al., 2023). Integrating AI with RS data will enable real-time monitoring, early detection of stress conditions, and predictive analytics for optimized crop management.

Policy and Regulatory Considerations

As RS technologies become more prevalent in agriculture, policy and regulatory considerations will be crucial in their adoption and implementation. Data privacy, ownership, and security issues must be addressed to ensure RS data's responsible and ethical use. Policies and guidelines for data sharing, interoperability standards, and data governance will facilitate collaboration and integration of RS with other agricultural technologies. Additionally, regulatory frameworks may be needed to ensure the responsible use of drones and UAVs in agrarian settings, addressing concerns related to safety, privacy, and environmental impact (Mohsan et al., 2023).

Multi-Source Data Fusion and Integration

The fusion and integration of data from multiple sources, including satellite imagery, UAV/drone data, ground-based sensor networks, and ancillary data like yield records and farmer knowledge, is an emerging trend (Barrile et al., 2022). This multi-source approach aims to leverage the strengths of different data sources, providing a comprehensive understanding of agricultural systems. It combines aerial observations, ground-level measurements, and contextual information for holistic crop condition assessment, yield prediction, and informed decision-making.

Citizen Science and Crowdsourcing Approaches

Citizen science and crowdsourcing approaches involve farmers, researchers, and the public in data collection, observation sharing, and collaborative problem-solving (Ebitu et al., 2021). These participatory methods leverage local knowledge and on-the-ground observations to validate RS data, improve accuracy, and foster inclusive and tailored agricultural monitoring. Crowdsourcing platforms, mobile apps, and user-friendly interfaces enable data sharing, knowledge exchange, and collective problem-solving, empowering local communities and promoting sustainable practices (Shakya et al., 2023).

In summary, the future of RS in agriculture is poised to be transformative. The convergence of advanced sensor technologies, IoT and smart farming systems, big data analytics, and AI/ML techniques will enable unprecedented precision, efficiency, and sustainability in agricultural practices. However, these advancements must be accompanied by robust policies and regulations to ensure responsible and ethical implementation, data privacy, and environmental stewardship. Collaborative efforts among researchers, technology developers, policymakers, and agricultural stakeholders will be crucial in addressing the challenges and unlocking the full potential of RS for agriculture. By embracing these emerging trends and future directions, we can pave the way for data-driven, sustainable, and resilient agricultural systems that meet the growing global demand for food while minimizing environmental impacts.

Conclusion

In conclusion, this chapter has comprehensively explored how RS technologies are revolutionizing agriculture. We began by laying the foundational principles of RS, elucidating the core concepts and mechanisms that underpin this powerful technology. Various RS platforms, including satellites and UAVs, were examined for their unique advantages and constraints in data acquisition. The discussion extended to different types of sensors, such as optical, thermal, and radar, each suited to specific agricultural applications. We also addressed the techniques and challenges of data acquisition, emphasizing the complexities involved in capturing high-quality RS data. Feature extraction and selection techniques were highlighted as critical processes for transforming raw data into meaningful insights. The diverse applications of RS in agriculture, from crop monitoring to soil health assessment, were explored alongside real-world case studies demonstrating these technologies' practical benefits and transformative potential. Despite the challenges, including technical limitations and data management issues, emerging trends like ML integration and hyperspectral imaging are paving the way for more sophisticated and precise agricultural mapping solutions. Overall, RS is poised to be a pivotal technology in modern agriculture, significantly enhancing productivity, sustainability, and efficiency and driving substantial advancements and improvements in the sector.

Declaration

Ethical Responsibilities of Authors

All authors have read, understood, and have complied as applicable with the statement on "Ethical responsibilities of Authors" as found in the Instructions for Authors.

Ethical Approval

Not applicable.

Use of AI Tool

The corresponding author declares that no AI tool has been used to prepare this manuscript.

Conflict of Interest

On behalf of all the authors the corresponding author states that there is no conflict of interest to disclose.

Competing Interests

Not applicable.

Authors' Contributions

Rufai Yusuf Zakari: Conceptualization, Writing - Original Draft, **Zaharaddeen Karami Lawal**: Writing - Review & Editing, Visualization, **Kassim Kalinaki**: Writing - Review & Editing, Visualization, Proof-reading, **Wasswa Shafik:** Writing - Review & Editing, Visualization.

Funding

No funding has been received.

References

Abdulraheem, M.I., Zhang, W., Li, S., Moshayedi, A.J., Farooque, A.A. and Hu, J. 2023. Advancement of remote sensing for soil measurements and applications: A comprehensive review. *Sustainability*, 15(21): 15444. https://doi.org/10.3390/SU152115444.

Abid, S.K., Sulaiman, N., Chan, S.W., Nazir, U., Abid, M., Han, H., Ariza-Montes, A. and Vega-Muñoz, A. 2021. Toward an integrated disaster management approach: How artificial intelligence can boost disaster management. In *Sustainability (Switzerland)* (Vol. 13, Issue 22). https://doi.org/10.3390/su132212560.

Afxentiou, N., Georgali, P.Z.M., Kylili, A. and Fokaides, P.A. 2021. Greenhouse agricultural plastic waste mapping database. *Data in Brief*, 34. https://doi.org/10.1016/j.dib.2020.106622.

Ahmad, U., Nasirahmadi, A., Hensel, O. and Marino, S. 2022. Technology and data fusion methods to enhance site-specific crop monitoring. *Agronomy*, 12(3): 555. https://doi.org/10.3390/agronomy12030555.

Alabdulatif, A., Thilakarathne, N.N., Lawal, Z.K., Fahim, K.E. and Zakari, R.Y. 2023. Internet of Nano-Things (IoNT): A comprehensive review from architecture to security and privacy challenges. In *Sensors* (Vol. 23, Issue 5). https://doi.org/10.3390/s23052807.

Anuradha, C.T. 2024. Soil quality analysis and mapping of various land uses using geospatial technology: a case study. *Applied Geomatics*, 16(1). https://doi.org/10.1007/s12518-023-00538-3.

Barbosa Júnior, M.R., Moreira, B.R. de A., Carreira, V. dos S., Brito Filho, A.L. de, Trentin, C., Souza, F.L.P. de, Tedesco, D., Setiyono, T., Flores, J.P., Ampatzidis, Y., Silva, R.P. da and Shiratsuchi, L.S. 2024. Precision agriculture in the United States: A comprehensive meta-review inspiring further research, innovation, and adoption. *Computers and Electronics in Agriculture*, 221: 108993. https://doi.org/10.1016/J.COMPAG.2024.108993.

Barrile, V., Simonetti, S., Citroni, R., Fotia, A. and Bilotta, G. 2022. Experimenting Agriculture 4.0 with sensors: A data fusion approach between remote sensing, UAVs and self-driving tractors. *Sensors*, 22(20): 7910. https://doi.org/10.3390/S22207910.

Bei, G., Zhang, S., Guo, Y., Yanli, L., Hu, N. and Liu, J. 2022. Study on meteorological disaster monitoring of field fruit industry by remote sensing data. In *Advances in Meteorology* (Vol. 2022). https://doi.org/10.1155/2022/1659053.

Benediktsson, J.A. and Wu, Z. 2021. Distributed computing for remotely sensed data processing. *Proceedings of the IEEE*, 109(8): 1278–1281. https://doi.org/10.1109/JPROC.2021.3094335.

Bennion, M., Fisher, J., Yesson, C. and Brodie, J. 2019. Remote sensing of Kelp (Laminariales, Ochrophyta): Monitoring tools and implications for wild harvesting. *Reviews in Fisheries Science & Aquaculture*, 27(2): 127–141. https://doi.org/10.1080/23308249.2018.1509056.

Bouguettaya, A., Zarzour, H., Kechida, A. and Taberkit, A.M. 2022. Deep learning techniques to classify agricultural crops through UAV imagery: A review. In *Neural Computing and Applications* (Vol. 34, Issue 12). https://doi.org/10.1007/s00521-022-07104-9.

Chouhan, S.S., Singh, U.P. and Jain, S. 2024. Introduction to computer vision and drone technology. In *Applications of Computer Vision and Drone Technology in Agriculture 4.0* (pp. 1–5). Springer Nature Singapore. https://doi.org/10.1007/978-981-99-8684-2_1.

Chouhan, S.S., Singh, U.P., Sharma, U. and Jain, S. 2021. Leaf disease segmentation and classification of *Jatropha Curcas* L. and *Pongamia Pinnata* L. biofuel plants using computer vision based approaches. *Measurement*, 171: 108796. https://doi.org/10.1016/J.MEASUREMENT.2020.108796.

Chouhan, S.S., Singh, U.P., Sharma, U. and Jain, S. 2024. Classification of different plant species using deep learning and machine learning algorithms. *Wireless Personal Communications*, 136(4): 2275–2298. https://doi.org/10.1007/S11277-024-11374-Y/FIGURES/9.

Chuphal, D.S., Kushwaha, A.P., Aadhar, S. and Mishra, V. 2024. Drought Atlas of India, 1901–2020. *Scientific Data*, 11(1). https://doi.org/10.1038/s41597-023-02856-y.

Dhal, S., Wyatt, B.M., Mahanta, S., Bhattarai, N., Sharma, S., Rout, T., Saud, P. and Acharya, B.S. 2024. Internet of Things (IoT) in digital agriculture: An overview. *Agronomy Journal*, 116(3): 1144–1163. https://doi.org/10.1002/AGJ2.21385.

Ebitu, L., Avery, H., Mourad, K.A. and Enyetu, J. 2021. Citizen science for sustainable agriculture—A systematic literature review. *Land Use Policy*, 103: 105326. https://doi.org/10.1016/J.LANDUSEPOL.2021.105326.

Escandón-Panchana, P., Herrera-Franco, G., Jaya-Montalvo, M. and Martínez-Cuevas, S. 2024. Geomatic tools used in the management of agricultural activities: A systematic review. In *Environment, Development and Sustainability*. https://doi.org/10.1007/s10668-024-04576-8.

Escandón-Panchana, P., Herrera-Franco, G., Martínez Cuevas, S. and Morante-Carballo, F. 2024. Prospects of UAVs in agricultural mapping. *Communications in Computer and Information Science*, 1874 CCIS. https://doi.org/10.1007/978-3-031-46813-1_21.

Grossiord, C. 2020 . Having the right neighbors: how tree species diversity modulates drought impacts on forests. In *New Phytologist* (Vol. 228, Issue 1). https://doi.org/10.1111/nph.15667.

Han, H., Liu, Z., Li, J. and Zeng, Z. 2024. Challenges in remote sensing based climate and crop monitoring: Navigating the complexities using AI. *Journal of Cloud Computing*, 13(1): 1–14. https://doi.org/10.1186/S13677-023-00583-8.

Hariri, M.A. and Goli, A. 2024. Issues and research challenges for implementing cyber-physical agricultural supply chains. In *Agri 4.0 and the Future of Cyber-Physical Agricultural Systems* (pp. 73–89). Elsevier. https://doi.org/10.1016/B978-0-443-13185-1.00005-8.

Hunt, D.A., Tabor, K., Hewson, J.H., Wood, M.A., Reymondin, L., Koenig, K., Schmitt-Harsh, M. and Follett, F. 2020. Review of remote sensing methods to map coffee production systems. In *Remote Sensing* (Vol. 12, Issue 12). https://doi.org/10.3390/rs12122041.

Jafarbiglu, H. and Pourreza, A. 2022. A comprehensive review of remote sensing platforms, sensors, and applications in nut crops. *Computers and Electronics in Agriculture*, 197: 106844. https://doi.org/10.1016/J.COMPAG.2022.106844.

Javaid, M., Haleem, A., Khan, I.H. and Suman, R. 2023. Understanding the potential applications of Artificial Intelligence in agriculture sector. *Advanced Agrochem*, 2(1). https://doi.org/10.1016/j.aac.2022.10.001.

Jiang, N., Li, P. and Feng, Z. 2022. Remote sensing of swidden agriculture in the tropics: A review. In *International Journal of Applied Earth Observation and Geoinformation* (Vol. 112). https://doi.org/10.1016/j.jag.2022.102876.

Kalinaki, K., Abdullatif, M., Nasser, S.A.-K., Nsubuga, R. and Kugonza, J. 2024a. Paving the path to a sustainable digital future with green cloud computing. *Emerging Trends in Cloud Computing Analytics, Scalability, and Service Models*, 44–66. https://doi.org/10.4018/979-8-3693-0900-1.CH002.

Kalinaki, K., Acaru, S.F., Kugonza, J. and Nsubuga, R. 2024b. Towards an intelligent tomorrow: Machine Learning enabling sustainable development. In *Methodologies, Frameworks, and Applications of Machine Learning* (pp. 66–89). https://doi.org/10.4018/979-8-3693-1062-5.CH004.

Kalinaki, K., Shafik, W., Namuwaya, S. and Namuwaya, S. 2024c. Perspectives, applications, challenges, and future trends of IoT-Based logistics. In *Navigating Cyber Threats and Cybersecurity in the Logistics Industry* (pp. 148–171). IGI Global. https://doi.org/10.4018/979-8-3693-3816-2.CH005.

Kalinaki, K., Yahya, U., Malik, O.A. and Lai, D.T.C. 2024d. A review of big data analytics and artificial intelligence in Industry 5.0 for smart decision-making. In *Human-Centered Approaches in Industry 5.0: Human-Machine Interaction, Virtual Reality Training, and Customer Sentiment Analysis* (pp. 24–47). https://doi.org/10.4018/979-8-3693-2647-3.ch002.

Kalinaki, K., Shafik, W., Gutu, T.J.L. and Malik, O.A. 2023. Computer vision and machine learning for smart farming and agriculture practices. In *Artificial Intelligence Tools and Technologies for Smart Farming and Agriculture Practices* (pp. 79–100). IGI Global. https://doi.org/10.4018/978-1-6684-8516-3.ch005.

Karnawat, M., Trivedi, S.K., Nagar, D. and Nagar, R. 2020. Future of AI in agriculture. *Biotica Research Today*, 2(9).

Kashyap, B. and Kumar, R. 2021a. Sensing methodologies in agriculture for monitoring biotic stress in plants due to pathogens and pests. *Inventions*, 6(2). https://doi.org/10.3390/INVENTIONS6020029.

Kashyap, B. and Kumar, R. 2021b. Sensing methodologies in agriculture for soil moisture and nutrient monitoring. *IEEE Access*, 9. https://doi.org/10.1109/ACCESS.2021.3052478.

Kattenborn, T., Leitloff, J., Schiefer, F. and Hinz, S. 2021. Review on Convolutional Neural Networks (CNN) in vegetation remote sensing. In *ISPRS Journal of Photogrammetry and Remote Sensing* (Vol. 173). https://doi.org/10.1016/j.isprsjprs.2020.12.010.

Khajuria, R., Sundanda, S. and Singh, S. 2018. A survey of existing leaf disease techniques using Artificial Neural Network. *International Journal of Computer Trends and Technology*, 59(1): 52–62. https://doi.org/10.14445/22312803/IJCTT-V59P109.

Khaki, S., Wang, L. and Archontoulis, S.V. 2020. A CNN-RNN framework for crop yield prediction. *Frontiers in Plant Science*, 10. https://doi.org/10.3389/fpls.2019.01750.

Khanal, S., Kushal, K.C., Fulton, J.P., Shearer, S. and Ozkan, E. 2020. Remote sensing in agriculture—Accomplishments, limitations, and opportunities. In *Remote Sensing* (Vol. 12, Issue 22). https://doi.org/10.3390/rs12223783.

Kluger, D.M., Wang, S. and Lobell, D.B. 2021. Two shifts for crop mapping: Leveraging aggregate crop statistics to improve satellite-based maps in new regions. *Remote Sensing of Environment*, 262. https://doi.org/10.1016/j.rse.2021.112488.

Kumalawati, R., Salamiah, Yuliarti, A. and Murliawan, K.H. 2021. Potential mapping agricultural commodities to mitigation of food problem in the future. In *Geojournal of Tourism and Geosites* (Vol. 33, Issue 4). https://doi.org/10.30892/gtg.334spl05-596.

Kumar, R., Kumar, A., Bhatia, K., Nisar, K.S., Chouhan, S.S., Maratha, P. and Tiwari, A.K. 2024a. Hybrid approach of cotton disease detection for enhanced crop health and yield. *IEEE Access*. https://doi.org/10.1109/ACCESS.2024.3419906.

Kumar, V., Sharma, K.V., Pham, Q.B., Srivastava, A.K., Bogireddy, C. and Yadav, S.M. 2024b. Advancements in drought using remote sensing: assessing progress, overcoming challenges, and exploring future opportunities. In *Theoretical and Applied Climatology*. https://doi.org/10.1007/s00704-024-04914-w.

Ladi, T., Jabalameli, S. and Sharifi, A. 2022. Applications of machine learning and deep learning methods for climate change mitigation and adaptation. *Environment and Planning B: Urban Analytics and City Science*, 49(4). https://doi.org/10.1177/23998083221085281.

Li, M., Shamshiri, R.R., Weltzien, C. and Schirrmann, M. 2022. Crop monitoring using Sentinel-2 and UAV multispectral imagery: A comparison case study in northeastern Germany. *Remote Sensing*, 14(17): 4426. https://doi.org/10.3390/RS14174426/S1.

Li, X., Li, S., Zhao, M., Guo, X. and Zhang, T. 2023. Mapping the shifting focus in remote sensing literature: Technology, methodology, and applications. *Processes*, 11(2). https://doi.org/10.3390/pr11020571.

Li, Y., Ma, J. and Zhang, Y. 2021. Image retrieval from remote sensing big data: A survey. *Information Fusion*, 67: 94–115. https://doi.org/10.1016/J.INFFUS.2020.10.008.

Liu, S.Y. 2020. Artificial Intelligence (AI) in Agriculture. In *IT Professional* (Vol. 22, Issue 3). https://doi.org/10.1109/MITP.2020.2986121.

Marinello, F., Zou, X., Liu, Z., Zhu, X., Zhang, W., Qian, Y., Li, Y., Karunathilake, E.M.B.M., Le, A.T., Heo, S., Chung, Y.S. and Mansoor, S. 2023. The path to smart farming: Innovations and opportunities in precision agriculture. *Agriculture*, 13(8): 1593. https://doi.org/10.3390/AGRICULTURE13081593.

Martos, V., Ahmad, A., Cartujo, P. and Ordoñez, J. 2021. Ensuring agricultural sustainability through remote sensing in the era of Agriculture 5.0. *Applied Sciences*, 11(13): 5911. https://doi.org/10.3390/APP11135911.

Mohsan, S.A.H., Othman, N.Q.H., Li, Y., Alsharif, M.H. and Khan, M.A. 2023. Unmanned aerial vehicles (UAVs): Practical aspects, applications, open challenges, security issues, and future trends. *Intelligent Service Robotics*. https://doi.org/10.1007/s11370-022-00452-4.

Nadeem, F., Jacobs, B. and Cordell, D. 2022. Mapping agricultural vulnerability to impacts of climate events of Punjab, Pakistan. *Regional Environmental Change*, 22(2). https://doi.org/10.1007/s10113-022-01918-y.

Nurdin, D.P., Karim, S. and Kurniadin, N. 2022. Geographic information system for mapping agricultural land in north samarinda district. *TEPIAN*, 3(4). https://doi.org/10.51967/tepian.v3i3.701.

Okengwu, U.A., Onyejegbu, L.N., Oghenekaro, L.U., Musa, M.O. and Ugbari, A.O. 2023. Environmental and ethical negative implications of AI in agriculture and proposed mitigation measures. *Scientia Africana*, 22(1). https://doi.org/10.4314/sa.v22i1.13.

Oymatov, R., Mamatkulov, Z., Maksudov, R., Bakhriyev, M., Aminova, G., Menlimuratova, I. and Esirgapov, A. 2023. Improving the methods of agricultural mapping using remote sensing data. *E3S Web of Conferences*, 386. https://doi.org/10.1051/e3sconf/202338604008.

Peters, A.J., Walter-Shea, E.A., Ji, L., Viña, A., Hayes, M. and Svoboda, M.D. 2002. Drought monitoring with NDVI-based Standardized Vegetation Index. *Photogrammetric Engineering and Remote Sensing*, 68(1).

Pungavi, R. and Praveenkumar, C. 2024. *Unmanned Aerial Vehicles (UAV) for Smart Agriculture*, 251–269. https://doi.org/10.1007/978-981-97-0341-8_13.

Qiao, Y., Wang, T., Huang, Q., Guo, H., Zhang, H., Xu, Q., Shen, Q. and Ling, N. 2024. Core species impact plant health by enhancing soil microbial cooperation and network complexity during community coalescence. *Soil Biology and Biochemistry*, 188. https://doi.org/10.1016/j.soilbio.2023.109231.

Rahimi, E., Barghjelveh, S., Dong, P., Pirlar, M.A. and Jahanbakhshian, M.M. 2021. PollMap: A software for crop pollination mapping in agricultural landscapes. *Journal of Ecology and Environment*, 45(1). https://doi.org/10.1186/s41610-021-00210-0.

Raj, E.F.I., Appadurai, M. and Athiappan, K. 2021. *Precision Farming in Modern Agriculture*. https://doi.org/10.1007/978-981-16-6124-2_4.

Rajak, P., Ganguly, A., Adhikary, S. and Bhattacharya, S. 2023. Internet of Things and smart sensors in agriculture: Scopes and challenges. *Journal of Agriculture and Food Research*, 14: 100776. https://doi.org/10.1016/J.JAFR.2023.100776.

Ram, B.G., Oduor, P., Igathinathane, C., Howatt, K. and Sun, X. 2024. A systematic review of hyperspectral imaging in precision agriculture: Analysis of its current state and future prospects. *Computers and Electronics in Agriculture*, 222: 109037. https://doi.org/10.1016/J.COMPAG.2024.109037.

Rao, M.S., Modi, S., Singh, R., Prasanna, K.L., Khan, S. and Ushapriya, C. 2023. Integration of cloud computing, IoT, and Big Data for the development of a novel smart agriculture model. *2023 3rd International Conference on Advance Computing and Innovative Technologies in Engineering, ICACITE 2023*, 2779–2783. https://doi.org/10.1109/ICACITE57410.2023.10182502.

Remote Sensing in Applications of Geoinformation. 2022. In *Remote Sensing in Applications of Geoinformation*. https://doi.org/10.3390/books978-3-0365-2326-2.

Rusňák, T., Kasanický, T., Malík, P., Mojžiš, J., Zelenka, J., Sviček, M., Abrahám, D. and Halabuk, A. 2023. Crop mapping without labels: Investigating temporal and spatial transferability of crop classification models using a 5-year sentinel-2 series and Machine Learning. *Remote Sensing*, 15(13): 3414. https://doi.org/10.3390/RS15133414.

Sachithra, V. and Subhashini, L.D.C.S. 2023. How artificial intelligence uses to achieve the agriculture sustainability: Systematic review. *Artificial Intelligence in Agriculture*, 8. https://doi.org/10.1016/j.aiia.2023.04.002.

San Emeterio de la Parte, M., Martínez-Ortega, J.F., Castillejo, P. and Lucas-Martínez, N. 2024. Spatio-temporal semantic data management systems for IoT in agriculture 5.0: Challenges and future directions. *Internet of Things*, 25: 101030. https://doi.org/10.1016/J.IOT.2023.101030.

Shafik, W. 2024. Navigating emerging challenges in robotics and artificial intelligence in Africa. In *Examining the Rapid Advance of Digital Technology in Africa*. https://doi.org/10.4018/978-1-6684-9962-7.ch007.

Shafik, W., Tufail, A., Namoun, A., De Silva, L.C. and Apong, R.A.A.H.M. 2023. A systematic literature review on plant disease detection: Motivations, classification techniques, datasets, challenges, and future trends. *IEEE Access*, 11. https://doi.org/10.1109/ACCESS.2023.3284760.

Shakya, A., Tripathi, G. and Ningombam, D.D. 2023. Crowdsourcing applications in agriculture. *Social Media and Crowdsourcing Application and Analytics*, 49–70. https://doi.org/10.1201/9781003346326-3.

Shanmugapriya, P., Rathika, S., Ramesh, T. and Janaki, P. 2019. Applications of remote sensing in agriculture—A review. *International Journal of Current Microbiology and Applied Sciences*, 8(01). https://doi.org/10.20546/ijcmas.2019.801.238.

Singh, G., Singh, S., Sethi, G. and Sood, V. 2022. Deep Learning in the mapping of agricultural land use using Sentinel-2 satellite data. *Geographies*, 2(4). https://doi.org/10.3390/geographies2040042.

Steven, M.D. and Clark, J.A. 1990. Applications of remote sensing in agriculture. *Applications of Remote Sensing in Agriculture*. https://doi.org/10.2478/gein-2014-0007.

Tang, L. and Shao, G. 2015. Drone remote sensing for forestry research and practices. In *Journal of Forestry Research* (Vol. 26, Issue 4). https://doi.org/10.1007/s11676-015-0088-y.

Tey, Y.S., Brindal, M., Wong, S.Y., Ardiansyah, Ibragimov, A. and Yusop, M.R. 2024. Evolution of precision agricultural technologies: A patent network analysis. *Precision Agriculture*, 25(1). https://doi.org/10.1007/s11119-023-10076-y.

Ullo, S.L. and Sinha, G.R. 2021. Advances in IoT and smart sensors for remote sensing and agriculture applications. *Remote Sensing*, 13(13). https://doi.org/10.3390/rs13132585.

van Klompenburg, T., Kassahun, A. and Catal, C. 2020. Crop yield prediction using machine learning: A systematic literature review. In *Computers and Electronics in Agriculture* (Vol. 177). https://doi.org/10.1016/j.compag.2020.105709.

Verma, B., Porwal, M., Jha, A.K., Vyshnavi, R.G., Rajpoot, A. and Nagar, A.K. 2023. Enhancing precision agriculture and environmental monitoring using proximal remote sensing. *Journal of Experimental Agriculture International*, 45(8). https://doi.org/10.9734/jeai/2023/v45i82168.

Victor, N., Maddikunta, P.K.R., Mary, D.R.K., Murugan, R., Chengoden, R., Gadekallu, T.R., Rakesh, N., Zhu, Y. and Paek, J. 2024. Remote sensing for agriculture in the era of Industry 5.0—A survey. *IEEE Journal of Selected Topics in Applied Earth Observations and Remote Sensing*, 17. https://doi.org/10.1109/JSTARS.2024.3370508.

Vyas, S., Dalhaus, T., Kropff, M., Aggarwal, P. and Meuwissen, M.P.M. 2021. Mapping global research on agricultural insurance. *Environmental Research Letters*, 16(10). https://doi.org/10.1088/1748-9326/ac263d.

Wang, R., Gamon, J.A., Moore, R., Zygielbaum, A.I., Arkebauer, T.J., Perk, R., Leavitt, B., Cogliati, S., Wardlow, B. and Qi, Y. 2021. Errors associated with atmospheric correction methods for airborne imaging spectroscopy: Implications for vegetation indices and plant traits. *Remote Sensing of Environment*, 265: 112663. https://doi.org/10.1016/J.RSE.2021.112663.

Wood, D., Rathnasabapathy, M., Stober, K.J. and Menon, P. 2024. Challenges and progress in applying space technology in support of the sustainable development goals. *Acta Astronautica*, 219: 678–692. https://doi.org/10.1016/J.ACTAASTRO.2024.03.064.

Wu, B., Zhang, M., Zeng, H., Tian, F., Potgieter, A.B., Qin, X., Yan, N., Chang, S., Zhao, Y., Dong, Q., Boken, V., Plotnikov, D., Guo, H., Wu, F., Zhao, H., Deronde, B., Tits, L. and Loupian, E. 2023. Challenges and opportunities in remote sensing-based crop monitoring: A review. *National Science Review*, 10(4). https://doi.org/10.1093/NSR/NWAC290.

Yan, C.S., Chen, Y.W., Yang, H.M. and Ahokas, E. 2023. Optical spectrum analyzers and typical applications in astronomy and remote sensing. In *Review of Scientific Instruments* (Vol. 94, Issue 8). https://doi.org/10.1063/5.0138963.

Zakari, R.Y., Shafik, W., Kalinaki, K. and Iheaturu, C.J. 2024. Internet of Forestry Things (IoFT) technologies and applications in forest management. In *Advanced IoT Technologies and Applications in the Industry 4.0 Digital Economy*. https://doi.org/10.1201/9781003434269-15.

Zhang, Q. 2023a. Opinion: AI in agriculture, researchable issues. In *Computers and Electronics in Agriculture* (Vol. 212). https://doi.org/10.1016/j.compag.2023.108110.

Zhang, X., Guo, Y., Tian, X. and Bai, Y. 2023b. Enhancing crop mapping precision through multi-temporal Sentinel-2 image and spatial-temporal neural networks in northern slopes of Tianshan mountain. *Agronomy*, 13(11). https://doi.org/10.3390/agronomy13112800.

Zhu, Z., Bo, Y., Sun, T., Zhang, X., Sun, M., Shen, A., Zhang, Y., Tang, J., Cao, M. and Wang, C. 2024. A downscaling-and-fusion framework for generating spatio-temporally complete and fine resolution remotely sensed surface soil moisture. *Agricultural and Forest Meteorology*, 352: 110044. https://doi.org/10.1016/J.AGRFORMET.2024.110044.

Chapter 4

Basil Crop Detection Using Computer Vision and Deep Learning Approach

***Rajneesh Kumar Patel,*[1] *Siddharth Singh Chouhan,*[1,*] *Thota Chandan*[1] and *Uday Pratap Singh*[2]**

Introduction

In recent years, applying deep learning models in agriculture has gained significant attention due to their potential to revolutionize farming practices. Precision farming maximizes crop yields while mitigating environmental repercussions and increasing economic returns by efficiently managing agricultural resources. Accurate detection and monitoring of crop health and growth are crucial for its success. Basil (Ocimum basilicum) is a widely cultivated herb with diverse culinary and medicinal uses (Sipos et al., 2021; Joshi and Bhavsar, 2023). The successful development and yield of basil plants depend on various factors, including adequate irrigation, nutrient supply, and pest management. Monitoring basil crops' health and growth stages is crucial for optimizing these factors and ensuring a successful harvest. Precision agriculture has become increasingly vital in modern farming practices to ensure optimal crop management, resource allocation, and sustainability (Abbasi et al., 2023; Farooqui et al., 2023). Traditional crop monitoring methods involve labor-intensive manual inspections or basic imaging techniques, which can be time-consuming and require more precision for effective decision-making (Chouhan et al., 2019a; Chouhan et al., 2019b). The advent of deep learning and computer vision technologies has

[1] School of Computing Science and Engineering, VIT Bhopal University, Sehore, 466114, India.

[2] Department of Mathematics, Central University of Jammu, Bagla Suchani, Jammu and Kashmir, 181143, India.

* Corresponding author: siddharth.smvdu@gmail.com

opened new avenues for automating crop monitoring tasks with higher accuracy and efficiency (Zhu et al., 2024; Singh et al., 2023). Object detection, in particular, enables the automated identification and localization of specific objects within digital images or video frames. YOLO (You Only Look Once) is a pioneering object detection architecture known for its real-time performance and accuracy (Sukhadeo Bajait et al., 2023). It is well-suited for agricultural applications where timely decision-making is critical. YOLOv8 represents an evolution of the YOLO architecture, incorporating advancements in model design, training techniques, and computational efficiency.

By leveraging a combination of deep convolutional neural networks and efficient inference strategies, YOLOv8 promises improved detection performance and reduced computational overhead compared to earlier iterations (Yang et al., 2023; Kunduracioglu and Pacal, 2024). Basil crops are susceptible to various environmental factors, pests, and diseases, making timely and accurate detection critical for maintaining crop health and maximizing yield. This study aims to develop a practical and efficient solution using YOLOv8 to detect basil crops in real-time, enabling farmers and agronomists to promptly make informed decisions and interventions (Fuentes-Pacheco et al., 2024; Mao et al., 2023). The Research Questionnaire (RQ) we developed to conduct this study are as follows:

RQ1: What is the reason for selecting the basil plant?

RQ2: What are the key advantages of using YOLOv8 compared to traditional methods?

RQ3: What improvements or additional features would you suggest for YOLOv8 to optimize its use in precision agriculture?

RQ4: What challenges have you encountered with existing methods for crop detection and monitoring methods?

RQ5: How effective are these technologies in terms of accuracy and efficiency?

Literature Review

Precision agriculture has witnessed substantial advancements driven by integrating deep learning and computer vision techniques. Object detection has been pivotal in automating tasks crucial for efficient crop management (Bouguettaya et al., 2022). In this context, YOLO (You Only Look Once) models have emerged as a prominent choice, given their real-time capabilities and accuracy (Badgujar et al., 2024). YOLOv8, a recent iteration in the YOLO series, stands out for its robustness and versatility. Numerous studies have showcased the potential of YOLOv8 in diverse applications, from pedestrian detection to vehicle counting. However, the deployment of YOLOv8 in the domain of precision agriculture, specifically for basil crop detection, remains a relatively unexplored avenue. The proposed research contributes to this burgeoning field by presenting a practical study that leverages YOLOv8 for basil crop detection, demonstrating its adaptability and efficacy in a high-value herb context. Wenjie et al. (Yang et al., 2023) introduce a pioneering strategy for swift cattle detection utilizing deformable convolution and coordinate attention. Dhingra et al. (2019) proposed a Basil leaves' disease classification and identification technique. Although not directly related to crop detection, this work highlights the significance of precise agricultural classification techniques. On-fire detection demonstrates the capability of deep learning models to handle object detection tasks, which has implications for precision agriculture. Dai et al. (2024; 2021) proposed a deep learning model for disease detection and tried to find the disease status with the help of deep learning techniques. Zhang et al. (2019) suggest a model for image processing based on multi-layer CNN using neighborhood optimization. Augusto et al. (Souza et al., 2022) delve into classifying cadmium-contaminated leafy vegetables using hyperspectral imaging and machine learning, which has relevance in crop health assessment.

Zhou et al. (2023) and Anand et al. (2022) discussed detecting nonstandard gear surface defects, a method demonstrating machine learning for defect detection relevant to crop quality assessment. Explore the application of artificial intelligence in plant disease detection, emphasizing the importance of AI in crop health assessment. Zhang et al. (2019) proposed a multilayer CNN based on a neighborhood optimization method for depth-measuring images. Sheng et al. (2018; 2021) proposed an image decomposition model for a deep learning-based model. Giakoumoglou et al. (2023) suggested that a synthetic dataset for object detection in agriculture has implications for training and improving models in precision agriculture. Using machine learning and deep learning, the state of leaf disease detection provides insights into crop health monitoring in precision agriculture (Sarkar et al., 2023; Saini and Nagpal, 2023). Instance segmentation for power lines and transmission towers with various practical applications. In detecting individual crops or objects in precision agriculture. The influence of starch-based bioplastic on basil plant growth, emphasizing the broader ecological context of crop cultivation in precision agriculture (Chodey and Shariff, 2022; Qin et al., 2024). Jiang et al. (2022) proposed a deep-learning model for extracting deep information from photographs. Lin et al. (2021) proposed a transformer-based attention pyramid model for image processing. Chen et al. (2023) discussed a neural network based on image processing, especially a one-shot neural network.

Li et al. (2021) proposed a waste detection model using the cascaded CNN model. Xie et al. (2021) suggested a higher-order feature interactions model based on attention graphs for image processing. Nazir et al. (2020) propose a model based on an end-to-end network for disease detection from blood vessels. Chen et al. (2020) proposed a deep learning model combining decomposition techniques. A Practical Study on Basil Crop Detection integrates insights from existing literature in computer vision, object detection, and agriculture. We leverage innovative methodologies and methods from referenced papers to extend the applicability of YOLOv8 to basil crop detection. By synthesizing and adapting best practices from these works, we contribute to advancing precision agriculture and crop monitoring. Our study demonstrates how we build upon existing research to push the boundaries of knowledge in our specific domain.

Proposed Work

In this section, the problem statement is discussed in detail followed by the proposed methodology along with the model description at the end.

Problem Statement

The realm of precision agriculture is undergoing a transformative evolution driven by the need to optimize resource utilization, maximize crop yield, and reduce environmental impact. In this context, accurate and timely crop health and growth monitoring is paramount. Basil, a high-value herb with diverse applications in culinary and medicinal fields, presents unique challenges in precision agriculture.

Challenges in Basil Crop Detection

The specific challenges in basil crop detection arise from the plant's growth characteristics and variability. Basil plants exhibit diverse appearances as they progress through different growth stages, and their leaves and foliage can vary significantly. Accurate basil crop detection can enable early identification of diseases, pests, and nutrient deficiencies, offering the potential for timely intervention to maintain crop health (Makri and Kintzios, 2008).

The Role of Deep Learning

Deep learning has emerged as a potent technology for object detection and holds significant promise in revolutionizing precision agriculture. Applying YOLOv8, a cutting-edge deep learning model is pivotal in addressing the basil crop detection problem. The problem statement involves the rigorous evaluation of the YOLOv8 model in the context of basil crop detection, assessing its capability to provide accurate and real-time monitoring.

Proposed Methodology

This paper investigates the application of YOLOv8, a state-of-the-art deep learning model, for basil crop detection in precision agriculture. It seeks to evaluate the model's utility and performance, advancing precision agriculture through practical experimentation and assessment. The cornerstone of this study is a meticulously curated dataset comprising 209 basil images. These images have been extensively pre-processed to ensure data quality and consistency. Furthermore, the dataset was thoughtfully partitioned into training (171 images) and validation (38 images) subsets to facilitate robust model training and evaluation. Our methodology begins with describing a comprehensive basil crop dataset, reflecting diverse growth stages and conditions crucial for model success. We then delve into the architecture of the YOLOv8 model, emphasizing optimizations tailored for basil crop detection. Transparent data pre-processing steps ensure the model receives clean, relevant data. Model training and fine-tuning, including crucial hyperparameter tuning, are highlighted for optimal performance. By conducting these steps, the research contributes to the baseline for basil crop image datasets and supports the reproducibility of the study. This research utilizes the YOLOv8 model, renowned for its accuracy and real-time object detection capabilities, to train and evaluate basil crop detection rigorously. Performance metrics such as precision, recall, and F1-score are comprehensively employed and supported by graphical representations for enhanced visualization of the model's efficacy.

Outline of the Proposed Methodology

Table 1 proposed methodology, trained Yolov8 model for Predicting Basil Crop.

Table 1. Proposed methodology.

Input: *Basil Crop Dataset*

Procedure:

Step 1: Start
Step 2: Resizing the Basil Crop Images and Annotations
Step 3: Normalizing the Image Annotation to the desired Yolov8 Format
Step 4: Divide the dataset into training and testing sets
Step 5: Build the Yolov8 Model
Step 6: Calculate the Model Performance using

- *Train Box loss, Train Cls loss, Train Dfl loss*
- *Val Box loss, Val Cls loss, Val Dfl loss*
- *Precision, Recall, F1 Score*
- *mAP50, mAP50 95*

Output: *Trained Yolov8 model for Predicting Basil Crop*

Dataset Description

The dataset utilized for training and validating our basil crop detection model is meticulously curated from the Rob flow platform, designated as version 4 (v4). It comprises 209 high-resolution images, uniformly sized at 640 × 640 pixels, ensuring consistency in training and evaluation. Annotations follow. A structured format, providing bounding box coordinates (x, y, height, width) are crucial for precise object detection, particularly in training the YOLOv8 model. In Fig. 1 the basil crop dataset

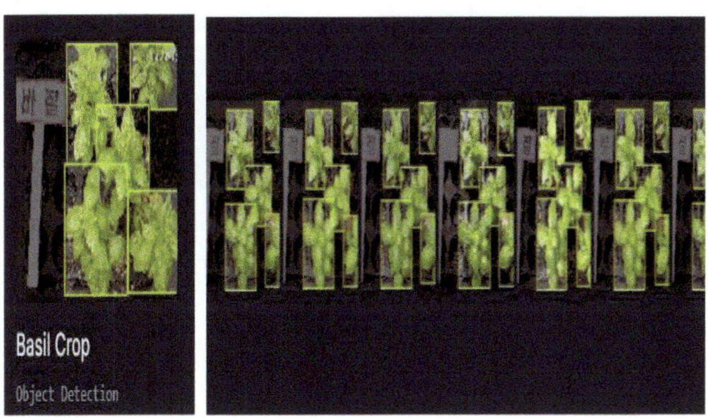

Fig. 1. Basil crop dataset.

employed in our research offers a comprehensive representation of basil plants, covering various growth stages and orientations encountered in real-world agricultural scenarios. This diversity enhances the adaptability of our YOLOv8 model, ensuring its proficiency in recognizing basil plants across different conditions.

Data Pre-processing

The robustness and adaptability of a deep learning model for object detection in agriculture are often attributed to the quality of the dataset and the effectiveness of data pre-processing steps. The first critical step in data pre-processing involves resizing the images and their corresponding annotations to uniform dimensions. Each image in the dataset, initially of varying sizes, is transformed into a consistent format of 640 × 480 pixels. This resizing operation not only ensures that the model is fed with standardized inputs but also enhances computational efficiency during training and evaluation. A uniform image size is essential for our YOLOv8 model to maintain consistency in its predictions across diverse scenarios and orientations in Fig. 2.

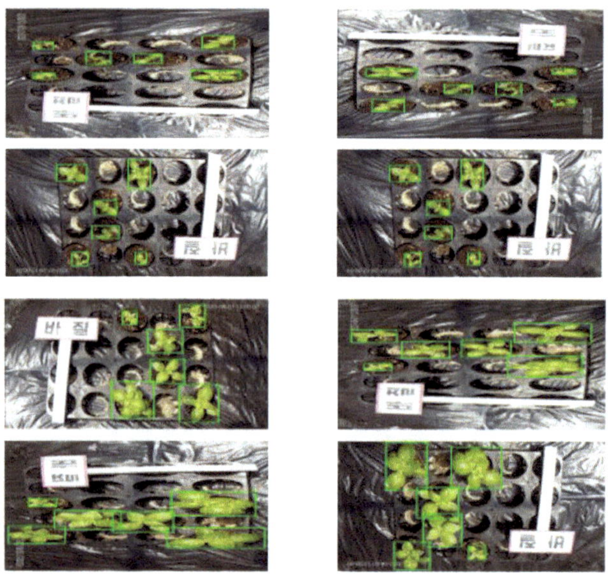

Fig. 2. Resized basil crop images and annotations.

This normalization facilitates the capability of the model to precisely predict and locate basil plants within the Fig. 3 images.

```
0 0.317 0.30354206008 0.114 0.173819742489
0 0.694 0.33726094420 0.156 0.23605150214
0 0.395 0.32257467811 0.13 0.195278969957
```

Fig. 3. Basil crop annotation format.

Yolov8 Model

In this section, we explore the core of our research, concentrating on the YOLOv8 model, which stands for "You Only Look Once Version 8." The Modern deep learning model YOLOv8 is famous for its effectiveness, quickness, and high accuracy in real-time object identification applications. It incorporates cutting-edge architectural advancements and optimization methods to expand upon the YOLO family's forerunners.

(i) Yolov8 Architecture Overview

The YOLOv8 architecture employs a single forward pass through a fully convolutional neural network, adhering to the single-shot object detection paradigm for efficiency. It divides the input image into a grid of cells, and the model directly predicts bounding boxes and probabilities for different classes. Maintaining high accuracy while being substantially faster than previous versions. YOLOv8 incorporates architectural improvements, such as the CSPDarknet53 backbone for feature extraction, which utilizes cross-stage feature fusion to enhance information flow and representation. Additionally, PANet (Path Aggregation Network) improves functionality at different scales, enabling better recognition of objects of various sizes through multi-scale feature fusion (Wang et al., 2024).

YOLOv8 uses various detection scales and a pyramid-like method for detecting the head. This enables the model to detect objects of different sizes efficiently. Additionally, it adds anchor-free object identification, eliminating the requirement for predefined anchor boxes and enhancing generalization to various object aspect ratios. Figure 4 represents Yolov8, an overview of architecture.

(ii) Training and Optimization

To train YOLOv8, a composite loss function that includes terms for objectless scores, bounding box coordinates, and class predictions must be optimized. The capacity of YOLOv8 to handle big datasets efficiently and learn from various examples to enhance generalization is a crucial component of its success.

(iii) Transfer Learning and Fine-Tuning

We leverage the power of transfer learning for our basil crop detection task. We initialized our YOLOv8 Model with pre-trained weights on an extensive dataset. This initialization provides our model with valuable prior knowledge of object detection, accelerating training convergence and enhancing performance.

(iv) Customization for Basil Crop Detection

While YOLOv8 is a versatile object detection model, the specifics of its performance depend on the choice of dataset and task. In our study, we fine-tune the model on our carefully curated basil image dataset, tailoring it to the nuances of basil crop detection. This fine-tuning process ensures that the model accurately identifies basil plants amidst diverse backgrounds and growth stages.

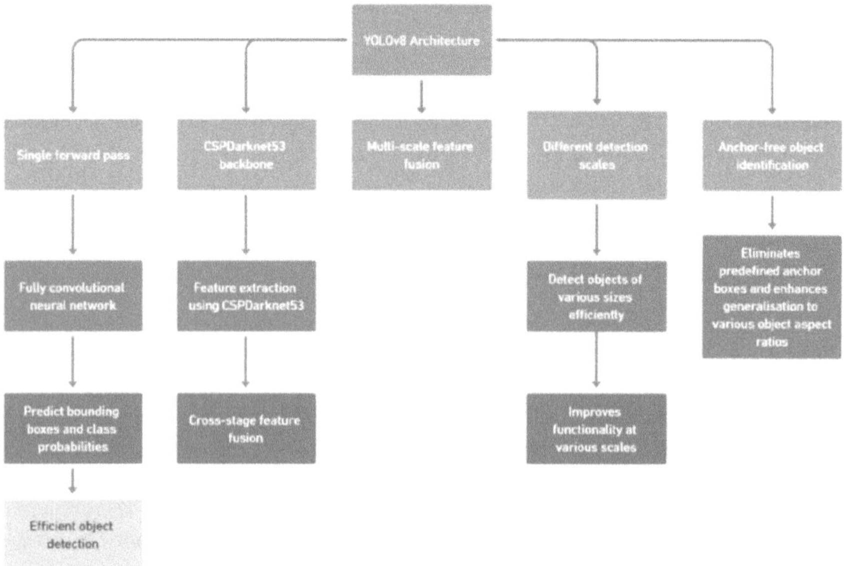

Fig. 4. Yolov8 architecture features.

Results

In this section, we unveil the outcomes of our meticulous efforts in basil crop detection utilizing the YOLOv8 model. Through rigorous dataset preparation, model training, and validation, we have gathered a wealth of data that encapsulates our deep learning solution's performance, strengths, and areas for improvement (Dai et al., 2021). The results presented here culminate extensive experimentation, representing a significant step towards revolutionizing precision agriculture. Our findings provide detailed insights into the model's accuracy and efficiency and form the basis for an in-depth discussion on the implications of our work in the broader context of agricultural technology and sustainable farming practices. The section begins with thoroughly presenting the model's training and testing results, detailing evaluation metrics like precision, recall, mean average precision (mAP), and others to quantitatively measure its performance in detecting basil crops across various conditions. This quantitative foundation is complemented by a qualitative analysis in the discussion, contextualizing the results within existing knowledge and comparing methodologies with prior research in crop detection and deep learning applications in agriculture. Real-world implications are explored, considering impacts on agricultural practices, resource optimization, and sustainability. The discussion doesn't solely focus on limitations and challenges but extends to broader implications, bridging theoretical advancements with practical applications. This comprehensive examination provides valuable insights, raises pertinent questions, and suggests future research avenues and innovative solutions in agricultural technology.

Evaluation Parameters

Assessing our YOLOv8 model's performance in basil crop detection is integral to comprehending its effectiveness and reliability in precision agriculture. We utilize a comprehensive set of evaluation parameters that provide quantitative and qualitative insights into the model's capabilities.

Mean Average Precision (mAP) is a fundamental metric in object detection tasks, representing the model's ability to locate objects with different degrees of precision accurately. We compute mAP at two primary thresholds: at an intersection over union (IoU) of 0.50 (mAP50) and across a broader range from 0.50 to 0.95 (mAP50-95). mAP50 assesses the model's accuracy in detecting basil crops with a moderate level of precision, while mAP50-95 extends this assessment across a wider spectrum of IoU values, offering a more comprehensive view of the model's performance.

Precision evaluates the model's ability to precisely identify basil crops by measuring the proportion of true positive predictions, while recall assesses its capability to locate all actual basil crops in the images among the total actual positives. The F1 score balances precision and recall, offering a single metric capturing the model's performance in both aspects. Training and validation loss metrics, including box loss and classification loss, reflect disparities between model predictions and actual annotations, with lower values indicating closer alignment and improved effectiveness in basil crop detection. These Evaluation parameters collectively offer a detailed understanding of the YOLOv8 model's accuracy and completeness in basil crop detection, showcasing its proficiency across diverse conditions and growth stages in precision agriculture.

Training the Yolov8 Model

Following the preprocessing of images and annotations, we embarked on the training of our YOLOv8 model. The Ultralytics package in Python served as our primary tool for model training. To facilitate this process, we compiled a 'Basil.yaml', a configuration file that defines crucial aspects of the training pipeline. This file specified the locations of the train and veil images and annotations folders, the number of classes, and the class names, enabling our model to recognize basil crops precisely.

Our training dataset comprised 171 images, and the validation dataset included 38 images. After comprehensive experimentation, the optimal batch size was determined to be 19. This configuration was crucial for achieving an optimal balance between computational efficiency and training effectiveness. The model was trained with these configurations for 100 epochs, encompassing numerous iterations where the model learned to detect basil crops within the images. The model refined its predictive capabilities during each epoch, gradually converging towards accurate detections. This process encompassed both the training dataset, used to teach the model, and the validation dataset, employed to assess its performance.

Evaluating the Yolov8 model

Once the model was successfully trained, we subjected it to rigorous evaluation using the validation dataset. The evaluation process involved running the model on the validation dataset and assessing its performance against the ground truth annotations (Li et al., 2020). The results were meticulously calculated and presented, offering insights into the model's precision, recall, f1 score, and other metrics that gauge its proficiency in basil crop detection. The concise Table 2 located below, summarizes the major validation metrics, highlighting the model's precision, recall, F1 Score, mAP50, and mAP50-95 results. These metrics collectively attest to the YOLOv8 model's exceptional performance in basil crop detection, underpinning the core findings of our research.

Table 2. Validation results on evaluation metrics.

Metric	Precision	Recall	F1 Score	mAP50	mAP50-95
Result	0.960	0.987	0.969	0.983	0.895

Table 3 succinctly summarizes the training and validation losses, which are vital indicators of your YOLOv8 model's performance. These results provide insights into the model's efficiency in making predictions that closely align with the ground truth annotations.

Table 3. Validation results on bounding boxes.

Metric	Train	Validation
Box Loss	0.428	0.435
Classification Loss	0.377	0.435
DFL Loss	0.891	0.778

In our study of basil crop detection with YOLOv8, we meticulously track box losses, classification losses, and deformable losses during training and validation. These loss metrics quantify various aspects of the model's performance: box losses assess its ability to localize basil crops accurately, classification losses evaluate its capacity to distinguish basil crops from background elements, and deformable losses add spatial adaptability for diverse basil crop orientations. Monitoring these losses provides nuanced insights into the model's precision and resilience, highlighting its suitability for precision agriculture. They are fundamental pillars in optimizing the YOLOv8 model for basil crop detection, refinement, and enhancing practical viability in real-world scenarios. The evaluation of our YOLOv8 model in basil crop detection includes a visual summary of true and predicted images with corresponding bounding box annotations. These annotated images showcase the model's precision and recall by accurately identifying basil crops across different growth stages and orientations

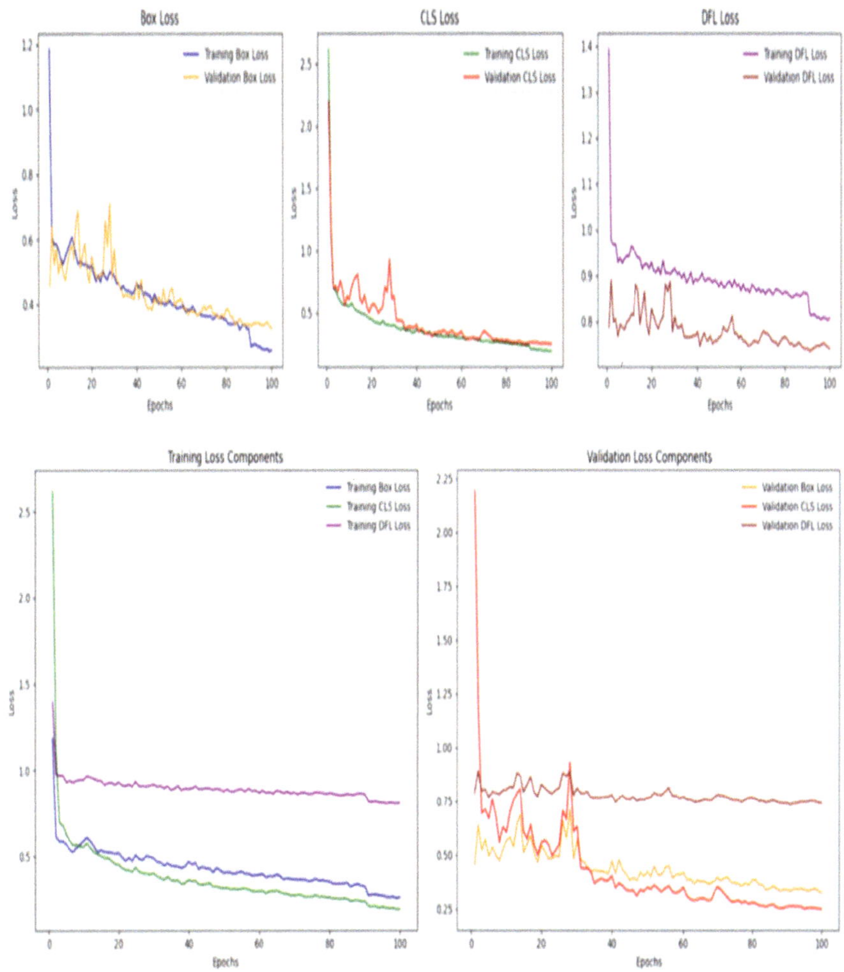

Fig. 5. Losses evaluation over epochs.

(Uddin et al., 2023). The visual representation enhances the empirical evidence of our findings and contributes to the transparency and interpretability of the research. Figure 5 shows loss evaluation over epochs, and Fig. 6 shows evaluation metrics over epochs. Figure 7 shows True and Predicted images that prove the model's performance. This addition ensures a thorough understanding of its practical implications in precision agriculture, making the research accessible and valuable to the scientific community and agricultural practitioners.

Scope and Limitations of the Proposed Model

Our research aims to advance precision agriculture, focusing on basil crop detection using the YOLOv8 deep learning model. It includes developing and evaluating an

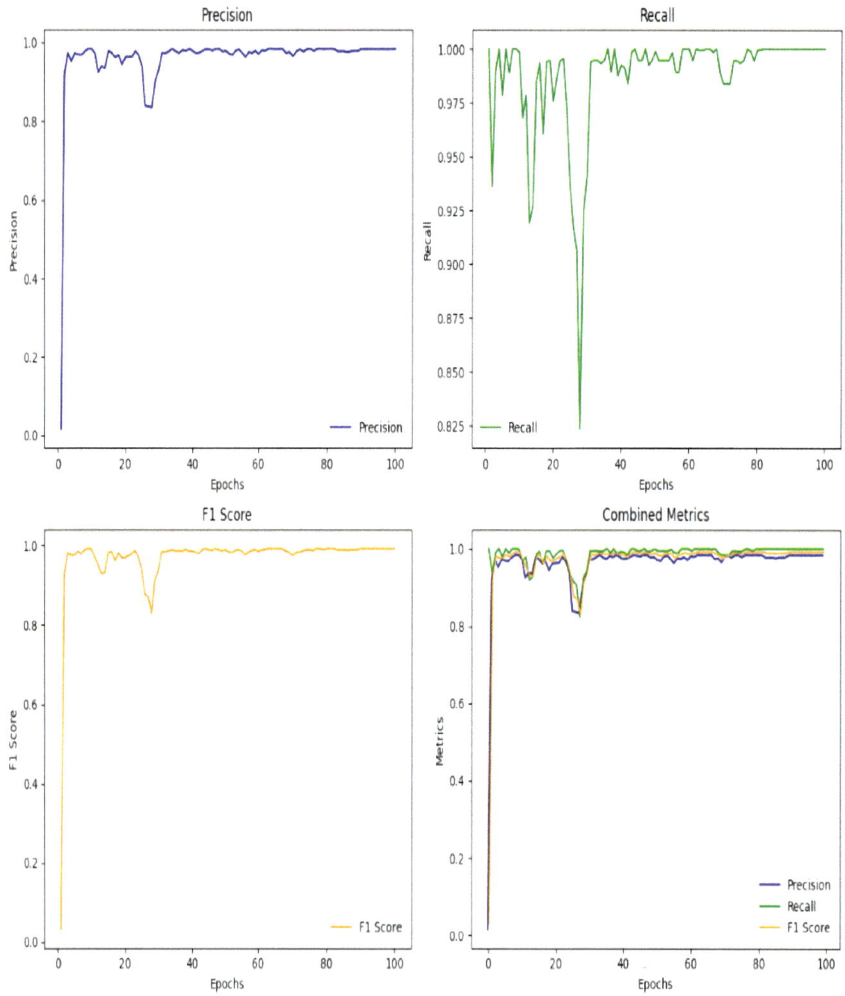

Fig. 6. Evaluation metrics over epochs.

accurate model adaptable to diverse growth stages and environmental conditions. We emphasize the model's configuration, training methodology, and dataset curation, highlighting its transformative potential in agriculture. Future enhancements could extend to other crops and applications like real-time monitoring and decision support systems. Acknowledging limitations, our dataset size is relatively small, and model generalization may not cover all variations. Real-world conditions like unpredictable environments may affect model performance (Gupta, 2016). GPU resources are needed for training, potentially limiting practicality, and broader societal implications should be extensively explored.

Fig. 7. True and predicted images.

Conclusion

This research combines deep learning technology, advanced agricultural practices, and machine vision systems to revolutionize precision agriculture, focusing on basil crop monitoring. The meticulously designed methodology, relying on curated datasets, rigorous model training, and validation, yielded promising results with the YOLOv8 model. The model demonstrated its adaptability and efficiency in basil crop detection by achieving impressive precision, recall, F1 score, and mean average precision metrics. Beyond empirical findings, the research highlights broader implications for enhancing crop management and sustainability practices. It serves as a foundation for future research and innovation in precision agriculture, emphasizing the potential of technology to optimize farming practices. Opportunities for future enhancements in deep learning capabilities and agricultural needs are recognized, signaling continued progress in this field.

References

Abbasi, R., Martinez, P. and Ahmad, R. 2023. Crop diagnostic system: A robust disease detection and management system for leafy green crops grown in an aquaponics facility. *Artificial Intelligence in Agriculture*, 10: 1–12, doi: 10.1016/J.AIIA.2023.09.001.

Anand, R., Mishra, R.K. and Khan, R. 2022. Plant diseases detection using artificial intelligence. *Application of Machine Learning in Agriculture*, pp. 173–190, doi: 10.1016/B978-0-323-90550-3.00007-2.

Badgujar, C.M., Poulose, A. and Gan, H. 2024. Agricultural Object Detection with You Look Only Once (YOLO) Algorithm: A Bibliometric and Systematic Literature Review, Accessed: May 01, 2024. [Online]. Available: http://arxiv.org/abs/2401.10379.

Bouguettaya, A., Zarzour, H., Kechida, A. Amine and Taberkit, M. 2022. A survey on deep learning-based identification of plant and crop diseases from UAV-based aerial images. *Cluster Computing*, 26(2): 1297–1317, doi: 10.1007/S10586-022-03627-X.

Chen, Z., Gao, T., Sheng, B., Li, P., Chen, C.L.P. and Li, P. 2020. Outdoor shadow estimating using multiclass geometric decomposition based on BLS. *IEEE Transactions on Cybernetics*, 50(5): 2020, doi: 10.1109/TCYB.2018.2875983.

Chen, Z., Qiu, G., Li, P., Zhu, L., Yang, X. and Sheng, B. 2023. Transactions on Pattern Analysis and Machine Intelligence, and undefined 2023. Mngnas: Distilling adaptive combination of multiple searched networks for one-shot neural architecture search. *IEEE Transactions on Pattern Analysis and Machine Intelligence*, Accessed: May 10, 2024. [Online]. Available: https://ieeexplore.ieee.org/abstract/document/10179167/.

Chodey, M.D. and Noorullah Shariff, C. 2022. Hybrid deep learning model for in-field pest detection on real-time field monitoring. *Journal of Plant Diseases and Protection*, 129(3): 635–650, doi: 10.1007/S41348-022-00584-W/METRICS.

Chouhan, S.S., Kaul, A. and Singh, U.P. 2019a. Plants leaf segmentation using bacterial foraging optimization algorithm. *2019 International Conference on Communication and Electronics Systems (ICCES)*, pp. 1500–1505, doi: 10.1109/ICCES45898.2019.9002039.

Chouhan, S.S., Kaul, A. and Singh, U.P. 2019b. Radial basis function neural network for the segmentation of plant leaf disease. *2019 4th International Conference on Information Systems and Computer Networks (ISCON)*, pp. 713–716, doi: 10.1109/ISCON47742.2019.9036299.

Dai, L. et al. 2024a. A deep learning system for predicting time to progression of diabetic retinopathy. *Nat. Med.*, 30(2): 584–594, doi: 10.1038/S41591-023-02702-Z.

Dai, L. et al. 2024b. A deep learning system for detecting diabetic retinopathy across the disease spectrum. *Nature Communications*, Accessed: May 10, 2024. [Online]. Available: https://www.nature.com/articles/s41467-021-23458-5.

Dhingra, G., Kumar, V. and Joshi, H.D. 2019. Basil leaves disease classification and identification by incorporating survival of fittest approach. *Chemometrics and Intelligent Laboratory Systems*, 186: 1–11, doi: 10.1016/J.CHEMOLAB.2019.01.006.

Farooqui, N.A., Mishra, A.K. and Mehra, R. 2023. Concatenated deep features with modified LSTM for enhanced crop disease classification. *Int. J. Intell. Robot. Appl.*, 7(3): 510–534, doi: 10.1007/S41315-022-00258-8/METRICS.

Fuentes-Pacheco et al. 2024. A curriculum learning approach to classify nitrogen concentration in greenhouse basil plants using a very small dataset and low-cost RGB images. *IEEE Access*, 12: 27411–27425, doi: 10.1109/ACCESS.2024.3367614.

Giakoumoglou, N., Pechlivani, D. and Tzovaras, D. 2023. Generate-Paste-Blend-Detect: Synthetic dataset for object detection in the agriculture domain. *Smart Agricultural Technology*, 5: 100258, doi: 10.1016/J.ATECH.2023.100258.

Gupta, A.K. 2016. Low contrast image enhancement technique by using fuzzy method. *International Journal of Engineering Research and General Science*, 4(2): 518–526.

Jiang, N., Sheng, B., Li, P. and Lee, T.Y. 2022. Transactions on Multimedia on, and undefined 2022. Photohelper: Portrait photographing guidance via deep feature retrieval and fusion. *IEEE Transactions on Multimedia*, Accessed: May 10, 2024. [Online]. Available: https://ieeexplore.ieee.org/abstract/document/9693256/.

Joshi, B.M. and Bhavsar, H. 2023. A nightshade crop leaf disease detection using enhance-nightshade-CNN for ground truth data. *The Visual Computer*, 2023: 1–20, doi: 10.1007/S00371-023-03127-Y.

Kunduracioglu, I. and Pacal, I. 2024. Advancements in deep learning for accurate classification of grape leaves and diagnosis of grape diseases. *Journal of Plant Diseases and Protection*, pp. 1–20, doi: 10.1007/S41348-024-00896-Z/TABLES/7.

Li, J., Chen, J., Sheng, B., Li, P., Yang, P., Feng, D.D. and Qi, J. 2021. Transactions on industrial informatics, and undefined 2021. Automatic detection and classification system of domestic waste via multimodel cascaded convolutional neural network. *IEEE Transactions on Industrial Informatics*, doi: 10.1109/TII.2021.3085669.

Li, M., Zhang, Z., Lei, L., Wang, X. and Guo, X. 2020. Sensors, and undefined 2020. Agricultural greenhouses detection in high-resolution satellite images based on convolutional neural networks: Comparison of faster R-CNN, YOLO v3 and mdpi.com, Accessed: May 01, 2024. [Online]. Available: https://www.mdpi.com/1424-8220/20/17/4938.

Lin, X., Sun, S., Huang, W., Sheng, B., Li., P. and Feng, D.D. 2021. Transactions on Multimedia on, and undefined 2021. EAPT: Efficient attention pyramid transformer for image processing. *IEEE Transactions on Multimedia*, Accessed: May 10, 2024. [Online]. Available: https://ieeexplore.ieee.org/abstract/document/9580642/?casa_token=-4ItzZScddgAAAAA:syEOoVm2z99GNwRrLmidmkxMoMARkAu5HB_NmW2li4SUIJx9jbSeYmvgcnfHERkFX3MI4KiL6t8.

Makri, O. and Kintzios, S. 2008. Ocimum sp. (Basil): Botany, cultivation, pharmaceutical properties, and biotechnology. *J. Herbs Spices Med. Plants*, 13(3): 123–150, doi: 10.1300/J044V13N03_10.

Mao, R. et al. 2023. DAE-Mask: A novel deep-learning-based automatic detection model for in-field wheat diseases. *Precision Agriculture*, 25(2): 785–810, doi: 10.1007/S11119-023-10093-X.

Nazir, A., Cheema, M.N., Sheng, B., Li, H., Li, P., Yang, P., Jung, Y., Qin, J., Kim, J. and Feng, D.D. 2020. OFF-eNET: An optimally fused fully end-to-end network for automatic dense volumetric 3D intracranial blood vessels segmentation. *IEEE Transactions on Image Processing*, doi: 10.1109/TIP.2020.2999854.

Qin, Y., Zhao, N., Yang, J., Pan, S., Sheng, B. and Lau, R.W.H. 2024. UrbanEvolver: Function-aware urban layout regeneration. *Int. J. Comput. Vis.*, doi: 10.1007/S11263-024-02030-W.

Saini, P. and Nagpal, B. 2023. PSO-CNN-Bi-LSTM: A hybrid optimization-enabled deep learning model for smart farming. *Environmental Modeling and Assessment*, 1: 1–18, doi: 10.1007/S10666-023-09920-2/FIGURES/9.

Sarkar, C., Gupta, D., Gupta, U. and Hazarika, B.B. 2023. Leaf disease detection using machine learning and deep learning: Review and challenges. *Appl. Soft Comput.*, 145: 110534, doi: 10.1016/J.ASOC.2023.110534.

Sheng, B., Li, P., Jin, Y., Tan, P. and Lee, T.Y. 2024. Transactions on Visualization and Computer Graphics on, and undefined 2018. Intrinsic image decomposition with step and drift shading separation. *IEEE Transactions on Visualization and Computer Graphics*, Accessed: May 10, 2024. [Online]. Available: https://ieeexplore.ieee.org/abstract/document/8457312/.

Sheng, B., Li, P., Ali, R. and Chen C.L.P. 2021. Transactions on Cybernetics on, and undefined 2021. Improving video temporal consistency via broad learning system. *IEEE Transactions on Cybernetics*, doi: 10.1109/TCYB.2021.3079311.

Singh, A., Kaur, J., Singh, K. and Singh, M.L. 2023. Deep transfer learning-based automated detection of blast disease in paddy crop. *Signal, Image and Video Processing*, 18(1): 569–577, doi: 10.1007/S11760-023-02735-4.

Sipos, L. et al. 2021. Optimization of basil (*Ocimum basilicum* L.) production in LED light environments—A review. *Sci. Hortic*, 289: 110486, doi: 10.1016/J.SCIENTA.2021.110486.

Souza, A., Rojas, M., Yang, Y. and Lee, L. 2022. Hoag and Heliyon, and undefined 2022. Classifying cadmium contaminated leafy vegetables using hyperspectral imaging and machine learning, doi: 10.1016/j.heliyon.2022.e12256.

Sukhadeo Bajait, V., Malarvizhi, N. and Schulz, M. 2023. Taylor Remora optimization enabled deep learning algorithms for percentage of pesticide detection in grapes. *Environmental Science and Pollution Research*, 1: 1–23, doi: 10.1007/S11356-023-30169-5.

Uddin, A., Chen, Y.L., Borkatullah, B., Khatun, M.S., Ferdous, J., Mahmud, P., Yang, J. and Ku, C.S. 2023. Mathematics, and undefined 2023. Deep-learning-based classification of Bangladeshi medicinal plants using neural ensemble models, Accessed: May 01, 2024. [Online]. Available: https://www.mdpi.com/2227-7390/11/16/3504.

Wang, G., Chen, Y., An, P., Hong, H., Hu, J. and Huang, T. 2023. UAV-YOLOv8: A small-object-detection model based on improved YOLOv8 for UAV aerial photography scenarios, Accessed: May 01, 2024. [Online]. Available: https://www.mdpi.com/1424-8220/23/16/7190.

Xie, Z., Zhang, W., Sheng, B., Li, P. and Chen, C.L.P. 2021. on, and undefined 2021. BaGFN: Broad attentive graph fusion network for high-order feature interactions. *IEEE Transactions on Neural Networks and Learning Systems*, Accessed: May 10, 2024. [Online]. Available: https://ieeexplore.ieee.org/abstract/document/9565146/.

Yang, G., Wang, J., Nie, Z., Yang, H. and Yu, S. 2023a. Agronomy, and undefined 2023. A lightweight YOLOv8 tomato detection algorithm combining feature enhancement and attention, Accessed: May 01, 2024. [Online]. Available: https://www.mdpi.com/2073-4395/13/7/1824.

Yang, W. et al. 2023b. Deformable convolution and coordinate attention for fast cattle detection. *Comput. Electron. Agric.*, 211: 108006, doi: 10.1016/J.COMPAG.2023.108006.

Zhang, B., Sheng, B., Li, P. and Lee, T.Y. 2019. Transactions on Visualization and Computer Graphics on, and undefined 2019. Depth of field rendering using multilayer-neighborhood optimization. *IEEE Transactions on Visualization and Computer Graphics*, Accessed: May 10, 2024. [Online]. Available: https://ieeexplore.ieee.org/abstract/document/8624466/.

Zhou, F., Chao, Y., Wang, C., Zhang, X., Li, H. and Song, X. 2023. A small sample nonstandard gear surface defect detection method. *Measurement*, 221: 113472, doi: 10.1016/J.MEASUREMENT.2023.113472.

Zhu, F., Qiao, X., Zhang, Y. and Jiang, J. 2024. Analysis and mitigation of illumination influences on canopy close-range hyperspectral imaging for the *in situ* detection of chlorophyll distribution of basil crops. *Comput. Electron. Agric.*, 217: 108553, doi: 10.1016/J.COMPAG.2023.108553.

Chapter 5

Remote Sensing for Agriculture Mapping

***Kurru Charitha,*[1,*] *Mohammed Umar Ali*[2] and *Chitteti Divya*[3]**

Introduction

"Remote sensing is a method of obtaining information about the properties of an object without coming into physical contact with it". It has become an indispensable tool in modern agriculture, offering insights into crucial aspects such as crop health, soil conditions, water availability, and environmental factors. This technology, utilizing sensors deployed on satellites, drones, and other platforms, has evolved significantly over time due to advancements in sensor technology, data processing algorithms, and satellite capabilities (Coppin and Bauer, 1996). Early satellite missions such as Landsat laid the groundwork by providing initial imagery for land cover and crop monitoring, marking the genesis of remote sensing's integration into agricultural practices. Subsequent developments in high-resolution optical sensors, thermal sensors, and radar sensors have enabled the collection of detailed data on various agricultural parameters, including crop health metrics, soil moisture levels, temperature fluctuations, and pest infestations (Wu et al., 2015).

This chapter aims to explore the diverse applications of remote sensing in agriculture mapping, with a particular emphasis on key areas such as crop health monitoring and soil analysis (Wu et al., 2015). By harnessing remote sensing data, farmers and agricultural stakeholders can make more informed decisions about

[1] Ph.D. Research Scholar, Department of Entomology, College of Agriculture, RVSKVV, Gwalior, Madhya Pradesh, India.

[2] M.Sc. Agricultural Science, University of Southern Queensland, Australia.

[3] PG Scholar, Department of Entomology, Naini Agricultural Institute, Allahabad, Uttar Pradesh, India.

* Corresponding author: Charithak2021@outlook.com

irrigation scheduling, fertilizer application, and pest management, resulting in increased productivity and optical resource utilization.

Moreover, this chapter seeks to highlight the transformative impact of remote sensing technology on sustainability within the agricultural sector. Through research studies and successful case examples, we will demonstrate how remote sensing tools have contributed to sustainable farming practices by reducing water consumption, minimizing chemical inputs, and fostering soil health (Nilsson et al., 2003; Elhag et al., 2019; Hansen et al., 2013). For instance, remote sensing techniques have been instrumental in estimating evapotranspiration rates, water use efficiency, and vegetation health, providing valuable insights for water resource management and environmental conservation initiatives (Wu et al., 2015).

Historical Overview

The integration of remote sensing technology into agriculture has a rich historical background marked by significant milestones and advancements. Early endeavours in remote sensing for agricultural purposes can be traced back to the 1960s and 1970s, with the launch of pioneering satellite missions such as Landsat. Landsat-1, launched in 1972, provided the first multispectral imagery of Earth's surface, laying the foundation for land cover mapping and crop monitoring (Campbell and Wynne, 2011). Subsequent Landsat missions, including Landsat-2 and Landsat-3, continued to contribute valuable data for agricultural applications, albeit with limited spectral and spatial resolution capabilities.

The 1980s witnessed a paradigm shift in remote sensing technology with the launch of more advanced satellites like SPOT (Satellite Pour Observation de la Terre) by France and the launch of the first synthetic aperture radar (SAR) satellite, Seasat, by NASA. These satellites offered higher spatial resolution and specialized sensors, enabling more detailed observations of agricultural landscapes, crop types, and vegetation health (Merchant, 2000).

The 1990s marked a period of significant technological advancement and diversification in remote sensing applications for agriculture. The launch of the Advanced Very High-Resolution Radiometer (AVHRR) on NOAA satellites provided global coverage for monitoring vegetation dynamics, land surface temperatures, and drought conditions (Justice et al., 1998). Additionally, the development of hyperspectral imaging sensors allowed for more precise spectral analysis of agricultural fields, leading to improved crop health assessments and disease detection capabilities (Thenkabail et al., 2000).

In the 21st century, the proliferation of commercial satellite providers, along with advancements in unmanned aerial vehicles (UAVs) or drones, revolutionized the accessibility and affordability of high-resolution remote sensing data for farmers and agricultural researchers. The launch of satellites with multispectral and hyperspectral sensors, such as Sentinel-2 and WorldView series, further enhanced the capabilities of remote sensing in agriculture, enabling detailed crop monitoring, soil analysis, and precision farming practices (Yang et al., 2018).

Overall, the historical evolution of remote sensing in agriculture reflects a trajectory of continuous innovation and technological refinement, from early satellite missions to the current era of high-resolution multispectral and hyperspectral imaging, empowering stakeholders in the agricultural sector with unprecedented insights for sustainable and efficient farming practices.

Principles of Remote Sensing in Agriculture

Remote sensing in agriculture relies on the fundamental principles of electromagnetic radiation, sensor technology, and data interpretation to gather and analyse information about agricultural landscapes. The following principles are key to understanding how remote sensing is applied in agricultural contexts:

1. **Electromagnetic Spectrum:** Remote sensing sensors detect electromagnetic energy produced or absorbed by entities on the globe's surfaces. Different wavelengths within the electromagnetic spectrum provide unique information about agricultural features such as vegetation health, soil properties, and water content (Merchant, 2000).
2. **Sensor Technology:** Remote sensing platforms, including satellites, drones, and airborne sensors, carry specialized instruments such as multispectral and hyperspectral cameras, thermal sensors, and radar systems. These sensors capture data across various spectral bands, allowing for the analysis of specific agricultural parameters (Thenkabail et al., 2000).
3. **Spectral Resolution:** The spectral resolution of remote sensing instruments refers to their ability to distinguish between different wavelengths of electromagnetic radiation. Higher spectral resolution enables detailed analysis of vegetation indices, soil composition, and other agricultural variables (Campbell and Wynne, 2011).
4. **Spatial Resolution:** Spatial accuracy enhances the amount of detail obtained in remote sensing images. High spatial resolution is crucial for identifying individual crops, monitoring field boundaries, and assessing crop health at a localized level (Merchant, 2000).
5. **Temporal Resolution:** Temporal resolution refers to the frequency at which remote sensing data is acquired over time. Regular data collection allows for monitoring seasonal changes, crop growth cycles, and dynamic agricultural processes (Campbell and Wynne, 2011).

How Remote Sensing Works

Satellite imagery is a powerful tool for acquiring data regarding entities or places from far away, usually by aviation or spacecraft. The process involves several key steps, each critical to acquiring, processing, and interpreting the data needed for various applications, including agriculture mapping. Here's an overview of how remote sensing works:

1. **Energy Source and Illumination:** Remote sensing relies on electromagnetic energy, usually sunlight or emitted radiation, as the primary source. The energy interacts with the Earth's surface and is either reflected, absorbed, or transmitted. Active remote sensing systems, such as radar and LiDAR, emit their energy and measure its reflection from the target (Yang, 2018).

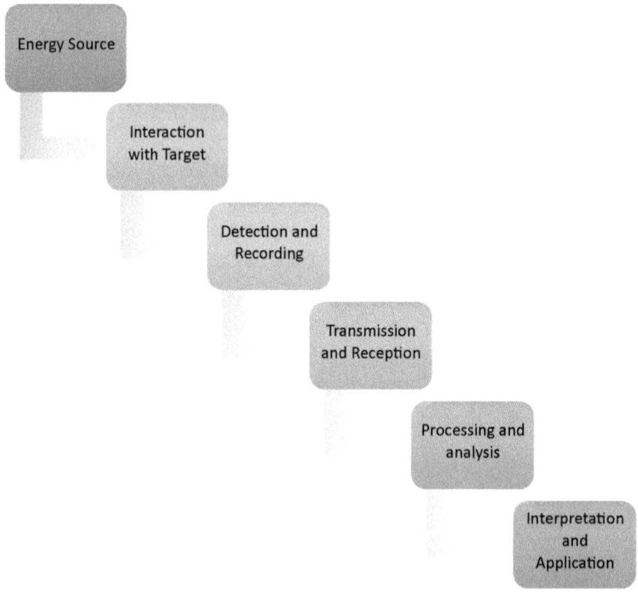

Fig. 1. Process of Remote Sensing.

2. **Interaction with the Target:** The energy interacts with the objects on the Earth's surface. Different materials reflect, absorb, and emit energy differently across various wavelengths of the electromagnetic spectrum. Vegetation, for example, has distinct reflectance properties in the visible and near-infrared regions, which are critical for assessing plant health (Thenkabail et al., 2011).
3. **Detection and Recording:** Sensors onboard satellites, aircraft, or drones detect the reflected or emitted energy. These sensors can be passive (measuring natural energy) or active (emitting energy and measuring the return). They record the energy in multiple spectral bands, creating multispectral or hyperspectral images (Campbell and Wynne, 2011).
4. **Transmission and Reception:** The collected data is transmitted to ground stations for processing. Modern satellites often use high-speed data links to ensure timely delivery of the large volumes of data they collect (Lillesand et al., 2015).
5. **Processing and Analysis:** The raw data received is processed to correct for distortions and extract meaningful information. This includes radiometric correction (to adjust for sensor noise and atmospheric effects), geometric correction (to align images with geographical coordinates), and spectral analysis (to identify specific features based on their spectral signatures) (Yang, 2018).
6. **Interpretation and Application:** The processed data is interpreted using various analytical techniques to derive actionable insights. In agriculture, this could involve generating vegetation indices like NDVI (Normalized Difference Vegetation Index) to assess crop health, mapping soil properties, or monitoring

water resources. Advanced machine learning algorithms and GIS tools are increasingly used to enhance the accuracy and utility of remote sensing data (Najafzadeh and Basirian, 2023).

Types of Remote Sensing Data

Remote sensing in agriculture utilizes various types of data collected through sensors deployed on satellites, drones, and other platforms. These data types provide valuable information about agricultural landscapes, crop health, soil conditions, and environmental factors. The key types of remote sensing data used in agriculture are as follows:

1. **Optical Imagery:** Optical sensors capture visible and near-infrared light, allowing for the analysis of vegetation health, crop types, and land cover characteristics. Multispectral and hyperspectral imagery provide detailed spectral information for vegetation indices and agricultural mapping (Thenkabail et al., 2000).
2. **Thermal Imagery:** Thermal sensors detect emitted infrared radiation, which correlates with surface temperatures. Thermal imagery is used to assess crop stress, water availability, and microclimate variations in agricultural fields (Kustas and Anderson, 2009).
3. **Radar Imagery:** Radar sensors transmit microwave pulses and measure their reflections, penetrating through cloud cover and vegetation. Radar imagery is valuable for monitoring soil moisture levels, land surface roughness, and crop structure in agricultural areas (McNairn and Shang, 2016).
4. **LiDAR Data:** LiDAR (Light Detection and Ranging) sensors emit laser pulses and measure their return time, providing detailed 3D elevation data and vegetation height profiles. LiDAR data are used for terrain modeling, canopy structure analysis, and precision agriculture applications (Pfeifer and Briese, 2007).
5. **Multisensor Fusion:** Integrating data from multiple sensors, such as optical, thermal, and radar, enables comprehensive analysis and enhanced understanding of agricultural landscapes. Multisensor fusion techniques improve accuracy in crop monitoring, soil mapping, and environmental assessments (Najafzadeh and Basirian, 2023).

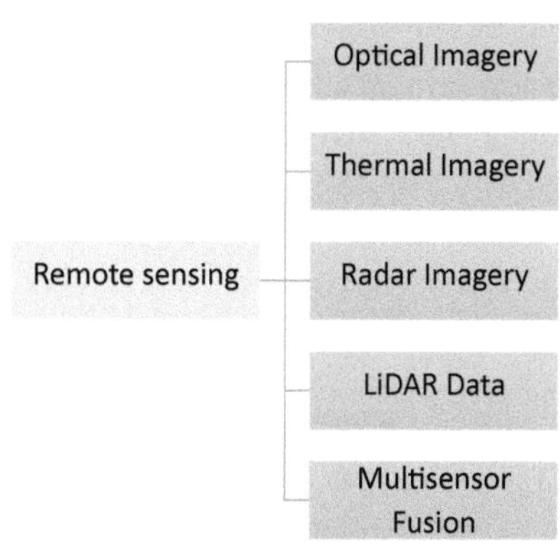

Fig. 2. Types of remote sensing data.

Applications of Remote Sensing in Agriculture Mapping

Remote sensing technology has numerous applications in agriculture mapping, offering valuable insights into crop health, soil conditions, water management, and land use planning. The following are key applications of remote sensing in agriculture mapping:

1. **Crop Health Monitoring:** Remote sensing enables the assessment of crop health and vigour by analysing vegetation metrics such as NDVI (Normalized Difference Vegetation Index) and NDWI (Normalized Difference Water Index). These indices help identify areas of stress, disease, or nutrient deficiencies in crops (Najafzadeh and Basirian, 2023).
2. **Soil Mapping and Analysis:** Remote sensing techniques are used to map soil properties, moisture content, and fertility levels across agricultural fields. Soil mapping aids in precision agriculture practices, optimizing irrigation and fertilization strategies (Thenkabail et al., 2000).
3. **Yield Prediction:** By analysing satellite imagery and historical data, remote sensing contributes to accurate yield prediction models. Yield prediction helps farmers optimize harvest planning, resource allocation, and marketing strategies (Yang, 2018).
4. **Pest and Disease Detection:** Early detection of pests and diseases is facilitated through remote sensing techniques. Monitoring changes in vegetation health and using spectral signatures help identify areas affected by pests or diseases, allowing for timely interventions (Elhag et al., 2019).
5. **Water Management:** Remote sensing data is utilized for efficient water management in agriculture. It helps monitor irrigation levels, detects water stress in crops, assesses water quality in rivers and reservoirs, and optimizes water resource allocation (Najafzadeh and Basirian, 2023).
6. **Land Use and Crop Zoning:** Remote sensing supports land use planning and crop zoning by identifying suitable areas for specific crops based on environmental conditions, soil suitability, and historical performance (Campbell and Wynne, 2011).

Challenges and Future Directions

While remote sensing has revolutionized agriculture mapping, several challenges remain, along with exciting future directions that promise further advancements. Here are some key challenges and future directions:

Challenges

1. **Data Interpretation Complexity:** Analysing large volumes of remote sensing data and extracting meaningful insights can be complex and time-consuming, requiring advanced data processing techniques and skilled personnel (Najafzadeh and Basirian, 2023).

2. **Cost Barriers:** Acquiring high-resolution remote sensing data and implementing sophisticated technologies can be costly, posing barriers for small-scale farmers and resource-constrained regions (Yang, 2018).
3. **Integration with Traditional Methods:** Integrating remote sensing data with traditional agricultural practices and farmer knowledge can be challenging, requiring effective communication and training initiatives (Elhag et al., 2019).
4. **Temporal and Spatial Resolution:** Balancing the trade-off between temporal and spatial resolution in remote sensing imagery to capture both detailed information and frequent updates remains a challenge (Campbell and Wynne, 2011).

Potential Instructions

1. **Machine Learning as well as AI Collaboration:** The combination of computational intelligence (AI) and machine learning techniques with remote sensing data holds immense potential for automated data analysis, pattern recognition, and predictive modelling in agriculture (Yang, 2018).
2. **Hyperspectral Imaging:** Advancements in hyperspectral imaging technology offer increased spectral resolution and finer spectral bands, enabling more precise analysis of agricultural parameters and vegetation characteristics (Najafzadeh and Basirian, 2023).
3. **Sensor Miniaturization and Cost Reduction:** Continued efforts in sensor miniaturization and cost reduction will make remote sensing technologies more accessible and affordable for a wider range of farmers and agricultural stakeholders (Yang, 2018).
4. **Precision Agriculture Applications:** Further development of precision agriculture techniques, such as variable rate technology (VRT) and automated farming systems, will enhance the efficiency and sustainability of agricultural practices (Campbell and Wynne, 2011).

By tackling these difficulties and looking ahead, remote sensing technology will continue playing a transformative role in agriculture mapping and encouraging sustainable farming methods.

Conclusion

In conclusion, remote sensing technology has revolutionized agriculture mapping by providing valuable insights into crop health, soil conditions, water management, and land use planning. Through applications such as crop health monitoring, soil mapping, yield prediction, pest detection, water management, and land use planning, remote sensing has significantly enhanced decision-making processes for farmers, agronomists, and policymakers.

Despite challenges such as data interpretation complexity, cost barriers, and integration with traditional methods, the future of remote sensing in agriculture looks promising. Advancements in artificial intelligence, machine learning, hyperspectral imaging, sensor miniaturization, and precision agriculture techniques offer exciting opportunities for automated data analysis, improved spectral resolution, cost reduction, and enhanced farming efficiency.

By addressing challenges and embracing future directions, remote sensing technology can continue to play an important role in promoting sustainable, efficient farming practices globally. The integration of remote sensing with traditional agricultural knowledge and practices, coupled with ongoing advancements in technology, will further propel the agriculture industry towards greater productivity, resource optimization, and environmental sustainability.

References

Campbell, J.B. and Wynne, R.H. 2011. *Introduction to Remote Sensing*. Guilford Press.

Coppin, P.R. and Bauer, M.E. 1996. Digital change detection in forest ecosystems with remote sensing imagery. *Remote Sensing Reviews*, 13(3-4): 207–234.

Elhag, M., Gitas, I., Othman, A., Bahrawi, J. and Gikas, P. 2019. Assessment of water quality parameters using temporal remote sensing spectral reflectance in arid environments, Saudi Arabia. *Water*, 11(3): 556.

Hansen, M.C., Potapov, P.V., Moore, R., Hancher, M., Turubanova, S.A., Tyukavina, A., Thau, D., Stehman, S.V., Goetz, S.J., Loveland, T.R., Kommareddy, A., Egorov, A., Chini, L., Justice, C.O. and Townshend, J.R. 2013. High-resolution global maps of 21st-century forest cover change. *Science*, 342(6160): 850–853.

Justice, C.O., Vermote, E., Townshend, J.R., Defries, R., Roy, D.P., Hall, D.K., Salomonson V.V., Privette, L., Riggs, G., Stragler, A., Lucht, W., Myneni, R.B., Knyazikhin Y., Running, S.W., Nemani, R.R., Wan, Z., Huete, A.R., Leeuwen, W.V., Wolfe, R.E., Giglio, L., Muller, J., Lewis and Barnsley, M.J. 1998. The Moderate Resolution Imaging Spectroradiometer (MODIS): Land remote sensing for global change research. *IEEE Transactions on Geoscience and Remote Sensing*, 36(4): 1228–1249.

Kustas, W. and Anderson, M. 2009. Advances in thermal infrared remote sensing for land surface modeling. *Agricultural and Forest Meteorology*, 149(12): 2071–2081.

Lillesand, T., Kiefer, R.W. and Chipman, J. 2015. *Remote Sensing and Image Interpretation*. John Wiley & Sons.

McNairn, H. and Shang, J. 2016. A review of multitemporal synthetic aperture radar (SAR) for crop monitoring. *Multitemporal Remote Sensing: Methods and Applications*, 317–340.

Merchant, J.W. 2000. Remote sensing of the environment: An earth resource perspective. *Cartography and Geographic Information Science*, 27(4): 311–311.

Najafzadeh, M. and Basirian, S. 2023. Evaluation of river water quality index using remote sensing and artificial intelligence models. *Remote Sensing*, 15(9): 2359.

Nilsson, C., Pizzuto, J.E., Moglen, G.E., Palmer, M.A., Stanley, E.H., Bockstael, N.E. and Thompson, L.C. 2003. Ecological forecasting and the urbanization of stream ecosystems: Challenges for economists, hydrologists, geomorphologists, and ecologists. *Ecosystems*, 6: 659–674.

Pfeifer, N. and Briese, C. 2007. Geometrical aspects of airborne laser scanning and terrestrial laser scanning. *International Archives of Photogrammetry, Remote Sensing and Spatial Information Sciences*, 36(3/W52): 311–319.

Thenkabail, P.S., Smith, R.B. and De Pauw, E. 2000. Hyperspectral vegetation indices and their relationships with agricultural crop characteristics. *Remote Sensing of Environment*, 71(2): 158–182.

Wu, X., Zhou, J., Wang, H., Li, Y. and Zhong, B. 2015. Evaluation of irrigation water use efficiency using remote sensing in the middle reach of the Heihe river, in the semi-arid Northwestern China. *Hydrological Processes*, 29(9): 2243–2257.

Yang, C. 2018. High resolution satellite imaging sensors for precision agriculture. *Frontiers of Agricultural Science and Engineering*, 5(4): 393–405.

Chapter 6

Advances of Remote Sensing Technologies in Agriculture

Current Progress and Future Perspectives

Bhabani Prasad Mondal,[1] ***Sumanta Das,***[2,4,*]
Suman Dutta[3] **and** ***Ali RA Moursy***[5]

Introduction

Agriculture, a pivotal driver of economic development in nations such as India, is presently experiencing its fourth revolution, commonly referred to as Agriculture 4.0, due to swift technological advancements. Remote sensing (RS) technologies have revolutionized various fields, including agriculture, by offering crucial information about crop conditions, soil health, and environmental parameters influencing agricultural productivity. Additionally, RS techniques can be combined with advanced technologies like global positioning systems (GPS), the Internet of Things (IoT), and artificial intelligence (AI) to create powerful tools for the precise management of agricultural inputs. Given the growing global population and the resulting food demand, there is an urgent need for innovative solutions to enhance

[1] Department of Soil Science and Agricultural Chemistry, Bihar Agricultural University, Sabour, Bhagalpur-813210.

[2] School of Environment and Disaster Management, Ramakrishna Mission Vivekananda Educational and Research Institute, Kolkata, West Bengal 700103, India.

[3] Dept. of Genetics and Plant Breeding, Ramakrishna Mission Vivekananda Educational and Research Institute, Kolkata, West Bengal 700103, India.

[4] The University of Queensland, St Lucia, Brisbane-4072 Queensland, Australia.

[5] Soil and Water Department, Faculty of Agriculture, Sohag University, 82524, Sohag, Egypt.

* Corresponding author: sumanta.das@uq.net.au

agricultural efficiency, sustainability, and resilience. These technologies together offer a powerful toolset to address these challenges by facilitating thorough monitoring and management of agricultural systems and by decreasing the expenses of cultivation and yield loss while concurrently enhancing production and economic gains (Elijah et al., 2018; Jha et al., 2019). Projections suggest that adopting precision agriculture practices could potentially boost global yield by as much as 67% by the year 2050 (Rosegrant, 2014). As per the 2018 report from the California Department of Food and Agriculture, a pivotal aspect of PA involves the extensive gathering of data on a large scale, primarily achieved through RS techniques, serving as the fundamental basis of the entire PA procedure.

The past decade has witnessed significant advancements in remote sensing technologies, ranging from traditional satellite imagery to emerging platforms such as unmanned aerial vehicles (UAVs) and hyperspectral sensors. Remote sensing involves collecting or measuring data about specific characteristics of a body or event using a recording device that does not have direct bodily contact or proximity to the subject being studied (ASPRS) (Estes et al., 2001). This process involves employing instrumentation from a distance. This RS technique throws electromagnetic radiation to capture diverse features of soil, plants, or other objects or phenomena (Jones and Vaughan, 2010). Various platforms such as ground-based, airborne, and spaceborne platforms carry a range of sensors of RS systems to collect data about remote objects or phenomena (Sishodia et al., 2020). These technological innovations have opened new avenues for agricultural research and management, facilitating the gathering of high-resolution spatial and temporal information at different scales. For instance, satellites equipped with multispectral sensors, such as Landsat and Sentinel, provide consistent and wide-area coverage, allowing for the monitoring of large-scale agricultural landscapes (Foody, 2019). On the other hand, UAVs offer flexibility and higher spatial resolution, enabling detailed assessments of crop health and variability within smaller agricultural fields (Colomina and Molina, 2014).

While numerous studies have discovered the usage of remote sensing in the domain of farming, there is a growing demand for comprehensive reviews that synthesize recent advances, identify current challenges, and outline future research directions. Such reviews play a crucial role in guiding researchers, policymakers, and practitioners in leveraging remote sensing technologies to address key agricultural issues, including yield optimization, resource management, and climate change adaptation. By critically evaluating the strengths and limitations of existing methodologies, these reviews pave the way for the progress of more efficient and robust remote sensing-based agricultural monitoring systems.

Furthermore, the complexity of agricultural systems necessitates interdisciplinary approaches that integrate remote sensing with other scientific disciplines, such as agronomy, ecology, and computer science. Researchers can obtain important insights into variables such as soil, plants, climate, and other environmental factors by integrating RS data with laboratory-based measurements and modeling techniques (Tucker et al., 2018). This interdisciplinary approach is essential for enhancing the accuracy and reliability of remote sensing-based agricultural assessments, ultimately supporting informed decision-making and sustainable agricultural practices.

Moreover, the need for sustainable agriculture has prompted increased interest in utilizing remote sensing technologies to monitor and mitigate environmental impacts associated with agricultural activities. From assessing soil erosion and nutrient runoff to monitoring deforestation and habitat loss, remote sensing offers valuable insights into the ecological footprint of agriculture (Turner et al., 2015). By combining environmental monitoring with agricultural management strategies, stakeholders can work towards enhancing productivity through improved land management practices while safeguarding natural environmental resources.

In this context, this chapter seeks to offer a complete summary of the most recent advances in RS technologies for farming, highlighting their applications, limitations, and prospects. By synthesizing findings from diverse research domains, this review will contribute to the collective understanding of how RS can be efficiently exploited to address the evolving challenges facing modern agriculture. Additionally, by identifying knowledge gaps and emerging trends, this review will guide future research efforts aimed at harnessing the full potential of remote sensing technologies to support sustainable and resilient agricultural systems.

The key research questions addressed in this review are:

- What are the latest advancements in remote sensing technologies for agriculture?
- How can RS technologies be effectively integrated with other advanced technologies for precision agriculture?
- What are the primary challenges and limitations of using RS in agricultural monitoring and management?
- How can RS contribute to sustainable agriculture practices and mitigate environmental impacts?
- What future research directions should be pursued to enhance the efficacy and adoption of RS technologies in agriculture?

Remote Sensing Data Acquisition—Platforms and Sensors

RS Platforms

The platforms employed in remote sensing, such as aircraft or satellites, are referred to as vehicles capable of transporting sensing devices, known as sensors, to conduct measurements from distant locations (Jafarbiglu and Pourreza, 2022). These platforms are constantly getting better in terms of operational speed, consistency, ease of use, and temporal resolution, which ultimately determines spatial resolution. As per Huang et al. (2018), RS systems possess the capability to produce extensive amounts of spectral data characterized by high resolutions, ranging from spatial to temporal resolutions which are particularly crucial for implementation in PA.

Space-based Platforms

Research efforts based on remote sensing began in the 1970s with the introduction of Landsat 1, which was designed to collect data from space-based remote sensing platforms. The rapid development of digital image processing has hastened the advancement of RS for various applications. Satellite imagery offers the advantage of extensive spatial coverage. Despite this advantage, its use in precision agriculture is still limited due to its coarse spatial resolution, which hinders its suitability for specific agricultural operations at the field level (Stafford, 2000). Another drawback of satellite-based measurements is their reliance on weather and atmospheric conditions, which can result in the attenuation of electromagnetic signals traveling from the object to the satellite sensor (Rees, 2013). Despite encountering numerous challenges, satellite imagery remains effective for large-scale mapping of diverse phenomena, including mapping of land uses, and land cover, mapping of soil erosion, desertification evaluation, regional yield estimation, and monitoring climate change (Jafarbiglu and Pourreza, 2022). This effectiveness stems from its ability to offer a comparatively low-priced "macro" perspective of the terrain (Barnes, 2018). Advanced satellite systems with high resolution include a red-edge band to monitor crop growth and are specifically designed for agricultural applications. RapidEye data has been used to evaluate nitrogen levels in wheat crops cultivated in pre-defined management zones (Basso et al., 2013; Basso et al., 2016).

Aerial Platforms

Unlike satellites, unmanned aerial systems (UAS) can capture and deliver high-resolution data, making them especially well-suited for agricultural operations in small and confined fields. However, the primary limitation lies in their restricted payload capacity and brief flight duration, hindering their extensive application. Conversely, manned aerial vehicles occupy an intermediate position, capable of covering larger areas than UAS and providing better resolution than satellites. In the 1920s, agrochemicals were applied to crops using manned aerial vehicles or piloted aircraft, marking the first use of airborne remote sensing in agriculture. As per Rees (2013), the commencement of analog aerial image capture and vegetation mapping began with the progress in aerial platforms such as airplanes and photographic equipment. During the 1970s, the National Aeronautics and Space Administration (NASA) in the United States emerged as a leading entity in utilizing aircraft for aerial remote sensing. They effectively mapped soil moisture for agricultural management by employing the P-3A aircraft, which was outfitted with various remote sensing sensors and operated at approximately 300 meters above ground level (AGL) (Jafarbiglu and Pourreza, 2022). Nevertheless, the utilization of aircraft remote sensing for research endeavors remains restricted, primarily owing to numerous drawbacks including the elevated expenses associated with sensors and platforms, requisite structure and skilled aviators, the operative intricacy of airlifts, and diminished reproducibility owing to the absence of detailed automatic pilot systems (Barnes, 2018). However, both private and governmental sectors have conducted research utilizing aerial platforms and sensors, yielding encouraging findings. While aerial photography provides excellent spatial resolution, its spectral resolution remains a concern due to the availability of only three basic bands: red (R), green (G), and blue (B). If imaging spectrometers, also known as hyperspectral sensors, were used in aerial platforms instead of multispectral sensors, they could enable the identification of plants' biochemical composition due to their numerous contiguous bands that capture the reflectance spectrum of each pixel (Asner and Martin, 2016). An essential aerial remote sensing platform is AVIRIS (Airborne Visible InfraRed Imaging Spectrometer), and its latest iteration, AVIRIS-NG (AVIRIS-Next Generation), has recently been utilized in diverse applications including geology, ecological research, and agriculture (Lundeen and Gowey, 2021). AVIRIS is capable of capturing 224 consecutive calibrated spectral bands (ranging from 400 to 2500 nm) at varying altitudes across regions encompassing the United States, Canada, and Europe. On the other hand, AVIRIS-NG is a hyperspectral sensor equipped with 425 consecutive hyperspectral bands spanning the visible to near-infrared spectrum (ranging from 380 to 2500 nm), with spatial resolution varying from 3 meters to 18 meters. Recently, the AVIRIS-NG sensor has been extensively utilized in India, covering an area of 22,840 square kilometers across 57 sites during the initial phase of an airborne science campaign. This campaign took place over 84 days, from December 16, 2015, to March 6, 2016, as part of a joint initiative between the Indian Space Research Organisation (ISRO) and NASA known as the HYperSpectral Imaging (HYSI) program. To back the campaign, organized fieldwork and on-site measurements were conducted at these sites, covering a range of topics such as crops, soil, forests, geology, coastal regions, oceans, river water, snow, urban areas, and beyond (Bhattacharya et al., 2019).

Given the drawbacks associated with manned aircraft, researchers are increasingly turning to unmanned aerial systems (UAS) for agricultural applications. Researchers have long documented the real strength of UAS for agrarian uses. Initially, UAS were primarily used for military purposes. Over time, their applications have expanded, and they are now widely utilized in the domain of precision farming. The use of Small UAS, commonly referred to as drones, weighing under 25 kg (55 lbs), began in agricultural operations in the early 2000s (Simpson et al., 2003). Its popularity surged rapidly due to its ability to provide high-resolution imagery, even at the centimeter scale. Eventually, in 2014, MIT's Technology Review acknowledged it as one of the top ten innovative technologies (MIT Technology Review, 2014). The Federal Aviation Administration (FAA) reported in March 2018 that by 2022, there will be 451,800 commercial drones, four times as many as there were 110,604 in 2017. The report also suggests that drone technologies are projected to generate services valued at $127 billion (Mazur et al., 2016). In addition to having high spatial resolution capabilities, farmers or growers can deploy drones as needed to survey fields regularly based on the intended temporal resolution, whether monthly, weekly, daily, or even hourly (Zhao, 2018). Moreover, drone-based remote sensing shows promise in offering comprehensive information on both inter and intra-field crop variability, aiding in the optimization of resource utilization based on the variability within fields. This includes assisting in irrigation scheduling, thereby enhancing water use efficiency (Barnes, 2018). The types of drones commonly used in agriculture include fixed-wing, single-rotor, and multirotor models. The vertical take-off and landing (VTOL) ability, along with the capability of staying in one place in the air without moving up or down rotary drones, have become highly popular in agriculture. However, concerns persist regarding flight endurance and energy efficiency. These challenges can be addressed by integrating fixed-wing and rotary drones, enabling the benefits of both VTOL capability and extended flight endurance to be enjoyed (Jafarbiglu and Pourreza,

2022). Moreover, the increased expense associated with sensors and cameras installed on drones restricts their feasible adoption among farmers and technicians (García-Tejero et al., 2018). In addition to the elevated costs, constrained payload capacity, and abbreviated flight durations, there are further obstacles such as the need for authorization to operate drones beyond visual line-of-sight (BVLOS) and to ascertain maximum flying altitudes. For instance, in India and the US, the maximum flying altitudes are restricted to 120 m and 122 m above ground level, respectively.

Notwithstanding the aforementioned drawbacks and limitations, the inclusion of a high-resolution camera on a drone has the potential to offer exceptionally detailed imagery at the level of 1 cm or finer, thereby rendering them suitable for extensive utilization (Hunt et al., 2010). Furthermore, significant endeavors have been undertaken to improve or extend restricted flight durations. These efforts involve implementing high-density fast-charging batteries, exploring other advanced battery technologies, and integrating solar panels with unmanned aerial systems (UAS) to extend or enhance flight times from 50 minutes to up to 5 hours (Manfreda et al., 2018). Recently, an academic chapter addressed the application of drone technology together with computer vision algorithms in the realm of precision agriculture to successful Agriculture 4.0 (Chouhan 2024).

Ground-based Platforms

Ground-based remote sensing platforms can be classified into three types: (i) handheld devices, (ii) standalone units positioned in the field, and (iii) those affixed to tractors or agricultural machinery, utilized primarily for precision farming purposes. These ground-based RS systems are referred to as 'proximal RS systems' because they collect necessary data close to the target object, such as soil or plants, in contrast to aerial or satellite-based. Active proximal sensors, often referred to as 'on-the-go sensors,' have enabled precision nutrient management (Barker and Sawyer, 2012). The fundamental process begins with diodes emitting light at various wavelengths to illuminate either the land surface or plant canopy. Sensors then capture the reflected light from the object, measuring the spectral reflectance across different wavelengths and recording them as the spectral signature of the object. Several notable instances of proximal sensors include spectroradiometers and Fourier-transform infrared spectrometers (FTIR), capable of application in both fieldwork and laboratory settings, facilitating spectral analysis of soil and crops (Roberts et al., 2010). The evaluation report highlights the significance of active proximal sensors, integral components of precision agriculture, which are affixed to the front of variable rate applicators. These sensors regulate the quantity of fertilizer dispensed, resulting in economic advantages and environmental benefits (van Evert et al., 2012). Alongside active proximal sensing, alternative methods using passive sensors can measure reflected solar radiation and the fluorescence of chlorophyll pigments (Tremblay et al., 2009). All these sensors can be installed on ground-based platforms and applicators that use variable rate technology for agricultural purposes. Farmers interested in adopting precision farming techniques need to prepare various spatial variability maps of soil and plants to assess nutrient variability in the soil and plant requirements for site-specific nutrient management (Hunt and Daughtry, 2018).

RS Sensors

Remote sensing sensors vary according to the spatial, spectral, radiometric, and temporal resolution they provide. The spatial resolution of an RS-based sensor is determined by the pixel size of an image, which corresponds to a specific part on the ground. Sensors characterized by high spatial resolution typically have smaller paths, while those with larger paths usually exhibit lower spatial resolution. The number of bands present within a specific region of the EMR spectrum defines the spectral resolution of an RS-based sensor (Nowatzki et al., 2004). While multispectral sensors provide information across an inadequate number of spectral bands, hyperspectral sensors provide an extensive array of contiguous bands, often with narrow widths (< 20 nm), ranging from 10 to over 400 bands, such as those found in AVIRIS-NG sensors, enabling detailed object analysis (Teke et al., 2013). Various machine learning and deep learning methods, such as random forest and neural networks, are employed to decrease the dimensionality of hyperspectral data while preserving its original variability. This aids in extracting valuable insights about soil and vegetation conditions for modeling purposes (Chang et al., 2020). Radiometric resolution in remote sensing pertains to the sensor's capability to detect and quantify minor variations in electromagnetic radiation intensity, indicating the level of detail it can discern in the captured data. Temporal resolution denotes a specific time interval in which the satellites can complete their movement in the orbital plane and revisit the same place after that certain period. Space-based platforms like satellites and airborne vehicles such as aircraft or UAVs commonly utilize passive sensors, meaning they rely on external light sources rather than generating their own energy. Nevertheless, certain satellites like ERS are equipped with active sensors, specifically the active microwave sensor. Green Seeker and Crop Circle are examples of ground-based platforms equipped with active sensors, and they are integral components of variable rate applicators. As fluctuations in daylight or sunlight intensity minimally affect the reflectance values measured by these active sensors, they can furnish more precise and consistent vegetation indices (VI) (Table 1) enhancing the assessment of crop nutrient status effectively (Sishodia et al., 2020).

In this section, we aim to represent a range of remote sensing sensors that are valuable for measuring various parameters relevant to agriculture.

Visual Digital or Panchromatic Sensors

A digital camera serves as an excellent illustration of a visual sensor, featuring a color filter array like the Bayer pattern, which overlays a chip made up of silicon. In this setup, each pixel is surrounded by two different colors from the filter array (Hunt Jr and Daughtry, 2018). To create an image, the raw image file needs to undergo demosaicing to extract the three primary spectral bands, specifically the RGB bands representing Red (R), Green (G), and Blue (B), i.e., the panchromatic bands. Typically, the camera implements several adjustments such as correcting sensitivity for light, balancing in white colour, removing noise from the dark coloured pixel, etc., to produce an image that is visually appealing to humanoid trichromatic visualization. According to optics principles, flying at a lower altitude with the same camera will produce smaller ground sample distances compared to flying at a higher altitude. A wide range of factors, including cost, lens quality, and features, significantly affect the performance of digital cameras and their visual sensors utilized in aerial remote sensing, spanning from simple point-and-shoot cameras for consumers to high-quality cameras for professionals offering extensive control over outcomes (Crusiol et al., 2017). The key determinant is the lens quality, which plays a crucial role in resolving distinctions between pixels and ensuring image quality. This suggests that the total number of pixels is not always a reliable indicator or determining factor for producing high-quality images with a small ground resolution (Candiago et al., 2015).

Multispectral Sensors

Multispectral sensors generally feature wide bands with large wavelength ranges, typically spanning from 50 to 100 nanometres in width. Multispectral remote sensing sensors employed in agriculture are specialized instruments designed to capture data across specific bands of the electromagnetic spectrum relevant to agricultural applications. These sensors typically collect information in several predetermined spectral bands, such as visible, near-infrared, and sometimes infrared wavelengths. By analyzing the data captured by these sensors, farmers and researchers can gather insights into various agricultural parameters, including crop health, vegetation vigor, moisture levels, and nutrient content of the soil. A multispectral sensor with narrower wavelength bands can offer an enhanced understanding of vegetation conditions by computing vegetation indices, such as the red-edge band responsible for chlorophyll absorption. This capability enables the multispectral sensor to provide reliable insights into the chlorophyll content of plants, similar to hyperspectral sensors (Hunt Jr and Daughtry, 2018). Some multispectral sensors such as Landsat, SPOT, and Sentinel-2 carried on satellite-based platforms have been employed in agricultural applications. After the year 2000, several other spacecraft like GeoEye-1 SkySat-2, and Superview-1 were launched in the year of 2008, 2014, and 2018 respectively. These are examples of nanosatellites and other spacecraft capable of capturing high-resolution multispectral images with a resolution of 2 meters or less. These satellites can be revisited daily or even less frequently.

Hyperspectral Sensors

Hyperspectral sensors are furnished with narrow width, contiguous spectral bands, typically with a spectral width of 10–20 nm. As a result, hyperspectral sensors like Hyperion and spectroradiometers (e.g., Headwall), which contain numerous or even hundreds of bands, are capable of capturing hyperspectral images and providing highly detailed spectral responses of various objects. This capability surpasses that of multispectral images, enabling hyperspectral sensors to more effectively detect detailed variations in surface covers and their changes over time. Therefore, it enables more precise and prompt identification of crop physiological conditions and vegetation parameters, including leaf area index (LAI) estimation, discrimination between crop types, retrieval of crop biomass, and assessment of leaf nitrogen content (Sun et al., 2017). Various platforms including space-based (satellite), airborne (UAS), and ground can be used as bases to install hyperspectral sensors can be installed, to effectively capture images having different resolutions including spatial and temporal resolutions. The space-based hyperspectral sensor Hyperion is predominantly used in agricultural applications. With its 30-meter spatial resolution and 10-nanometer spectral resolution, Hyperion collects data across a broad spectrum, including visible, near-infrared, and shortwave-infrared regions, making it highly effective for agricultural applications (Lu et al., 2020). The first airborne hyperspectral sensor, the Airborne Visible/Infrared Imaging Spectrometer (AVIRIS), introduced in 1987, had 224 bands spanning from visible to shortwave infrared. It was used for various agricultural applications, including assessing vegetation properties like canopy area, chlorophyll, and water content, as well as analyzing soil characteristics (Estep et al., 2004). Compact hyperspectral imaging sensors such as the Headwall company's Micro and Nano Hyperspec VIS-NIR sensors, UHD 185-Firefly, and HySpex VNIR, which feature hundreds of bands covering a broad spectral range (VIS-NIR), can be mounted on UAVs for use in agricultural and forestry tasks (Saari et al., 2011). In recent times, close-range hyperspectral imaging sensors (HySpex hyperspectral camera, hyperspectral imaging scanner of Pika XC, etc.), comprising ground or laboratory-based systems, have become increasingly popular due to their ability to capture hyperspectral imagery at super-high spatial resolutions, reaching levels as fine as centimeters or even sub-centimeters. This capability enables the investigation of fine-scale vegetation features, such as those at the canopy stages, thereby significantly helping to perceive initial symptoms of vegetation stress conditions, such as disease incidence, weed infestation, or nutritional deficits (Malmir et al., 2019).

Thermal and Microwave Sensors

Nowadays, advanced satellites are equipped with two additional sensors—thermal infrared sensors and microwave sensors—whose use in agriculture is increasing daily because of their longer wavelengths. Thermal infrared sensors can detect the thermal energy emitted by target objects such as crops or soils. The temperature data collected by these sensors can then be used to estimate various parameters, including water stress conditions in both soil and plants, the rate of evapotranspiration (ET), and the water requirements for irrigation (Khanal et al., 2017). Thermal sensors yield data comprising a blend of canopy, shadow, and bare soil temperatures. To assess water stress, compare the canopy temperature of crops with that of well-irrigated crops (Hunt Jr and Daughtry, 2018). However, thermal sensors are typically either precise yet costly or imprecise yet affordable. As a result, their application in agriculture is relatively limited compared to other types of sensors.

Microwave sensors function like thermal sensors in measuring the emitted energy and/or backpropagated pulse from the land surface, albeit using longer microwave wavelengths. Microwave sensors are primarily utilized for estimating soil moisture levels and crop water usage across extensive regions. One notable advantage they offer over other sensor types is their ability to operate day and night and under any weather conditions, as microwaves can penetrate through clouds (Palazzi et al., 2019).

Here is a compilation showcasing a range of remote sensing sensors, complete with their specifications and how they are utilized in precision framing (Table 1).

Table 1. List of agriculturally important satellite-based, airborne, and ground-based sensors.

Year of Launch	Name of Sensors	Type of Sensors	Spatial Resolution	Temporal Resolution	Agricultural Applications	References
Satellite based sensors						
1972	Landsat-1	Multispectral Scanner System (MSS)	80 m	18 days	Crop growth monitoring	Leslie et al. (2017)
1975	Landsat-2	MSS	80 m	18 days	Crop growth monitoring	Leslie et al. (2017)
1978	Landsat-3	Return Beam Vidicon (RBV) and MSS	40 m	18 days	Crop health monitoring	Seelan et al. (2003)
1979	NOAA-AVHRR (TIROS-N satellite)	MSS	1100 m	1 day	Soil moisture monitoring, nutrient management	Worsley and Bowler (2001)
1984	Landsat 5	MS and Thematic Mapper (TM)	120 m	16 days	Crop biomass monitoring	Scudiero et al. (2016)
1986	SPOT-1	MSS	20 m	2–6 days	Water management, Flood damage assessment	Worsley and Bowler (2001)
1988	IRS-1A	MSS	72 m	22 days	Water and nutrient management	Mondal and Basu (2009)
1995	LIDAR	Visible	0.1 m	–	Topographical map, digital elevation model preparation, post-harvest growth detection	Koenig et al. (2015)
1995	RADARSAT-1	Microwave (C Band-Synthetic Aperture Radar; i.e., SAR)	30 m	106 days	Mapping crop types and monitoring crop conditions	McNairn et al. (2002)
1999	IKONOS	MSS	3.2 m	3 days	Spatial variability evaluation of soil properties	Sullivan et al. (2005)
1999	MODIS-Terra	MSS	250 m	1–2 days	Crop yield forecasting	Filippi et al. (2019)

1999	Landsat-7	Enhanced Thematic Mapper (ETM)	60 m	16 days	Crop growth monitoring	Venancio et al. (2019)
2000	EO-1-Hyperion	Hyperspectral Sensor System (HSS)	30 m	16–30 days	Crop disease detection	Apan et al. (2004)
2001	Quick Bird	MSS	2.44 m	1–3.5 days	Oil palm disease identification	Santoso et al. (2011)
2001	PROBA-CHRIS	HSS	36 m	8 days	Leaf area index (LAI) estimation, vegetation cover assessment	Verger et al. (2011)
2002	SPOT-5	MSS	10 m (near-infrared, NIR), 20 m (shortwave infrared, SWIR)	2–3 days	Crop yield estimation	Yang et al. (2009)
2003	ResourceSat-1	MSS	5.6 (visible, V), 23.5 (SWIR)	5 days	Crop discrimination	Sai and Rao (2008)
2006	KOMPSAT-2	MSS	4 m	5.5 days	Predicting crop yield and its connection to vegetation indices	Lee et al. (2011)
2008	RapidEye	MSS	6.5 m	1–5.5 days	Spatial variability mapping of crop growth conditions	Shang et al. (2015)
2008	HJ-1A	Hyperspectral imager	20 m		Wheat grain yield estimation	Wang et al. (2014)
2009	WorldView-2	MSS	1.4 m	1.1 days	LAI mapping	Tian et al. (2017)
2012	SPOT-6	MSS	6 m	1 day	Crop disease (powdery mildew) mapping	Yuan et al. (2016)
2013	Landsat-8	Operational Land Imager (OLI) and Thermal Infrared (TI) Sensor	100 m	16 days	Crop yield estimation	Dong et al. (2016)
2014	Worldview-3	Super Spectral System (SSS)	1.24 m	Less than 1 day	Agricultural landscape mapping	Sidike et al. (2018)

Table 1 contd. ...

...Table 1 contd.

Year of Launch	Name of Sensors	Type of Sensors	Spatial Resolution	Temporal Resolution	Agricultural Applications	References
Satellite based sensors						
2015	Sentinel-2	MSS	10 m (V & NIR), 20 m (SWIR and Red Edge)	2–5 days	Barley yield prediction	Martínez-Casasnovas et al. (2019)
2015	SMAP	Microwave	1–3 km (L-band SAR)	2–3 days	Irrigation water management	Hao et al. (2019)
2017	TROPOMI (onboard Sentinel-5)	MSS with imaging spectrometers and radiometers	5.5 × 3.5 km	Less than 1 day	Crop phenology identification	de Gouw et al. (2020)
2018	ECOSTRESS-PHyTIR	Thermal sensors	38 × 69 m	1–5 days	Evapotranspiration estimation	Fisher et al. (2020)
2019	CARTOSAT-3	MSS	1.12 m	More than 1 day	Land use land cover mapping of coastal area	Bharteey et al. (2019)
Airborne sensors						
Released Year	**Name of Airborne Sensors**	**Types of Sensors**	**Spatial Resolution**	**No. of Spectral Bands**	**Agricultural Applications**	**References**
1989	CASI	HSS (380-1-50 nm)	1–20 m	288	Weed classification, chlorophyll estimation	Haboudane et al. (2002)
1993	AVIRIS	HSS (400-2500 nm)	7.1 m	224	Water status monitoring in nut crops	Cheng et al. (2014)
1999	HyMap	HSS (440-2500 nm)	5 m	128	Crop biophysical and biochemical parameters estimation	Jarmer (2013)
2006	mX-SIGHT	Thermal	35 cm	–	Water status monitoring	Gonzalez-Dugo et al. (2015)

2013	CERES Imaging	MSS	20 cm	6	Floral phenology identification	Chen et al., (2019)
2014	AISA-Eagle	HSS (400-970 nm)		244	Crop nitrogen estimation	Cilia et al. (2014)
2016	Phantom IV	RGB	3.5 cm	3	Crop height estimation	Erfanifard et al. 2019)
Ground-based sensors						
Launch Year	**Name of Ground-based Sensors**	**Types of Sensors**	**Spectral Resolution**	**No. of Spectral Bands**	**Agricultural Applications**	**References**
1974	FTIR	HSS (2500-25000 nm)	2 cm^{-1}	22500	Soil properties estimation, soil function assessment	Ng et al. (2022)
1993	Spectroradiometer	HSS (350-2500 nm)	3–10 nm	2150	Soil nutrients such as organic carbon prediction	Das et al. (2023)
2003	HySpex	HSS (400-2500 nm)	2.7 nm	400	Crop disease monitoring	Morel et al. (2018)

Applications of RS in Precision Agriculture: Current Progress

The following section includes the exploration of the diverse applications of RS technologies in agriculture, spanning from field planning to harvest. These applications encompass tasks such as generating field topographical maps, mapping soil moisture levels, managing water resources, monitoring pests and diseases, predicting crop yields, managing crop residues, and numerous other applications.

Field Topographical Maps Generation

Topographical characteristics of a field, including slope and elevation, significantly influence preseason management decisions. These decisions involve determining the optimal timing of field activities and practices, selecting appropriate crop types, and calculating rates of seed planting. Hence, accurately defining field topography using Digital Elevation Models (DEMs) is essential. This delineation can be achieved through grade surveying by utilizing the Global Positioning System (GPS) instrument, which is integrated into farm machinery, and remote sensing (RS) imagery (Khanal et al., 2020).

Soil Moisture Map Preparation

Preparation of a high-resolution field map of soil moisture is highly useful for taking management choices including crop sowing or transplanting, and particularly in irrigation scheduling for the survival of crops. Optical, thermal, and microwave types of remote sensing data are extensively utilized worldwide for determining soil moisture (Verstraeten et al., 2008). Optical and thermal data commonly employ methodologies like the "triangle" or "trapezoid" method, as well as the land surface temperature-vegetation index (LST-VI) approach. These approaches rely on the physical contact of temperature on the land surface (also latent heat fluxes) and characteristics of vegetation areas to estimate and map soil moisture and evapotranspiration (ET). Microwave sensors gauge the dielectric characteristics of soil for estimating soil moisture levels, relying on either land surface emissivity or scattering. Multiple satellites equipped with microwave active and passive sensors, for instance, 'Advanced Microwave Scanning Radiometer-Earth Observing System (AMSR-E)'; 'Soil Moisture and Ocean Salinity (SMOS), Soil Moisture Active Passive (SMAP)'; 'and Sentinel-1'. These examples of microwave sensors are utilized to monitor soil moisture purposes (Babaeian et al., 2018).

Irrigation Water Management

Utilizing a suitable quantity of irrigation water at a suitable time is crucial in alleviating the water stress in growing crops and ensuring optimal growth and yield of these crops. Many farmers rely on their experience for field irrigation, while commercial farms may use soil moisture sensors to irrigate crops. However, neither of these methods accounts for the variability in soil moisture levels across the field during irrigation (Thompson et al., 2007; Holt et al., 2019). The RS data assist in identifying differents at the ground, enabling the implementation of variable-rate irrigation using conventional irrigation systems like center pivots. Remote sensing images, gathered at multiple intervals throughout a developing time, are utilized to assess different indicators related to the water consumption of the crops, including evapotranspiration, levels of moisture levels, and drought stress conditions of crops. Those factors are then employed in accurately calculation the demand for water and accordingly irrigation tabulating for the crops.

Crop Nutrient Management

Promoting healthy crop development and gains while avoiding environmental hazards by preventing nutrient loss and contamination of both surface and groundwater resources requires making sure that fertilizer nutrients are applied to crops in suitable quantities and at the right times. However, soil nutrient levels exhibit variability in both spatial and temporal dimensions, as well as in terms of crop fertilizer requirements, owing to variations in soil conditions, topography, hydrological conditions, and activities of management (Hendricks et al., 2019). Mapping of spatial as well as temporal variations of those nutrients using conventional tools poses significant challenges for their applications in PA. Hence, the mapping of RS based on various indexes like difference vegetation index (DVI) and NDVI, holds promise for comprehending and delineating the spatial variability of nutrients. This is due to the notable correlation between vegetation indices and parameters such as plant chlorophyll, activity of the photosynthetic process, and the productivity of the growing crops. Certain commercially available remote sensors, like Green Seeker, Yara N-sensor, and Crop Circle, whether handheld or tractor-mounted, hold promise in measuring crop and soil spectral reflectance values. Subsequently, this data can aid in implementing spatially variable and real-time fertilizer applications (Ali et al., 2017). Remote sensing technology also assists in generating spatially variable or digital maps of the different nutrients of the soil such as the soil organic carbon, and available macronutrients (nitrogen, and phosphorus), facilitating tailored nutrient management at specific sites (Blasch et al., 2015; Kalambukattu et al., 2018). Malmir et al. (2019) utilized laboratory-based hyperspectral imagery acquired through a Pika XC2 hyperspectral camera mounted on a linear stage to accurately predict the soil macronutrients as well as the micronutrients, enabling site-specific soil nutrient management.

Crop Disease Detection

Given the widely recognized detrimental impact of pest and disease infestations on crop yields and agricultural profits, early identification of diseases is essential to resist the spreading of these diseases and maximize economic returns. The conventional approach to disease monitoring, such as field scouting, proves ineffective in detecting pest infestations until visible symptoms appear on plant parts. Furthermore, this traditional method consumes time, is laborious, needs experienced hands, and is economically inefficient (Ehsani and Maja, 2013). Remote sensing technology effectively circumvents these limitations and enables the earlier detection and monitoring of disease infestations compared to conventional methods. The alteration in leaf color resulting from pathogen attacks leads to a significant decline in plant photosynthetic activity. Remote sensing sensors pick up these changes in visible and near-infrared (Vis-NIR) portions of the electromagnetic spectra, which causes alterations in plants' spectral characteristics (West et al., 2003). The stressed plants generally show lower spectral reflectance values at the NIR area of EM spectra as compared to healthy vegetation, and thus such spectral response can able to differentiate healthy plants from stressed plants (Govaerts and Verhulst, 2010). In practical applications, utilizing disease-specific spectral disease indices (SDI) derived from remote sensing (RS) proves more effective for the accurate detection of plant diseases compared to conventional indexes of the vegetation (i.e., NDVI, DVI, etc.) and others (AL-Saddik et al., 2017). While remote sensing (RS) technology employing visual (RGB), multispectral, and hyperspectral sensors has demonstrated effectiveness in disease monitoring, achieving early disease detection remains challenging. Consequently, experiments have revealed that thermal and fluorescence-based RS hold promise, particularly in early disease detection when symptoms are not yet apparent in plant parts. This potential is attributed to the tendency of diseased plants to exhibit elevated temperatures compared to healthy ones (Mahlein, 2016).

Crop Yield Prediction

In the current context, crop yield is typically assessed and mapped using a yield monitor attached to a combine harvester, which usually consolidates yield data from numerous crop rows during harvesting (Bern et al., 2019). However, low resolution is a defining feature of these yield maps. Hence, remote sensing imagery integrated with various machine learning-based models has been utilized to create high-resolution yield maps, allowing for the evaluation of yield variations within fields and across landscapes (Khanal et al., 2018). This enables a more comprehensive understanding of the field's overall agricultural production. The RS data is commonly employed to predict crucial crop biological and physical factors (i.e., Leaf Area Index 'LAI'; biomass; etc.) which serve as essential indicators of crop health and development across various crops (Ali et al., 2020). These data are integrated into yield forecasting models to facilitate timely and accurate yield prediction.

Crop Residue Management

Efficient management of crop residues holds significant importance in different regions such as India, whereas residue burning is common across small to large-scale farm holdings. This practice aims to enhance soil quality and crop productivity while mitigating the adverse environmental effects caused by excessive burning of crop residues in the fields (Shah et al., 2017). Many studies have illustrated the capability of both multispectral and hyperspectral RS techniques to accurately map the spatial distribution of plants' residues in the site (Galloza et al., 2013; Sharma et al., 2016). Sharma and colleagues 2016 developed RS indexes (i.e., normalized difference tillage index 'NDTI' which is a multispectral index that can be applied to Landsat ETM+ and OLI imageries). Their findings indicated that NDTI could effectively distinguish the degree of residue cover associated with several practices of the tillage process in various crop fields located in a part of the USA (Sharma et al., 2016). Galloza and colleagues found that hyperspectral indices outperformed multispectral indices in distinguishing crop residue covers in the field (Galloza et al., 2013). For example, compared to Vis-NIR imagery, Thermal Infrared (TIR) from NASA's Airborne Terrestrial Application Sensor (ATLAS) is more useful in identifying variations in temperature differences among crop residues and soil systems (Sullivan et al., 2004). In recent times, scientists have turned their attention to site-specific tillage practices for managing residues, utilizing drone technology to gain a deeper insight into the impact of managing the crop residues through the water, nutrients, and soil dynamics (Khanal et al., 2020).

Identification of Plant Species

Different plant species can also be identified by integrating remote sensing technology with machine learning and deep learning techniques. A team of researchers employed a convolutional neural network (CNN), a deep learning algorithm, to identify six different plants based on their morphological characteristics (Chouhan et al., 2024).

Challenges and Future Direction

Remote sensing technologies and data-driven approaches have emerged as powerful tools in modern agriculture, offering unprecedented insights into crop health, soil conditions, and environmental factors. However, their implementation faces several challenges and technological constraints that need to be addressed for their widespread adoption and effective utilization.

Based on research conducted by The World Bank, it has been identified that around 65% of individuals working in agriculture are situated below the poverty threshold, equating to less than US $3.10/day (Castaneda et al., 2016). This creates a significant barrier for many smallholder farmers to embrace advanced technologies like satellite technologies, UAVs for precision agriculture on their farms, as the considerable cost disparity between one agricultural UAV and its financial benefits was unfeasible for them to afford such technology (Fernández et al., 2019). For small farm owners in places like India and Africa, even if they could buy such a device, they would still encounter problems with costs because the services and working costs for these instruments are already quite costly. Additionally, research indicates that previously utilized PA techniques like mapping of yield; GPS; and variable rate mechanisms had minimal influence on total returns. Yield mapping led to approximately a 2.5% increase, GPS guidance systems to a 2% increase, and variable-rate systems to a 1.1% increase in yield (Schimmelpfennig, 2016). Therefore, for many farmers, particularly those who own smaller-scale farms, it may be challenging to justify investing a substantial amount of time in setting up the initial capital required for these technologies. Bridging this technological gap requires innovative pricing models, capacity-building initiatives, and government support to make remote sensing technologies more affordable and user-friendly for farmers worldwide.

The next strong struggle is the complexity and variability of agricultural landscapes. Different regions exhibit diverse topographies, soil types, and crop varieties, making it challenging to develop universal remote sensing models that accurately capture agricultural dynamics across various environments (Hosseini et al., 2020). This variability requires the customization of algorithms and data processing techniques to suit local conditions, increasing the RS application complexity in agriculture.

Moreover, the spatiotemporal resolution of RS data poses a constraint on its utility in agriculture. While satellite imagery provides valuable insights into large-scale agricultural patterns, its spatial resolution is insufficient for detecting subtle variations in vegetation health or filed soil moisture (Kussul et al., 2017). Similarly, the temporal resolution of satellite data may not always align with the rapid changes occurring in agricultural systems, such as sudden weather events or pest outbreaks, limiting the timeliness of decision-making based on remote sensing observations.

Another critical challenge is the integration of data obtained from the RS and different sources of agricultural observations (i.e., ground-based sensors and agronomic knowledge). While the RS offers valuable insights from a bird's eye view, ground-based sensors provide field-scale high-resolution data, capturing localized soil moisture variability, nutrient levels, as well as crop health (Tao et al., 2018). Integrating these heterogeneous data sources requires sophisticated data fusion techniques and interoperable platforms, which remain a technological challenge in agricultural remote sensing applications.

As we navigate towards a future marked by an increase in population, the change in climate, as well as the shortage in resources, an integration of RS technologies becomes increasingly imperative. However, several avenues warrant further exploration to harness their full potential and address existing challenges.

Firstly, advancing sensor capabilities is crucial to enhance the precision and resolution of data collected. Research efforts should focus on developing sensors capable of capturing multi-spectral and hyperspectral images in higher spatiotemporal resolutions. Additionally, an integration between emerging technologies like LiDAR with a detecting and ranging light; and SAR can provide complementary data layers for comprehensive agricultural assessments (Gómez-Candón et al., 2018).

Secondly, developing robust data analytics methodologies tailored to agricultural applications is essential. Machine learning algorithms, particularly deep learning models, hold promise in analyzing vast amounts of remote sensing data for predictive modeling, anomaly detection, and crop classification tasks (Atzberger, 2013). However, further research is needed to refine these algorithms for specific crop types, geographical regions, and environmental conditions, ensuring their scalability and accuracy.

Thirdly, integrating RS data and other data sources such as meteorological data, soil properties, and agronomic management practices can provide holistic insights into agricultural systems. Interdisciplinary research efforts are required to develop integrated frameworks that leverage diverse datasets to optimize crop yields, minimize resource inputs, and mitigate environmental impacts (Lobell et al., 2019).

Moreover, fostering collaboration between researchers, policymakers, and agricultural stakeholders is imperative to facilitate the adoption of remote sensing technologies at scale. Interdisciplinary research consortia and public-private partnerships can accelerate technology transfer, promote knowledge exchange, and address barriers to adoption such as cost, infrastructure, and data access (Fritz et al., 2019).

Conclusion

In this chapter, we have examined current developments in remote sensing technology and how they apply to agriculture. The integration of remote sensing techniques has significantly transformed agricultural practices, offering new insights and solutions to address various challenges faced by the agricultural sector. From the comprehensive analysis of current developments, it is evident that remote sensing technologies, including space-based satellite imagery, airborne imageries, and hyperspectral imaging, have revolutionized crop monitoring, yield prediction, and precision agriculture. These technologies have empowered farmers to make well-informed decisions, optimize resource use, and boost productivity while reducing environmental impact. Additionally, combining remote sensing with new technologies like data analytics and machine learning has opened up new ways to interpret data, resulting in more accurate and timely decisions in agriculture. The integration of these technologies has facilitated the development of innovative solutions for pest and disease management, soil mapping, and irrigation optimization.

However, despite the remarkable progress, several challenges remain to be addressed. These include difficulty in data interpretation, restricted access to modern technologies in rural places, and the need for continuous research and development to enhance the capabilities of remote sensing systems. Additionally, there is a growing need for capacity building and educational programs to equip farmers with the technical knowhow and skills required for effective utilization of advanced technologies.

Looking ahead, the future of remote sensing in agriculture holds great promise. As technology continues to evolve, we anticipate the emergence of more advanced sensors, platforms, and analytical tools that will improve the accuracy, efficiency, and sustainability of farming methods. Collaboration between stakeholders, including researchers, policymakers, industry players, and farmers, will be essential to drive innovation and ensure the widespread adoption of remote sensing technologies in agriculture.

In summary, using remote sensing technologies marks a major change in agriculture, providing new opportunities to tackle global food security issues, mitigate climate change impacts, and promote sustainable agricultural development. With continued investment, innovation, and collaboration, remote sensing is poised to play a central role in shaping the future of agriculture, ultimately leading to more resilient and productive agricultural systems.

References

Ali, M.M., Al-Ani, A., Eamus, D. and Tan, D.K. 2017. Leaf nitrogen determination using non-destructive techniques—A review. *J. Plant Nutr.*, 40: 928–953.

Ali, A. and Imran, M. 2020. Evaluating the potential of red edge position (REP) of hyperspectral remote sensing data for real time estimation of LAI & chlorophyll content of kinnow mandarin (Citrus reticulata) fruit orchards. *Sci. Hortic.*, 267: 109326.

Apan, A., Held, A., Phinn, S. and Markley, J. 2004. Detecting sugarcane 'orange rust'disease using EO-1 Hyperion hyperspectral imagery. *Int. J. Remote Sens.*, 25: 489–498.

Al-Saddik, H., Simon, J.C. and Cointault, F. 2017. Development of spectral disease indices for 'Flavescence Dorée'grapevine disease identification. *Sensors*, 17: 2772.

Asner, G.P. and Martin, R.E. 2016. Spectranomics: Emerging science and conservation opportunities at the interface of biodiversity and remote sensing. *Glob. Ecol. Conserv.*, 8: 212–219.

Atzberger, C. 2013. Advances in remote sensing of agriculture: Context description, existing operational monitoring systems and major information needs. *Remote Sens.*, 5: 949–981.

Babaeian, E., Sadeghi, M., Franz, T.E., Jones, S. and Tuller, M. 2018. Mapping soil moisture with the OPtical TRApezoid Model (OPTRAM) based on long-term MODIS observations. *Remote Sens. Environ.*, 211: 425–440.

Barker, D.W. and Sawyer, J.E. 2012. Using active canopy sensing to adjust nitrogen application rate in corn. *Agron. J.*, 104: 926–933.

Barnes, J. 2018. *Drones vs Satellites*: Competitive or Complementary? Commercial UAV News.

Bern, C.J., Quick, G. and Herum, F.L. 2019. *Harvesting and Postharvest Management. In*: Corn (pp. 109–145). AACC International Press.

Basso, B., Cammarano, D., Fiorentino, C. and Ritchie, J.T. 2013. Wheat yield response to spatially variable nitrogen fertilizer in Mediterranean environment. *Eur. J. Agron.*, 51: 65–70.

Basso, B., Fiorentino, C., Cammarano, D. and Schulthess, U. 2016. Variable rate nitrogen fertilizer response in wheat using remote sensing. *Precis. Agric.*, 17: 168–182.

Bhattacharya, B.K., Green, R.O., Rao, S., Saxena, M., Sharma, S., Kumar, K.A., Srinivasulu, P., Sharma, S., Dhar, D., Bandyopadhyay, S. and Bhatwadekar, S. 2019. An overview of AVIRIS-NG airborne hyperspectral science campaign over India. *Curr. Sci.*, 7: 1082–1088.

Bharteey, P.K., Deka, B., Dutta, M., Parit, R.K. and Maurya, P.K. 2019. Remote sensing application in precision agriculture: A review. *Remote Sens.*, 6: 233–241.

Blasch, G., Spengler, D., Hohmann, C., Neumann, C., Itzerott, S. and Kaufmann, H. 2015. Multitemporal soil pattern analysis with multispectral remote sensing data at the field-scale. *Comput. Electron. Agric.*, 113: 1–13.

Candiago, S., Remondino, F., De Giglio, M., Dubbini, M. and Gattelli, M. 2015. Evaluating multispectral images and vegetation indices for precision farming applications from UAV images. *Remote Sens.*, 7: 4026–4047.

Castaneda, R., Doan, D., Newhouse, D.L., Nguyen, M., Uematsu, H. and Azevedo, J.P. 2016. Who are the poor in the developing world?. World Bank Policy Research Working Paper, 7844.

Chang, C.Y., Zhou, R., Kira, O., Marri, S., Skovira, J., Gu, L. and Sun, Y. 2020. An Unmanned Aerial System (UAS) for concurrent measurements of solar-induced chlorophyll fluorescence and hyperspectral reflectance toward improving crop monitoring. *Agric. For. Meteorol.*, 294: 108145.

Chen, S., She, D., Zhang, L., Guo, M. and Liu, X. 2019. Spatial downscaling methods of soil moisture based on multisource remote sensing data and its application. *Water*, 11: 1401.

Cheng, T., Riaño, D. and Ustin, S.L. 2014. Detecting diurnal and seasonal variation in canopy water content of nut tree orchards from airborne imaging spectroscopy data using continuous wavelet analysis. *Remote Sens. of Environ.*, 143: 39–53.

Chouhan, S.S., Singh, U.P. and Jain, S. 2024. Introduction to computer vision and drone technology. pp. 1–5. In *Applications of Computer Vision and Drone Technology in Agriculture 4.0*. Singapore: Springer Nature Singapore.

Chouhan, S.S., Singh, U.P., Sharma, U. and Jain, S. 2024. Classification of different plant species using deep learning and machine learning algorithms. *Wirel. Pers. Commun.*, 1–24.

Cilia, C., Panigada, C., Rossini, M., Meroni, M., Busetto, L., Amaducci, S., Boschetti, M., Picchi, V. and Colombo, R. 2014. Nitrogen status assessment for variable rate fertilization in maize through hyperspectral imagery. *Remote Sens.*, 6: 6549–6565.

Colomina, I. and Molina, P. 2014. Unmanned aerial systems for photogrammetry and remote sensing: A review. *ISPRS J. Photogramm. Remote Sens.*, 92: 79–97.

Crusiol, L.G.T., Nanni, M.R., Silva, G.F.C., Furlanetto, R.H., da Silva Gualberto, A.A., Gasparotto, A.D.C. and De Paula, M.N. 2017. Semi professional digital camera calibration techniques for Vis/NIR spectral data acquisition from an unmanned aerial vehicle. *Int. J. Remote Sens.*, 38: 2717–2736.

Das, B., Chakraborty, D., Singh, V.K., Das, D., Sahoo, R.N., Aggarwal, P., Murgaokar, D. and Mondal, B.P. 2023. Partial least square regression based machine learning models for soil organic carbon prediction using visible–near infrared spectroscopy. *Geoderma Reg.*, 33: 628.

De Gouw, J.A., Veefkind, J.P., Roosenbrand, E., Dix, B., Lin, J.C., Landgraf, J. and Levelt, P.F. 2020. Daily satellite observations of methane from oil and gas production regions in the United States. *Sci. Rep.*, 10: 1379.

Dong, T., Liu, J., Qian, B., Zhao, T., Jing, Q., Geng, X., Wang, J., Huffman, T. and Shang, J. 2016. Estimating winter wheat biomass by assimilating leaf area index derived from fusion of Landsat-8 and MODIS data. *Int. J. Appl. Earth Obs. Geoinf.*, 49: 63–74.

Ehsani, R. and Maja, J.M. 2013. The rise of small UAVs in precision agriculture. *Resource Magazine*, 20: 18–19.

Elijah, O., Rahman, T.A., Orikumhi, I., Leow, C.Y. and Hindia, M.N. 2018. An overview of Internet of Things (IoT) and data analytics in agriculture: Benefits and challenges. *IEEE Internet Things J.*, 5: 3758–3773.

Erfanifard, S.Y., Chenari, A., Dehghani, M. and Amiraslani, F. 2019. Effect of spatial resolution of UAV aerial images on height estimation of wild pistachio (Pistacia atlantica Desf.) trees. *Iranian Journal of Forest and Poplar Research*, 27: 169–181.

Estep, L., Terrie, G. and Davis, B. 2004. Crop stress detection using AVIRIS hyperspectral imagery and artificial neural networks.

Estes, J., Kline, K. and Collins, E. 2001. Remote sensing. pp. 13144–13150. *In*: Smelser, N.J. and Baltes, P.B. (eds.). *International Encyclopedia of the Social & Behavioral Sciences*. Pergamon, Oxford.

Fernández, N., Velayos, C., Díaz-Guerra, L. and Sanz-Cañada, J. 2019. Satellite remote sensing for precision agriculture: A review of application examples in Spain. *Remote Sens.*, 11: 424.

Filippi, P., Jones, E.J., Wimalathunge, N.S., Somarathna, P.D., Pozza, L.E., Ugbaje, S.U., Jephcott, T.G., Paterson, S.E., Whelan, B.M. and Bishop, T.F. 2019. An approach to forecast grain crop yield using multi-layered, multi-farm data sets and machine learning. *Precis. Agric.*, 20: 1015–1029.

Fisher, J.B., Lee, B., Purdy, A.J., Halverson, G.H., Dohlen, M.B., Cawse-Nicholson, K., Wang, A., Anderson, R.G., Aragon, B., Arain, M.A. and Baldocchi, D.D. 2020. ECOSTRESS: NASA's next generation mission to measure evapotranspiration from the international space station. *Water Resour. Res.*, 56: 6058.

Foody, G.M. 2019. Earth observation for agricultural applications: Making the case for a systematic approach. *Int. J. Appl. Earth Obs. Geoinf.*, 77: 158–165.

Fritz, S., See, L., Rembold, F., Coppola, A. and Di Leginio, L. 2019. Citizen science and the United Nations Sustainable Development Goals: Connecting citizens and decision makers to address complex problems. *WIREs Climate Change*, 10: 590.

Galloza, M.S., Crawford, M.M. and Heathman, G.C. 2013. Crop residue modeling and mapping using Landsat, ALI, Hyperion and airborne remote sensing data. *IEEE J. Sel. Top. Appl. Earth Obs. Remote Sens.*, 6: 446–456.

García-Tejero, I.F., Ortega-Arévalo, C.J., Iglesias-Contreras, M., Moreno, J.M., Souza, L., Tavira, S.C. and Durán-Zuazo, V.H. 2018. Assessing the crop-water status in almond (Prunus dulcis mill.) trees via thermal imaging camera connected to smartphone. *Sensors*, 18: 1050.

Gómez-Candón, D., López-Granados, F., Caballero-Novella, J.J. and García-Ferrer, A. 2018. Assessment of a LiDAR-based canopy height model for monitoring crop growth: A comparison with high-resolution UAV photogrammetry. *Precis. Agric.*, 19: 650–669.

Gonzalez-Dugo, V., Hernandez, P., Solis, I. and Zarco-Tejada, P.J. 2015. Using high-resolution hyperspectral and thermal airborne imagery to assess physiological condition in the context of wheat phenotyping. *Remote Sens.*, 7: 13586–13605.

Govaerts, B. and Verhulst, N. 2010. The normalized difference vegetation index (NDVI) Greenseeker (TM) handheld sensor: Toward the integrated evaluation of crop management part A: concepts and case studies.

Günal, E., Wang, X., Kılıc, O.M., Budak, M., Al Obaid, S., Ansari, M.J. and Brestic, M. 2021. Potential of Landsat 8 OLI for mapping and monitoring of soil salinity in an arid region: A case study in Dushak, Turkmenistan. *Plos One*, 16: 695.

Haboudane, D., Miller, J.R., Tremblay, N., Zarco-Tejada, P.J. and Dextraze, L. 2002. Integrated narrow-band vegetation indices for prediction of crop chlorophyll content for application to precision agriculture. *Remote Sens. Environ.*, 81: 416–426.

Hao, Z., Zhao, H., Zhang, C., Wang, H. and Jiang, Y. 2019. Detecting winter wheat irrigation signals using SMAP gridded soil moisture data. *Remote Sens.*, 11: 2390.

Hendricks, G.S., Shukla, S., Roka, F.M., Sishodia, R.P., Obreza, T.A., Hochmuth, G.J. and Colee, J. 2019. Economic and environmental consequences of overfertilization under extreme weather conditions. *J. Soil Water Conserv.*, 74: 160–171.

Holt, N., Sishodia, R.P., Shukla, S. and Hansen, K.M. 2019. Improved water and economic sustainability with low-input compact bed plasticulture and precision irrigation. *J. Irrig. Drain. Eng.*, 145: 04019013.

Hosseini, M., Zhang, Q. and Golian, S. 2020. Challenges in remote sensing applications for precision agriculture: A review. *Agric. Water Manag.*, 231: 105987.

Huang, Y., Chen, Z.X., Tao, Y.U., Huang, X.Z. and Gu, X.F. 2018. Agricultural remote sensing big data: Management and applications. *J. Integr. Agric.*, 17: 1915–1931.

Hunt Jr, E.R., Hively, W.D., Fujikawa, S.J., Linden, D.S., Daughtry, C.S. and McCarty, G.W. 2010. Acquisition of NIR-green-blue digital photographs from unmanned aircraft for crop monitoring. *Remote Sens.*, 2: 290–305.

Hunt Jr, E.R. and Daughtry, C.S. 2018. What good are unmanned aircraft systems for agricultural remote sensing and precision agriculture? *Int. J. Remote Sens.*, 39: 5345–5376.

Jafarbiglu, H. and Pourreza, A. 2022. A comprehensive review of remote sensing platforms, sensors, and applications in nut crops. *Comput. Electron. Agric.*, 197: 106844.

Jarmer, T. 2013. Spectroscopy and hyperspectral imagery for monitoring summer barley. *Int. J. Remote Sens.*, 34: 6067–6078.

Jha, K., Doshi, A., Patel, P. and Shah, M. 2019. A comprehensive review on automation in agriculture using artificial intelligence. *Artificial Intelligence in Agriculture*, 2: 1–12.

Jones, H.G. and Vaughan, R.A. 2010. *Remote Sensing of Vegetation: Principles, Techniques, and Applications*. Oxford University Press, USA.

Kalambukattu, J.G., Kumar, S. and Arya Raj, R. 2018. Digital soil mapping in a Himalayan watershed using remote sensing and terrain parameters employing artificial neural network model. *Environ. Earth Sci.*, 77: 1–14.

Khanal, S., Fulton, J. and Shearer, S. 2017. An overview of current and potential applications of thermal remote sensing in precision agriculture. *Comput. Electron. Agric.*, 139: 22–32.

Khanal, S., Fulton, J., Klopfenstein, A., Douridas, N. and Shearer, S. 2018. Integration of high resolution remotely sensed data and machine learning techniques for spatial prediction of soil properties and corn yield. *Comput. Electron. Agric.*, 153: 213–225.

Khanal, S., Kc, K., Fulton, J.P., Shearer, S. and Ozkan, E. 2020. Remote sensing in agriculture—Accomplishments, limitations, and opportunities. *Remote Sens.*, 12: 3783.

Koenig, K., Höfle, B., Hämmerle, M., Jarmer, T., Siegmann, B. and Lilienthal, H. 2015. Comparative classification analysis of post-harvest growth detection from terrestrial LiDAR point clouds in precision agriculture. *ISPRS J. Photogramm. Remote Sens.*, 104: 112–125.

Kussul, N., Lavreniuk, M. and Skakun, S. 2017. Deep learning classification of land cover and crop types using remote sensing data. *IEEE Geosci Remote S.*, 14: 778–782.

Lee, J.W., Park, G., Joh, H.K., Lee, K.H., Na, S.I., Park, J.H. and Kim, S.J. 2011. Analysis of relationship between vegetation indices and crop yield using KOMPSAT (KoreaMulti-Purpose SATellite)-2 imagery and field investigation data. *Journal of the Korean Society of Agricultural Engineers*, 53: 75–82.

Leslie, C.R., Serbina, L.O. and Miller, H.M. 2017. Landsat and agriculture—Case studies on the uses and benefits of Landsat imagery in agricultural monitoring and production (No. 2017–1034). US Geological Survey.

Lobell, D.B., Ortiz-Monasterio, J.I., Asner, G.P. and Matson, P.A. 2019. Integrating airborne remote sensing and ground census data to track the effects of agricultural intensification on aboveground biomass production in the Yucatan Peninsula, Mexico. *Environ. Res. Lett.*, 14: 054010.

Lu, B., Dao, P.D., Liu, J., He, Y. and Shang, J. 2020. Recent advances of hyperspectral imaging technology and applications in agriculture. *Remote Sens.*, 12: 2659.

Lundeen, S. and Gowey, K. 2021. AVIRIS-Airborne Visible/Infrared Imaging Spectrometer-Data Processing.

Mahlein, A.K. 2016. Plant disease detection by imaging sensors–parallels and specific demands for precision agriculture and plant phenotyping. *Plant Dis.*, 100: 241–251.

Malmir, M., Tahmasbian, I., Xu, Z., Farrar, M.B. and Bai, S.H. 2019. Prediction of soil macro-and micro-elements in sieved and ground air-dried soils using laboratory-based hyperspectral imaging technique. *Geoderma*, 340: 70–80.

Manfreda, S., McCabe, M.F., Miller, P.E., Lucas, R., Pajuelo Madrigal, V., Mallinis, G., Ben Dor, E., Helman, D., Estes, L., Ciraolo, G. and Müllerová, J. 2018. On the use of unmanned aerial systems for environmental monitoring. *Remote Sens.*, 10: 641.

Martínez-Casasnovas, J.A., Uribeetxebarría, A., Escolà, A. and Arnó, J. 2019. Sentinel-2 vegetation indices and apparent electrical conductivity to predict barley (*Hordeum vulgare L.*) yield. *Precis. Agric.*, 19: 415–421. Wageningen Academic Publishers.

Mazur, M., Wisniewski, A. and McMillan, J. 2016. Clarity from above: PwC global report on the commercial applications of drone technology. Warsaw: Drone Powered Solutions, PriceWater house Coopers.

McNairn, H., Ellis, J., Van Der Sanden, J.J., Hirose, T. and Brown, R.J. 2002. Providing crop information using RADARSAT-1 and satellite optical imagery. *International Journal of Remote Sens.*, 23: 851–870.

MIT Technology Review. 2014. 10 Most Important Technology Milestones for 2014. MIT Technology Review. https://www.technologyreview.com/lists/technologies/2014/.

Mondal, P. and Basu, M. 2009. Adoption of precision agriculture technologies in India and in some developing countries: Scope, present status and strategies. *Prog. Nat. Sci.*, 19: 659–666.

Morel, J., Jay, S., Féret, J.B., Bakache, A., Bendoula, R., Carreel, F. and Gorretta, N. 2018. Exploring the potential of PROCOSINE and close-range hyperspectral imaging to study the effects of fungal diseases on leaf physiology. *Sci. Rep.*, 8: 15933.

Ng, W., Minasny, B., Jeon, S.H. and McBratney, A. 2022. Mid-infrared spectroscopy for accurate measurement of an extensive set of soil properties for assessing soil functions. *Soil Security*, 6: 100043.

Nowatzki, J., Andres, R. and Kyllo, K. 2004. Agricultural remote sensing basics.

Palazzi, V., Bonafoni, S., Alimenti, F., Mezzanotte, P. and Roselli, L. 2019. Feeding the world with microwaves: How remote and wireless sensing can help precision agriculture. *IEEE Microwave Magazine*, 20: 72–86.

Rees, G. 2013. *Physical Principles of Remote Sensing*. Cambridge University Press.

Roberts, D.F., Kitchen, N.R., Scharf, P.C. and Sudduth, K.A. 2010. Will variable-rate nitrogen fertilization using corn canopy reflectance sensing deliver environmental benefits?. *Agron J.*, 102: 85–95.

Rosegrant, M.W., Koo, J., Cenacchi, N., Ringler, C., Robertson, R.D., Fisher, M., Cox, C.M., Garrett, K., Perez, N.D. and Sabbagh, P. 2014. Food security in a world of natural resource scarcity: The role of agricultural technologies. *Intl. Food Policy Res. Inst.*

Saari, H., Pellikka, I., Pesonen, L., Tuominen, S., Heikkilä, J., Holmlund, C., Mäkynen, J., Ojala, K. and Antila, T. 2011. Unmanned Aerial Vehicle (UAV) operated spectral camera system for forest and agriculture applications. *Remote Sensing for Agriculture, Ecosystems, and Hydrology*, 8174: 170–184.

Sai, M.S. and Rao, P.N. 2008. Utilization of Resourcesat-1 data for improved crop discrimination. *Int. J. Appl. Earth Obs. Geoinf.*, 10: 206–210.

Santoso, H., Gunawan, T., Jatmiko, R.H., Darmosarkoro, W. and Minasny, B. 2011. Mapping and identifying basal stem rot disease in oil palms in North Sumatra with QuickBird imagery. *Precis. Agric.*, 12: 233–248.

Schimmelpfennig, D. 2016. Farm profits and adoption of precision agriculture (No. 1477-2016-121190).

Scudiero, E., Corwin, D.L., Wienhold, B.J., Bosley, B., Shanahan, J.F. and Johnson, C.K. 2016. Downscaling Landsat 7 canopy reflectance employing a multi-soil sensor platform. *Precis. Agric.*, 17: 53–73.

Seelan, S.K., Laguette, S., Casady, G.M. and Seielstad, G.A. 2003. Remote sensing applications for precision agriculture: A learning community approach. *Remote Sens. Environ.*, 88: 157–169.

Shah, A., Darr, M., Khanal, S. and Lal, R. 2017. A techno-environmental overview of a corn stover biomass feedstock supply chain for cellulosic biorefineries. *Biofuels*, 8: 59–69.

Shang, J., Liu, J., Ma, B., Zhao, T., Jiao, X., Geng, X., Huffman, T., Kovacs, J.M. and Walters, D. 2015. Mapping spatial variability of crop growth conditions using RapidEye data in Northern Ontario, Canada. *Remote Sens. Environ.*, 168: 113–125.

Sharma, V., Irmak, S., Kilic, A., Sharma, V., Gilley, J.E., Meyer, G.E., Knezevic, S.Z. and Marx, D. 2016. Quantification and mapping of surface residue cover for maize and soybean fields in south central Nebraska. *Transactions of the ASABE*, 59: 925–939.

Sidike, P., Sagan, V., Maimaitijiang, M., Maimaitiyiming, M., Shakoor, N., Burken, J., Mockler, T. and Fritschi, F.B. 2019. dPEN: Deep Progressively Expanded Network for mapping heterogeneous agricultural landscape using WorldView-3 satellite imagery. *Remote Sens. Environ.*, 221: 756–772.

Simpson, A., Stombaugh, T., Wells, L. and Jacob, J. 2003. Imaging techniques and applications for UAV's in agriculture. pp. 1. *In*: ASAE Annual Meeting J. ASABE.

Sishodia, R.P., Ray, R.L. and Singh, S.K. 2020. Applications of remote sensing in precision agriculture: A review. *Remote Sens.*, 12: 3136.

Stafford, J.V. 2000. Implementing precision agriculture in the 21st century. *Res. Agric. Eng.*, 76: 267–275.

Sullivan, D.G., Shaw, J.N., Mask, P.L., Rickman, D., Guertal, E.A., Luvall, J. and Wersinger, J.M. 2004. Evaluation of multispectral data for rapid assessment of wheat straw residue cover. *Soil Sci. Soc. Am. J.*, 68: 2007–2013.

Sullivan, D.G., Shaw, J.N. and Rickman, D. 2005. IKONOS imagery to estimate surface soil property variability in two Alabama physiographies. *Soil Sci. Soc. Am. J.*, 69: 1789–1798.

Sun, J., Yang, J., Shi, S., Chen, B., Du, L., Gong, W. and Song, S. 2017. Estimating rice leaf nitrogen concentration: Influence of regression algorithms based on passive and active leaf reflectance. *Remote Sens.*, 9: 951.

Tao, H., Jin, X., Li, X. and Feng, W. 2018. The role of remote sensing in precision agriculture: A review. *Journal of Plant Nutrition*, 41: 744–749.

Teke, M., Deveci, H.S., Haliloğlu, O., Gürbüz, S.Z. and Sakarya, U. 2013. A short survey of hyperspectral remote sensing applications in agriculture. pp. 171–176. *In*: *2013 6th International Conference on Recent Advances in Space Technologies (RAST)*. IEEE.

Thompson, R.B., Gallardo, M., Valdez, L.C. and Fernández, M.D. 2007. Using plant water status to define threshold values for irrigation management of vegetable crops using soil moisture sensors. *Agric. Water Manag.*, 88: 147–158.

Tian, J., Wang, L., Li, X., Gong, H., Shi, C., Zhong, R. and Liu, X. 2017. Comparison of UAV and WorldView-2 imagery for mapping leaf area index of mangrove forest. *Int. J. Appl. Earth Obs. Geoinf.*, 61: 22–31.

Tremblay, N., Wang, Z., Ma, B.L., Belec, C. and Vigneault, P. 2009. A comparison of crop data measured by two commercial sensors for variable-rate nitrogen application. *Precis. Agric.*, 10: 145–161.

Tucker, C.J., Pinzon, J.E., Brown, M.E., Slayback, D.A., Pak, E.W., Mahoney, R. and El Saleous, N. 2018. An extended AVHRR 8-km NDVI dataset compatible with MODIS and SPOT vegetation NDVI data. *Int. J. Remote Sens.*, 39: 5137–5143.

Turner, W., Spector, S., Gardiner, N., Fladeland, M., Sterling, E. and Steininger, M. 2015. Remote sensing for biodiversity science and conservation. *Trends Ecol. Evol.*, 30: 84–91.

van Evert, F.K., Booij, R., ukema, J.N., ten Berge, H.F., Uenk, D., Meurs, E.B., van Geel, W.C. and Wijnholds, K.H. 2012. Using crop reflectance to determine sidedress N rate in potato saves N and maintains yield. *Eur. J. Agron.*, 43: 58–67.

Venancio, L.P., Mantovani, E.C., do Amaral, C.H., Neale, C.M.U., Gonçalves, I.Z., Filgueiras, R. and Campos, I. 2019. Forecasting corn yield at the farm level in Brazil based on the FAO-66 approach and soil-adjusted vegetation index (SAVI). *Agric. Water Manag.*, 225: 105779.

Verger, A., Baret, F. and Camacho, F. 2011. Optimal modalities for radiative transfer-neural network estimation of canopy biophysical characteristics: Evaluation over an agricultural area with CHRIS/PROBA observations. *Remote Sens. Environ.*, 115: 415–426.

Verstraeten, W.W., Veroustraete, F. and Feyen, J. 2008. Assessment of evapotranspiration and soil moisture content across different scales of observation. *Sensors*, 8: 70–117.

Wang, L., Tian, Y., Yao, X., Zhu, Y. and Cao, W. 2014. Predicting grain yield and protein content in wheat by fusing multi-sensor and multi-temporal remote-sensing images. *Field Crops Res.*, 164: 178–188.

West, J.S., Bravo, C., Oberti, R., Lemaire, D., Moshou, D. and McCartney, H.A. 2003. The potential of optical canopy measurement for targeted control of field crop diseases. *Annu. Rev. Phytopathol.*, 41: 593–614.

Worsley, P. and Bowler, J. 2001. Assessing flood damage using SPOT and NOAA AVHRR data. *Geospat. Inf. Agric.*, 2–7.

Yang, C., Everitt, J.H. and Bradford, J.M. 2009. Evaluating high resolution SPOT 5 satellite imagery to estimate crop yield. *Precis. Agric.*, 10: 292–303.

Yuan, L., Pu, R., Zhang, J., Wang, J. and Yang, H. 2016. Using high spatial resolution satellite imagery for mapping powdery mildew at a regional scale. *Precis. Agric.*, 17: 332–348.

Zhao, T. 2018. *Remote Sensing of Water Stress in Almond Trees using Unmanned Aerial Vehicles*. University of California, Merced.

Chapter 7

Remote Sensing for Sustainable Agriculture

A Machine Learning Approach to Optimizing Farm Yield and Economic Returns

Harshit Mishra,[1,*] *Supriya*[1] and *Shripati Dwivedi*[2]

Introduction

In the twenty-first century, agricultural production is confronted with numerous difficulties. An expanding world population needs more food to be produced with as little negative influence on the environment as possible. To do this, sustainable agricultural methods are essential. When combined with machine learning approaches, remote sensing provides a potent means of maximizing agricultural yield and financial gains (Lin et al., 2021). Farmers have historically made decisions regarding crop health, irrigation, and resource allocation based on field observations and experience. These techniques, however, are frequently laborious, subjective, and have a constrained scope in terms of both time and space. For getting beyond these restrictions, remote sensing offers a useful tool. It serves as a way of learning about an object or part of the Earth's surface without embarking there in person (Bharadiya et al., 2023; Sharma et al., 2020). This information is collected by sensors mounted on satellites, unmanned aerial vehicles (UAVs), or ground-based platforms. Sensors

[1] Department of Agricultural Economics, College of Agriculture, Acharya Narendra Deva University of Agriculture and Technology, Kumarganj, Ayodhya (U.P.) - 224 229, India.
[2] Department of Agricultural Economics, Post Graduate College of Agriculture, Dr. Rajendra Prasad Central Agricultural University, Pusa, Samastipur (Bihar) - 848 125, India.
* Corresponding author: wehars@gmail.com

detect and record electromagnetic radiation emitted or reflected from the Earth's surface, providing valuable insights into various agricultural parameters (Martos et al., 2021).

Different types of remote sensing data are available, each with its advantages and limitations. Satellite imagery offers wide-area coverage at regular intervals, making it suitable for regional and national-level monitoring. UAVs, also known as drones, provide high-resolution images and can be deployed on-demand for targeted monitoring of specific fields. Ground-based sensors provide continuous data collection for highly localized areas (Diaz-Gonzalez et al., 2022). Hyperspectral imaging captures data across a large number of spectral bands, allowing for detailed analysis of crop health and nutrient status. To utilize remote sensing data effectively, data acquisition and preprocessing are crucial steps. Radiometric calibration compensates for differences in sensor response, whereas geometric correction provides precise spatial representation. Data fusion combines information from multiple sensors or sources to generate a more comprehensive view of the agricultural landscape (Weiss et al., 2020; Salaria and Suresh, 1990).

Once the data is prepared, image classification techniques are employed to extract meaningful information. Supervised classification uses labelled training data to categorize pixels into specific classes, such as different crop types, soil types, or areas of water stress. Unsupervised classification identifies natural groupings within the data without predefined labels. Object-based classification analyzes groups of pixels representing specific things, such as trees or crop rows, for a more comprehensive description (Benami et al., 2021; Saxena et al., 2024). By leveraging machine learning algorithms, remote sensing data can be used for a variety of applications in crop monitoring and management. Crop type classification enables farmers to map their fields and track changes in planting patterns over time. Crop health monitoring allows for early detection of stress caused by nutrient deficiencies, pests, or diseases, facilitating timely interventions. Remote sensing data also plays a vital role in detecting and mapping the spread of pests and diseases, enabling targeted application of pesticides and minimizing environmental impact (Tiwari et al., 2024; Li et al., 2022).

Yield prediction models are another key application of remote sensing in agriculture. Machine learning algorithms can analyze historical yield data along with remote sensing data on crop health, weather conditions, and soil moisture to predict future crop yields. This information empowers farmers to make informed decisions about resource allocation, fertilizer application, and harvest planning, potentially increasing overall yield and economic returns. Water management and irrigation optimization are critical aspects of sustainable agriculture (Benos et al., 2021; Solanki et al., 2024). Remote sensing data helps estimate soil moisture content, enabling farmers to identify areas requiring irrigation and optimize water use efficiency. By monitoring water stress levels in crops, farmers can schedule irrigation more effectively, reducing water waste and conserving precious resources (Sishodia et al., 2020).

Soil mapping and fertility assessment are crucial for targeted nutrient management. Remote sensing data enables the identification and mapping of different soil types across a field. By analyzing spectral reflectance patterns, it is possible to assess soil fertility and identify areas deficient in specific nutrients. This information allows for the development of targeted nutrient management strategies, optimizing fertilizer application and minimizing environmental pollution. Weed detection and management are significant challenges in agriculture. Remote sensing data can be used to classify weed species and map their distribution across a field. This information facilitates the development of precision herbicide application strategies, targeting only areas with weeds and minimizing herbicide use (Chlingaryan et al., 2018).

Economic analysis and decision support systems are essential for translating the benefits of remote sensing into tangible gains for farmers. Cost-benefit analyzes assess the economic viability of adopting remote sensing technologies, considering factors like initial investment, operational costs, and potential yield improvements. Economic impact assessments evaluate the broader economic benefits of remote sensing for agriculture, such as increased food production and improved resource efficiency (Khanal et al., 2018; Meraj et al., 2022). Finally, the development of decision support systems integrates remote sensing data with other relevant information sources to provide farmers with user-friendly tools for making informed decisions about their agricultural practices. Looking towards the future, emerging technologies like advanced sensors with higher resolution and wider spectral coverage hold immense promise for even more refined data collection (Khanal et al., 2020). The integration of AI and computer vision with remote sensing data analysis will further enhance the capabilities of these technologies.

However, addressing challenges like scalability, data accessibility for small-scale farmers, and ensuring ethical data practices remain crucial for maximizing the positive impact of remote sensing on sustainable agriculture. By overcoming these hurdles, remote sensing and machine learning have the potential to revolutionize agricultural practices, leading to a future of increased food production, improved economic returns for farmers, and a more sustainable approach to agriculture that safeguards our planet for generations to come (Elavarasan and Vincent, 2020; Bali et al., 2023). This chapter explores the application of remote sensing and machine learning for sustainable agriculture. It first explains the fundamental principles and various technologies used in remote sensing for agricultural purposes. Next, it outlines the essential workflow for agricultural remote sensing, including data acquisition, preprocessing, and image classification techniques. Finally, the chapter explores how machine learning algorithms can be used to predict crop yield, optimize water management, and assess the economic viability of agricultural practices (Tiwari and Mishra, 2024; Solanki et al., 2023).

Fundamentals of Remote Sensing

Remote sensing, a vital tool in modern agriculture, encompasses the acquisition and interpretation of data from a distance to understand and manage the resources of this earth. This section delves into the foundational principles governing remote sensing techniques, the various types of data sources employed, and the significance of spectral bands in agricultural applications.

Principles of Remote Sensing

The basis of remote sensing is the interaction of electromagnetic radiation with the surface of the Earth. Fundamental to it all is the idea that various electromagnetic spectrum wavelengths are absorbed, reflected, and emitted by surface-level things on Earth. The ability to record and analyze these interactions underpins the basic idea of remote sensing, making it easier to extract important data about the Earth's surface (Machwitz et al., 2021). Some principles governing remote sensing include:

- *Interaction of Electromagnetic Radiation*: Objects interact differently with electromagnetic radiation based on their composition, structure, and surface properties. This interaction results in the absorption, transmission, or reflection of energy at specific wavelengths.
- *Spectral Signatures*: Objects exhibit unique spectral signatures characterized by their reflectance properties across different wavelengths. These signatures enable the identification and differentiation of features on Earth's surface, such as vegetation, water bodies, and soil types.
- *Spatial and Temporal Resolution*: Remote sensing systems vary in their spatial and temporal resolutions, influencing their ability to capture fine-scale details and monitor changes over time. High-resolution sensors offer detailed spatial information but may have limited coverage, whereas low-resolution sensors provide broader coverage but with less detail.

Understanding these principles is crucial for effectively utilizing remote sensing techniques in agriculture to monitor crop health, assess environmental conditions, and optimize management practices.

Types of Remote Sensing Data

Remote sensing data can be obtained from a variety of platforms, each offering unique advantages and capabilities. The primary types of remote sensing platforms include:

- *Satellite-based Remote Sensing*: Satellites orbiting Earth capture imagery using various sensors, such as optical, thermal, and radar systems. Satellite data offer wide spatial coverage and are suitable for regional and global-scale monitoring. However, they may have limitations in spatial resolution and revisit frequency.
- *Unmanned Aerial Vehicles (UAVs)*: UAVs, or drones, are increasingly used for agricultural remote sensing due to their flexibility and high spatial resolution capabilities. UAVs can capture imagery at customizable flight altitudes, providing detailed information about small-scale features such as individual plants or field variations.
- *Aircraft-based Remote Sensing*: Manned aircraft equipped with remote sensing instruments are employed for specialized applications requiring high spatial resolution and real-time data acquisition. Aircraft-based platforms offer flexibility in sensor configuration and flight planning, making them suitable for research and precision agriculture applications.

The choice of a remote sensing platform depends on factors such as spatial resolution requirements, area of interest, budget constraints, and data accessibility (Johnson et al., 2016; Solanki et al., 2023).

Spectral Bands and their Significance

Spectral bands represent specific ranges of wavelengths within the electromagnetic spectrum and play a critical role in remote sensing applications. Different spectral bands are sensitive to distinct surface properties and facilitate the extraction of valuable information about the Earth's surface. The significance of spectral bands in agricultural remote sensing includes:

- *Visible and Near-Infrared (VNIR) Bands*: VNIR bands are sensitive to vegetation properties such as chlorophyll content, leaf structure, and biomass. These bands are commonly used for vegetation monitoring, crop health assessment, and yield estimation. The near-infrared (NIR) region, in particular, is highly reflective for healthy vegetation due to chlorophyll absorption, making it valuable for detecting vegetation stress and assessing crop vigour.
- *Shortwave Infrared (SWIR) Bands*: SWIR bands penetrate vegetation canopy and are sensitive to moisture content, soil properties, and biochemical composition. SWIR imagery can help identify soil types, detect water stress in crops, and assess soil moisture levels, providing valuable information for irrigation management and soil fertility mapping.
- *Thermal Infrared (TIR) Bands*: TIR bands measure the emitted thermal radiation from the Earth's surface, providing information about surface temperature variations. TIR imagery is used to monitor crop water stress, detect disease outbreaks, and assess crop water requirements, aiding in precision irrigation scheduling and crop management decisions.

Choosing the right sensors and refining data processing methods for agricultural applications requires an understanding of the spectral properties of various bands. A thorough examination of crop health, environmental factors, and soil characteristics is made possible by integrating spectral data over several bands, which helps decision-makers adopt sustainable agricultural methods (Tripathi et al., 2022).

Remote Sensing Technologies in Agriculture

Effective management of agricultural land is essential for ensuring food security and economic prosperity. Remote sensing technologies provide farmers and other agricultural professionals with an invaluable tool that allows them to have a thorough, non-invasive picture of their land. By collecting data remotely, these technologies enable the monitoring of crop health, soil conditions, and other critical factors that impact yield and profitability. The various remote sensing technologies utilized in sustainable agriculture are examined in this section, along with their unique characteristics and methods of increasing agricultural output.

Satellite Remote Sensing

Satellite remote sensing collects data over large geographic areas using orbiting platforms fitted with cutting-edge sensors. These sensors record visible, infrared, and microwave electromagnetic radiation that is reflected off the earth's surface. Crop varieties may be identified, crop health can be evaluated, and anomalies like insect infestations or water stress can be found through the analysis of this data (Eugenio et al., 2020). We must take a closer look at the advantages and limits of satellite remote sensing.

Advantages:

- Provides wide-area coverage, enabling the monitoring of large farms and regional agricultural trends.
- Offers consistent data collection over time, allowing for the tracking of crop growth patterns and changes in field conditions.
- Freely available satellite data from government agencies exists for basic applications.

Limitations:

- Spatial resolution might be limited, making it challenging to identify features on the ground smaller than a few meters.
- Cloud cover can significantly hinder data collection and necessitate repeat imaging.
- Advanced data analysis often requires specialized software and expertise.

Unmanned Aerial Vehicles (UAVs)

UAVs, also known as drones, have emerged as a valuable tool for high-resolution remote sensing in agriculture. These miniature aerial platforms equipped with sensors fly autonomously or remotely controlled, capturing detailed data over specific fields or areas of interest.

UAVs may be more advantageous than conventional satellite images.

- *High-resolution imaging*: They can capture images with pixel sizes of a few centimetres, enabling the detection of finer details like weed patches or localized disease outbreaks.
- *Flexibility*: UAV flights can be tailored to specific needs, providing on-demand data collection over targeted areas.
- *Cost-effectiveness*: Compared to manned aircraft, UAV data acquisition is typically less expensive.

However, UAVs also have limitations to consider:

- *Limited range and flight time*: Battery life restricts the area a single UAV can cover in a single flight.
- *Weather dependency*: Wind and precipitation can impede safe operation and data quality.
- *Regulations*: Operating UAVs for commercial purposes often require adherence to specific regulations set by aviation authorities (Kumar et al., 2015).

Ground-Based Sensors

Ground-based sensors provide *in-situ* data collection, offering highly detailed measurements of various environmental parameters crucial for crop growth. These sensors are typically deployed within fields and continuously collect data throughout the growing season.

Some common types of ground-based sensors in agriculture include:

- *Soil moisture sensors*: Measure the water content in the soil, aiding in irrigation management and water use efficiency.
- *Temperature sensors*: Monitor air and soil temperature, providing insights into frost risk and plant growth conditions.
- *Nutrient sensors*: Assess the levels of essential nutrients in the soil, enabling targeted fertilizer application.

The primary advantages of ground-based sensors include:

- *Highly accurate data collection*: They provide precise measurements of specific parameters at the point of deployment.
- *Real-time monitoring*: Data can be continuously collected and transmitted, allowing for immediate response to changes in field conditions.

However, limitations exist for ground-based sensors as well:

- *Limited spatial coverage*: Data is only collected from the immediate vicinity of the sensor, requiring a network of sensors for comprehensive field monitoring.
- *Maintenance requirements*: Sensors need regular calibration and maintenance to ensure data accuracy.
- *Installation costs*: The initial investment in purchasing and installing a network of ground-based sensors can be significant.

Hyperspectral Imaging

Hyperspectral imaging is a sophisticated remote sensing technology that captures data across a very large number of contiguous spectral bands. This allows for the identification of specific elements and compounds based on their unique spectral signatures. In agriculture, hyperspectral imaging is used for:

- *Crop health assessment*: By analyzing the reflectance properties of leaves, it can detect subtle variations in plant health, indicating stress due to nutrient deficiencies, pests, or diseases.
- *Soil analysis*: Hyperspectral data can be used to map soil types and identify areas with specific mineral compositions.
- *Weed identification*: The unique spectral signatures of different weed species can be identified and targeted for selective herbicide application.

The advantages of hyperspectral imaging include:

- *Highly detailed spectral information*: Provides a wealth of data for advanced analysis and feature identification.
- *Ability to differentiate subtle variations*: Enables the detection of early signs of stress in crops before visible symptoms appear, allowing for timely intervention and improved crop management.
- *Precision agriculture*: Hyperspectral data can be integrated with other geospatial information to create variable rate application maps. This allows for the targeted application of fertilizers, pesticides, and irrigation water, optimizing resource use and minimizing environmental impact.

Limitations of hyperspectral imaging:

- *High data volume*: The vast amount of data collected by hyperspectral sensors requires specialized software and expertise for processing and analysis.
- *Cost*: Hyperspectral sensors are typically more expensive than other remote sensing technologies.
- *Data interpretation*: Extracting meaningful information from hyperspectral data often requires advanced chemometric techniques (Nishad et al., 2024; Schwalbert et al., 2020).

Data Acquisition and Preprocessing

To effectively use remote sensing for sustainable agriculture, careful consideration of data collection and preprocessing is essential. This section outlines the key methods and approaches used in this field to guarantee the accuracy and value of collected data.

Image Acquisition Techniques

Image acquisition encompasses various methods to capture data from different sensors mounted on satellites, drones, or other platforms. In agricultural settings, multispectral and hyperspectral imaging are commonly used. Multispectral sensors capture data in specific bands (e.g., red, green, blue, near-infrared), whereas hyperspectral sensors provide more detailed spectral information across a broader range of wavelengths. Thermal imaging is also valuable for monitoring crop health and water stress.

The choice of sensor and acquisition parameters (such as spatial resolution, spectral resolution, and temporal frequency) depends on the specific objectives of the agricultural application. For instance, high-resolution satellite imagery might be preferred for detailed crop monitoring and precision farming, while drones can offer flexibility in capturing data at different stages of crop growth (Rasid et al., 2021).

Preprocessing Steps

The raw data acquired from diverse remote sensing platforms undergo a series of preprocessing steps to rectify inherent imperfections, standardize formats, and enhance interpretability. These preprocessing steps encompass radiometric calibration, geometric correction, atmospheric correction, and normalization procedures, ensuring the consistency and accuracy of subsequent analyzes.

- *Radiometric calibration*: Radiometric calibration involves correcting sensor-specific radiometric distortions such as sensor noise, gain variations, and atmospheric effects. By calibrating raw digital numbers to absolute radiance or reflectance values, radiometric calibration facilitates quantitative analysis of surface properties and enables comparison across different sensors and acquisitions.
- *Geometric correction*: Geometric correction rectifies spatial distortions induced by sensor platform motion, Earth's topography, and atmospheric refraction. Through geometric transformation techniques such as orthorectification and georeferencing, raw imagery is projected onto a standardized coordinate system, ensuring spatial accuracy and alignment with ground truth data.
- *Atmospheric correction*: Atmospheric correction mitigates the influence of atmospheric constituents on remote sensing measurements, particularly in the presence of aerosols, water vapor, and ozone. Utilizing radiative transfer models and atmospheric correction algorithms, atmospheric effects are removed or minimized, enhancing the accuracy of quantitative analysis and interpretation.
- *Normalization procedures*: Normalization procedures standardize remote sensing data across different acquisitions, sensors, and environmental conditions. Techniques such as normalization by vegetation index or terrain flattening ensure consistency in spectral signatures and facilitate temporal comparison of imagery over extended periods.

Assuring the precision and dependability of a future analysis depends on each preprocessing stage, especially when combining data from various platforms or sensors (such as satellite and drone photography) (Sharma et al., 2020).

Data Fusion Techniques

Data fusion techniques amalgamate heterogeneous data from multiple sources and sensors to capitalize on their complementary strengths and enhance the overall information content. By integrating optical, radar, thermal, and ancillary data layers, data fusion facilitates the comprehensive characterization of agricultural landscapes and fosters synergistic insights beyond the capabilities of individual sensors.

- *Multi-sensor fusion*: Multi-sensor fusion combines data from diverse remote sensing platforms, such as optical, radar, and LiDAR sensors, to exploit their complementary spectral, spatial, and temporal resolutions. Fusion of optical and radar data, for instance, enables the derivation of biophysical parameters under varying weather and illumination conditions, enhancing crop monitoring and yield estimation accuracy.
- *Temporal fusion*: Temporal fusion integrates time-series data acquired over different growing seasons to capture seasonal dynamics and long-term trends in agricultural phenomena. By combining multi-temporal imagery from satellite archives or UAV surveys, temporal fusion facilitates trend analysis, anomaly

detection, and predictive modelling of crop growth cycles and phenological stages.

- *Spatial fusion*: Spatial fusion merges data at different spatial resolutions and scales to enhance spatial detail and context awareness. Techniques such as super-resolution imaging and spatial sharpening integrate high-resolution imagery with coarser-resolution datasets to derive fine-scale information on crop health, field boundaries, and land use patterns.

Proficient data acquisition and preprocessing are foundational pillars in harnessing remote sensing for sustainable agriculture. By employing advanced imaging techniques, meticulous preprocessing procedures, and innovative data fusion strategies, researchers and practitioners can unlock the full potential of remote sensing data to optimize farm yield, mitigate environmental impacts, and foster economic sustainability in agriculture (Sharifi, 2021; Virnodkar et al., 2020).

Image Classification Techniques

Using remote sensing data to optimize agricultural practices requires an understanding of the nuances of picture categorization algorithms, which we explore in this section. Image classification serves as a cornerstone in the analysis of remotely sensed data, facilitating the identification and categorization of various features within the imagery. This section delineates three primary methodologies employed in image classification: Supervised classification, Unsupervised classification, and Object-based classification. Each technique offers distinct advantages and is tailored to specific contexts, contributing significantly to the enhancement of agricultural practices through the utilization of remote sensing data.

Supervised Classification

Supervised classification stands as a fundamental approach in image analysis, relying on a priori knowledge of spectral signatures to categorize pixels within an image. The process entails the utilization of training data, wherein samples representing different land cover classes are identified and associated with their corresponding spectral characteristics. These spectral signatures serve as reference points for the classification algorithm, enabling it to assign pixels in the image to the predefined classes based on spectral similarity (Delgado et al., 2019). The workflow of supervised classification encompasses several key stages, commencing with the selection and acquisition of training data representative of the various land cover classes under consideration. This step necessitates meticulous attention to ensure the adequacy and representativeness of the training samples, thereby enhancing the accuracy and reliability of the classification outcomes. Subsequently, feature extraction techniques may be employed to extract pertinent information from the spectral bands, facilitating the discrimination of distinct land cover classes (Nay et al., 2018).

After obtaining training data, supervised learning approaches are employed to train a classification algorithm, which teaches it to distinguish between various land cover classes according to their spectral signatures. The most popular algorithms are Random Forest (RF), Support Vector Machines (SVM), and Maximum Likelihood

Classification (MLC); each has a distinct advantage over the others in terms of computational efficiency, noise resilience, and classification accuracy (Han et al., 2020; Balducci et al., 2018). The classification method is trained on the whole image and then used to categorize each pixel into a predetermined class of land cover. Subsequently, post-classification techniques like accuracy evaluation and error matrix analysis are carried out to appraise the classification's performance and make any required adjustments to the outcomes. Supervised classification is essential to remote sensing for sustainable agriculture to maximize farm productivity and financial returns by making it simpler to define and track the different land cover classes accurately (Zhang et al., 2021).

Unsupervised Classification

Unsupervised classification represents an alternative approach to image analysis, particularly valuable in scenarios where a priori knowledge of land cover classes may be limited or unavailable. Unlike supervised classification, unsupervised classification does not require predefined training samples; instead, it autonomously categorizes pixels within the image into distinct clusters based on their spectral similarities. The workflow of unsupervised classification entails the utilization of clustering algorithms such as K-means clustering, ISODATA (Iterative Self-Organizing Data Analysis Technique), and Hierarchical clustering. These algorithms iteratively partition the multivariate spectral space into clusters, with each cluster representing a distinct land cover class. The number of clusters, a critical parameter in unsupervised classification, may be determined empirically or through techniques such as the Elbow method or silhouette analysis (Mishra et al., 2024; Khan et al., 2022).

Once the clustering algorithm has been applied, post-processing steps such as cluster labelling and classification refinement may be employed to interpret the resultant clusters and assign them to specific land cover classes. While unsupervised classification offers the advantage of autonomy and adaptability to diverse datasets, it may be susceptible to misclassification errors and require additional human intervention for result interpretation and refinement (Ju et al., 2021). Despite these limitations, unsupervised classification remains a valuable tool in remote sensing for sustainable agriculture, particularly in scenarios where prior knowledge of land cover classes is limited or dynamic environmental conditions necessitate adaptive classification approaches.

Object-Based Classification

Object-based classification represents an innovative approach to image analysis that transcends traditional pixel-based classification methods by considering spatial contextual information in addition to spectral characteristics. Unlike pixel-based classification, which treats each pixel as an independent unit, object-based classification groups adjacent pixels into meaningful spatial objects or segments based on similarity in spectral, spatial, and contextual attributes (Gomez et al., 2019; Wei et al., 2020). The workflow of object-based classification encompasses several key stages, commencing with image segmentation, wherein the image is partitioned into homogeneous and meaningful objects or segments based on criteria such as spectral similarity, texture, and shape. This step aims to delineate objects corresponding to distinct land cover classes while minimizing intra-class spectral variability and inter-class spectral confusion (Kumar et al., 2022).

Following segmentation, feature extraction techniques are employed to derive pertinent attributes or descriptors for each object, capturing spectral, spatial, and contextual information essential for classification. These features may include spectral indices, texture measures, shape parameters, and contextual relationships with neighbouring objects. Subsequently, a classification algorithm is applied to the extracted features to assign each object to one of the predefined land cover classes (Tian et al., 2021). Commonly utilized algorithms in object-based classification include Decision Trees, Random Forests, and Support Vector Machines, each offering unique advantages in terms of classification accuracy, scalability, and computational efficiency.

Post-classification procedures such as accuracy assessment and error matrix analysis are then conducted to evaluate the performance of the classification and refine the results if necessary. Object-based classification thus offers a holistic and contextually informed approach to image analysis, leveraging spatial information to improve classification accuracy and enhance the interpretation of remotely sensed data for sustainable agriculture. Image classification techniques play a pivotal role in leveraging remote sensing data for sustainable agriculture, facilitating the precise delineation and monitoring of land cover classes essential for optimizing farm yield and economic returns. Supervised, unsupervised, and object-based classification methodologies each offer distinct advantages and are tailored to specific contexts, contributing significantly to the advancement of agricultural practices through the integration of remote sensing technology (Zhang et al., 2019).

Crop Monitoring and Management

Crop monitoring and management are integral components of modern agricultural practices aimed at ensuring optimal crop health and maximizing yield. In recent years, the integration of remote sensing technologies with machine learning algorithms has revolutionized these processes, enabling more efficient and precise monitoring and management strategies.

Crop Type Classification

Crop type classification involves the identification and categorization of different crops grown within a given agricultural area. Remote sensing techniques, particularly satellite imagery and aerial photography, offer valuable data sources for this task. Machine learning algorithms, such as SVM, random forests, and convolutional neural networks (CNN), are commonly employed for the classification of crop types based on spectral and spatial characteristics extracted from remote sensing data. The process of crop type classification begins with the acquisition of multispectral or hyperspectral imagery, which captures information across different wavelengths of the electromagnetic spectrum. Preprocessing techniques, including radiometric and atmospheric corrections, are applied to enhance the quality of the imagery and remove artifacts. Feature extraction methods are then utilized to extract relevant spectral signatures and spatial features from the imagery (Abbas et al., 2020).

Labelled training data—each sample connected to a particular crop type—is used to train machine learning algorithms. Based on the attributes that are retrieved, these models are trained to categorize individual pixels or image patches into various crop classes. Metrics including overall accuracy, producer accuracy, and user accuracy are used to assess the efficacy of the categorization model. Farmers and other agricultural stakeholders can learn more about the spatial distribution of various crops within their fields by using accurate crop type classification (Hara et al., 2021). With the use of this data, customized management strategies that are based on the unique needs of certain crop types can be implemented, such as variable rate treatments of pesticides, fertilizers, and irrigation. Furthermore, the creation of agricultural policies, crop rotation choices, and land use planning can all benefit from the usage of crop type maps generated from remote sensing data (Nguyen et al., 2021; Di et al., 2022).

Crop Health Monitoring

Crop health monitoring involves the assessment of the physiological condition and vigour of crops throughout the growing season. Early detection of stress factors, such as nutrient deficiencies, water stress, pests, and diseases, is crucial for implementing timely intervention measures and mitigating yield losses. Remote sensing technologies offer a non-destructive and cost-effective means of monitoring crop health at various spatial and temporal scales. The spectral reflectance properties of crops, as captured by remote sensing imagery, provide valuable information about their health status. Healthy plants exhibit characteristic spectral signatures associated with high reflectance in the NIR region and low reflectance in the visible (VIS) and SWIR regions. Conversely, stressed or diseased plants may exhibit alterations in their spectral signatures due to changes in chlorophyll content, leaf morphology, and cellular structure (Mishra and Mishra, 2024; Mhango et al., 2021).

To analyze remote sensing data and monitor crop health, machine learning techniques are essential. Spectral indices (e.g., normalized difference vegetation index, or NDVI) and machine learning classifiers (e.g., decision trees, support vector machines) are two popular supervised classification techniques used to pinpoint stressed areas in agricultural fields. Tracking crop health dynamics over time through temporal analysis of remote sensing data makes it possible to identify trends and anomalies linked to stressors (Radocaj et al., 2022). Integration with ground-based sensor networks and UAVs further enhances the spatial resolution and temporal frequency of crop health monitoring efforts. Crop health monitoring facilitates proactive decision-making by farmers and agronomists, enabling targeted interventions such as the precision application of fertilizers, irrigation scheduling, and pest management strategies. By identifying areas of concern early in the growing season, farmers can optimize resource allocation and minimize yield losses, ultimately contributing to sustainable agricultural practices (Paudel et al., 2021).

Detection of Pests and Diseases

The detection of pests and diseases poses significant challenges to crop production, as infestations can spread rapidly and cause substantial economic losses if left unchecked. Remote sensing technologies offer a valuable tool for early detection and monitoring of pest and disease outbreaks, enabling prompt response measures to mitigate their impact on crop yields. Pests and diseases often manifest as subtle changes in crop physiology and canopy structure, which can be detected using remote sensing imagery. Hyperspectral sensors, capable of capturing narrow spectral bands across the electromagnetic spectrum, are particularly well-suited for discriminating between healthy and stressed vegetation.

Automated detection and categorization of pest and disease symptoms in remote sensing data is achieved through the use of machine learning algorithms that have been trained on labelled datasets that contain spectral signatures of both healthy and diseased crops. Using spectral indices and textural data extracted from the images, supervised classification methods like random forests and support vector machines are frequently applied for this purpose (Wang et al., 2022). By tracking the evolution of pests and diseases across time, temporal analysis of remote sensing data enables early warning systems and epidemic dynamics prediction modelling. The accuracy and dependability of pest and disease detection activities are further improved by integration with meteorological data, crop growth models, and historical pest incidence records (Cai et al., 2019).

To limit the spread of pests and diseases and reduce production losses, farmers can employ targeted control techniques, such as precision pesticide treatments and cultural practices, as soon as outbreaks are identified. Furthermore, spatially explicit data from remote sensing can help prioritize management and surveillance activities, maximize resource allocation, and support environmentally friendly pest control techniques (Xu et al., 2021). The potential for improving agricultural crop monitoring and management practices through the combination of machine learning and remote sensing technology is enormous. Through the provision of precise and timely data on crop varieties, health conditions, and pest/disease incidents, these technologies enable farmers and other agricultural stakeholders to take proactive measures to maximize farm productivity and financial gains, all the while fostering sustainability in agricultural production.

Yield Prediction Models

In sustainable agriculture, the capacity to forecast crop yields with sufficient accuracy is essential for maximizing farm operations and guaranteeing financial gains. This section investigates the use of machine learning (ML) algorithms for yield prediction, looks at how remote sensing data can be integrated with these models, and notes the drawbacks and difficulties that come with using this method.

ML Algorithms for Yield Prediction

In recent years, machine learning algorithms have emerged as potent tools for predicting agricultural yields with remarkable accuracy. Leveraging the power of computational analysis and statistical learning, ML algorithms offer a sophisticated means of processing vast datasets derived from remote sensing technologies, weather patterns, soil characteristics, and historical yield records. The integration of these diverse data sources enables the development of robust predictive models capable of capturing the complex interplay of factors influencing crop growth and productivity (Shankar et al., 2020).

Example: Let us consider an example to elucidate the application of ML algorithms for yield prediction in agriculture. Suppose a farm situated in a region characterized by varying soil types, climatic conditions, and crop varieties aims to forecast its maize yield for the upcoming growing season.

Data Collection and Preprocessing: Initially, a comprehensive dataset is compiled, encompassing relevant variables such as soil moisture levels, temperature fluctuations, precipitation patterns, spectral reflectance data from satellite imagery, and past yield records. Before model development, this dataset undergoes meticulous preprocessing steps, including data cleaning, normalization, and feature engineering to enhance its suitability for machine learning analysis.

Model Selection: With the pre-processed dataset at hand, the next step involves the selection of an appropriate machine learning algorithm tailored to the specific task of yield prediction. Several algorithms exhibit promise in this domain, including but not limited to:

- *Random Forest*: A versatile ensemble learning technique capable of handling non-linear relationships and high-dimensional data effectively.
- *SVM:* Suited for both classification and regression tasks, SVMs excel in capturing complex decision boundaries.
- *Gradient Boosting Machines (GBM)*: By sequentially minimizing the loss function, GBM constructs a powerful ensemble of weak learners, thereby enhancing predictive accuracy.

Model Training and Evaluation: The Random Forest regression model was selected for its robustness and flexibility; the next step is to train the model on the prepared dataset. By repeatedly optimizing the selected objective function, the algorithm learns to map the input features—such as soil characteristics and meteorological variables—to the target variable—maize yield—during the training phase.

The trained model is then rigorously assessed using relevant metrics, including coefficient of determination (R^2), mean absolute error (MAE), and root mean squared error (RMSE). In this evaluation step, the predictive performance of the model is evaluated, and any areas that require improvement or fine-tuning are identified (Saikanth et al., 2024).

Model Deployment and Forecasting: Upon satisfactory validation of the model's performance, it is deployed to generate yield forecasts for the forthcoming growing season. By inputting real-time data acquired through remote sensing platforms and weather stations, the deployed model generates predictions about maize yield under varying environmental conditions and agronomic practices.

Continual Refinement and Adaptation: It is imperative to acknowledge that agricultural systems are inherently dynamic, and characterized by evolving environmental conditions and technological advancements. As such, the predictive model must undergo continual refinement and adaptation to remain attuned to changing circumstances. This iterative process may involve the incorporation of additional data sources, fine-tuning of model hyperparameters, or adoption of novel ML techniques to enhance predictive accuracy and reliability.

Integration of Remote Sensing Data with Yield Models

Precision agriculture is advancing with the use of remote sensing data integration with yield prediction algorithms. Drones and satellite photos are two examples of remote sensing technology that provide an abundance of information on crop health, soil moisture content, and environmental factors. Farmers can improve crop production forecasts by adding this data into yield prediction models, which will increase forecast accuracy and dependability (Kayad et al., 2019).

Example: Consider a case where a soybean farmer aims to predict the yield for the upcoming harvest season. To build an effective yield prediction model, the farmer collects remote sensing data using satellite imagery throughout the growth stages of the soybean crop. This data includes vegetation indices such as NDVI, which provides insights into the crop's health and vigour.

Next, the farmer integrates this remote sensing data with a machine learning model, such as a random forest regressor. The model is trained using historical yield data, along with features derived from the remote sensing imagery (e.g., NDVI values, temperature, precipitation). During the training phase, the model learns to capture the complex relationships between these variables and the final crop yield.

Once trained, the model can be used to predict the soybean yield for the current season. By inputting real-time remote sensing data into the model (e.g., updated NDVI readings), the farmer obtains dynamic yield forecasts that account for changing environmental conditions and crop health indicators.

The integration of remote sensing data with yield prediction models involves many important steps:

1. *Data Collection and Preprocessing*: Remote sensing data is collected using satellites or drones at regular intervals throughout the growing season. This data includes spectral bands capturing information on crop reflectance, moisture levels, and other environmental parameters.
2. *Feature Extraction*: From the raw remote sensing data, relevant features are extracted, such as vegetation indices (NDVI, EVI), thermal images, and soil moisture content. These features serve as input variables for the yield prediction model.
3. *Model Training*: Historical yield data, along with the extracted remote sensing features, are used to train the machine learning model. Various algorithms like random forests or neural networks are employed to learn the complex relationships between these features and crop yields.
4. *Model Validation and Tuning*: The trained model is validated using independent datasets to assess its predictive performance. Model parameters are fine-tuned to optimize accuracy and robustness.

5. *Real-Time Prediction*: Once validated, the model is deployed for real-time yield prediction. New remote sensing data collected during the season is fed into the model to generate updated yield forecasts (Ruan et al., 2022).

Challenges and Limitations

Despite their promising capabilities, yield prediction models encounter many challenges and limitations that merit careful consideration. Below, we explore these factors in detail, shedding light on the complexities inherent in implementing such models.

Data Quality and Availability

- One of the foremost challenges in developing accurate yield prediction models is the quality and availability of data. Remote sensing data, while abundant, may vary in resolution, accuracy, and consistency, posing significant challenges for model training and validation.
- Furthermore, accessing ground truth data for model calibration and validation purposes can be arduous, especially in remote or resource-constrained regions. Discrepancies between ground truth data and remote sensing observations can undermine the reliability of yield predictions.

Spatial and Temporal Variability

- Agricultural landscapes exhibit considerable spatial and temporal variability, influenced by factors such as soil composition, weather patterns, and agronomic practices. Capturing this variability accurately is essential for robust yield predictions.
- However, modelling such variability poses significant challenges, as it requires sophisticated algorithms capable of integrating heterogeneous data sources and accounting for spatial and temporal dynamics effectively (Muniasamy, 2020).

Model Complexity and Interpretability

- As yield prediction models become increasingly sophisticated, they also tend to grow in complexity, incorporating numerous variables and intricate algorithms. While such complexity enhances predictive accuracy, it may hinder model interpretability and transparency.
- Farmers and stakeholders often require interpretable models that elucidate the underlying factors driving yield predictions. Balancing model complexity with interpretability is thus a critical consideration in model development.

Transferability and Generalization

- Deploying yield prediction models across diverse geographical regions and cropping systems necessitates robust transferability and generalization capabilities. Models trained on data from one region or crop type may exhibit limited performance when applied elsewhere.

- Achieving transferable models requires careful attention to domain adaptation techniques, feature selection, and model architecture to ensure broad applicability and robust performance under varying conditions (Mishra and Mishra, 2023; Prodhan et al., 2021).

Uncertainty Quantification

- Yield prediction models inherently involve uncertainty stemming from sources such as data noise, model parameterization, and environmental variability. Effectively quantifying and communicating this uncertainty is crucial for informing decision-making and risk management.
- However, traditional machine learning approaches often provide point predictions without accounting for uncertainty, thereby potentially misleading stakeholders. Integrating probabilistic methods and uncertainty quantification techniques into yield prediction models is essential for enhancing their reliability and utility.

Scalability and Computational Resources

- Scalability and computational efficiency are paramount considerations in deploying yield prediction models at scale, particularly in the context of large agricultural landscapes and high-resolution remote sensing data.
- Resource-intensive model training and inference processes can pose practical challenges, especially for stakeholders with limited access to computational resources or expertise. Developing scalable algorithms and leveraging cloud computing infrastructure can help mitigate these challenges and facilitate the widespread adoption of yield prediction models (Shaikh et al., 2022).

Water Management and Irrigation Optimization

Water is a vital resource for agriculture, and sustainable farming methods depend heavily on how it is managed. In this section, we'll discuss how remote sensing and machine learning techniques can be used to optimize irrigation and water management strategies, increasing agricultural output while preserving water resources.

Estimation of Soil Moisture

Calculating the exact moisture content of the soil is crucial for assessing plant water availability and determining irrigation requirements. Remote sensing is a non-invasive method of gathering temporal and geographical data on soil moisture levels. A range of sensors, such as active microwave sensors and passive optical sensors, are used to track the moisture content of soil at various depths. Two examples of machine learning techniques that are used to analyze the data that was identified remotely and generate precise estimates of soil moisture are support vector machines and neural networks.

To guarantee accuracy and dependability in forecasting soil moisture levels across a variety of agricultural landscapes, these models are calibrated using ground-truth measurements and trained on historical data (Mekonnen et al., 2019).

Monitoring Water Stress

Water stress, caused by inadequate water availability or excessive transpiration, can significantly impact crop health and yield. Remote sensing techniques enable the monitoring of vegetation conditions, facilitating the detection and assessment of water stress in crops. Multispectral and hyperspectral sensors capture spectral reflectance patterns of crops, which are indicative of their physiological status, including water stress. Machine learning algorithms are employed to analyze these spectral signatures and classify vegetation into different stress levels. By leveraging advanced classification techniques such as random forests and CNNs, remote sensing data can effectively identify areas experiencing water stress, enabling timely intervention measures to mitigate crop losses and optimize water use efficiency.

Irrigation Scheduling Using Remote Sensing Data

Efficient irrigation scheduling is paramount for optimizing water usage and maximizing crop yields. Remote sensing technologies offer valuable insights into crop water requirements by monitoring various parameters, including evapotranspiration rates, crop coefficients, and soil moisture levels. By integrating remote sensing data with weather forecasts and crop growth models, machine learning algorithms can generate dynamic irrigation schedules tailored to specific crop types and environmental conditions (Tiwari et al., 2023). These models utilize historical data on crop response to irrigation treatments to optimize water allocation and timing, thereby minimizing water wastage and enhancing overall agricultural productivity. Additionally, the integration of remote sensing-based irrigation scheduling systems with precision irrigation technologies, such as drip and sprinkler systems, further enhances water efficiency and resource conservation in farm operations (De Rosa et al., 2021).

The integration of remote sensing with machine learning techniques offers a powerful toolkit for water management and irrigation optimization in agriculture. By providing timely and accurate information on soil moisture, water stress, and irrigation scheduling, these technologies enable farmers to make informed decisions that promote sustainable farming practices, mitigate water scarcity challenges, and enhance economic returns in agricultural production systems.

Soil Mapping and Fertility Assessment

One of the fundamental pillars of sustainable agriculture lies in understanding and optimizing soil health. This section will discuss how remote sensing, coupled with machine learning, revolutionizes soil mapping and fertility assessment, ultimately contributing to maximizing farm yield and economic returns.

Soil Type Classification

Understanding soil types is essential for efficient land management and crop production in modern agriculture. Sorting soil according to its physical, chemical, and biological characteristics is known as soil type classification. Farmers can now classify soil types more accurately by combining machine learning techniques with remote sensing. This allows farmers to make well-informed decisions about fertilization, irrigation, and crop selection (Escalante et al., 2019). Multispectral photographs of agricultural areas are taken by remote sensing platforms like satellites and drones, and these images include important information about the properties of the soil. Machine learning techniques are used to process these photos, analyzing spectral characteristics to distinguish between different types of soil. Distinct patterns linked to various soil types, including clayey, sandy, loamy, or silty soils, can be found using spectral analysis (Shendryk et al., 2021).

SVM, random forests, and neural networks are a few of the machine learning models that are trained on labelled data sets that contain spectral information and related soil types. With a high degree of accuracy, these algorithms identify patterns in the spectral data and classify pixels or regions into different soil types. There are multiple phases involved in classifying soil types: preprocessing the data, extracting features, training the model, and validating the model. The correctness and consistency of the spectrum data are guaranteed by preprocessing methods as an atmospheric and radiometric correction (Canata et al., 2021). To implement input features for machine learning models, feature extraction techniques gather pertinent data from the spectral bands, such as organic matter concentration, soil moisture content, and vegetation indices.

Once trained, the machine learning models can classify soil types across large agricultural landscapes with high spatial resolution. This information is invaluable for farmers in understanding the distribution of different soil types within their fields and optimizing management practices accordingly. By tailoring agronomic practices to specific soil types, such as adjusting irrigation schedules or selecting suitable crop varieties, farmers can maximize crop productivity while minimizing input costs and environmental impact (Bian et al., 2022).

Soil Fertility Mapping

To ensure sustainable crop production and optimize nutrient management strategies, soil fertility mapping is crucial. By monitoring soil fertility levels across agricultural landscapes using remote sensing and machine learning, farmers may increase overall farm output by more effectively targeting inputs. Platforms for remote sensing gather spectral data about soil characteristics such as the amount of organic matter, the concentration of nutrients, and the moisture content of the soil. This data is processed by machine learning algorithms, which produce comprehensive maps of soil fertility characteristics and important insights about the availability of nutrients and the health of the soil (Darwin et al., 2021).

Soil fertility indicators are proxied by a range of spectral indices that are obtained from remote sensing data. The NDVI, for instance, is frequently used to calculate vegetation vigor, which is associated with soil fertility. Additional indices that can offer important details regarding the condition of soil fertility include the chlorophyll vegetation index (CVI) and the soil-adjusted vegetation index (SAVI). Spectral data and matching ground-truth measurements of soil fertility metrics, like potassium, phosphorus, and nitrogen levels, are used to train machine learning models. These models can forecast soil fertility values across the terrain with accuracy by learning to recognize patterns in the spectral data that correspond to varying levels of fertility (Mishra, 2024; Teodoro et al., 2021).

The resulting soil fertility maps enable farmers to implement site-specific nutrient management strategies, tailoring fertilizer applications to the specific needs of different areas within their fields. By applying fertilizers more precisely, farmers can optimize nutrient use efficiency, reduce input costs, and minimize environmental impacts such as nutrient runoff and leaching. Furthermore, soil fertility mapping can help identify areas of nutrient deficiencies or excesses, allowing farmers to take corrective actions such as soil amendments or cover cropping to improve soil health and productivity over the long term. By integrating remote sensing technology with machine learning algorithms, farmers can make informed decisions about nutrient management, leading to more sustainable and profitable agricultural systems (Al-Adhaileh and Aldhyani, 2022).

Nutrient Management Strategies

Reducing environmental impact and raising crop yields depend on effective fertilizer management. With the use of machine learning and remote sensing, farmers can now optimize their nutrient management strategies and make data-driven decisions on crop rotation, soil amendments, and fertilizer application. By using machine learning algorithms to evaluate remote sensing data, farmers can discover areas of nutrient surpluses or shortfalls in their fields by creating comprehensive maps of soil nutrient levels. Through the overlay of agronomic prescriptions and crop yield data on these maps, farmers may create customized nutrient management plans to meet the unique nutrient requirements of their soil and maximize crop performance.

Farmers may apply inputs only where they are required thanks to variable rate technology (VRT), which enables accurate application of fertilizers and soil amendments based on real-time soil nutrient maps. This minimizes environmental effects including nutrient runoff and leaching and increases nutrient use efficiency while also lowering input costs (Saikai et al., 2020). Apart from conventional fertilizers, the efficacy of substitute nutrient sources such as agricultural leftovers, cover crops, and organic amendments can also be evaluated using remote sensing data. By analyzing spectral signatures linked to various soil additions and forecasting their effects on crop yield and soil fertility, machine learning models assist farmers in choosing the best nutrient management strategies for their unique circumstances. (Abioye et al., 2022).

Moreover, during the growing season, remote sensing technology can track crop growth and development, giving farmers the ability to instantly modify their nutrient management plans in response to changing crop requirements and environmental circumstances. Farmers can enhance nutrient management strategies to maximize yields, profitability, and environmental sustainability by combining remote sensing data with agronomic models and decision support systems (Cao et al., 2021; De Lara et al., 2023). An essential part of sustainable agriculture is soil mapping and fertility assessment, which help farmers reduce their environmental impact, maximize crop output, and optimize nutrient management techniques. To enable farmers to make educated decisions about nutrient management strategies and ultimately create more profitable and sustainable farming systems, remote sensing and machine learning are combined to provide powerful tools for accurately mapping soil properties and predicting nutrient levels across agricultural landscapes (Gupta et al., 2022).

Weed Detection and Management

A major challenge to agricultural productivity is weed infestation, which raises production costs and results in yield losses. Recently, promising solutions for efficient weed detection and management have emerged: remote sensing technology combined with machine learning algorithms. This section explores the nuances of using these methods to address different weed detection and management issues, including the classification of weed species, mapping weed densities, and applying herbicides precisely.

Weed Species Classification

Classifying weed species is an essential part of managing weeds since different species have varied growth patterns, morphologies, and reactions to control methods. Accurate weed species discrimination among crop plants is made possible by utilizing powerful machine learning algorithms in conjunction with remote sensing data, specifically multispectral or hyperspectral photography. To this end, feature extraction is essential, since it allows the identification and use of spectral signatures specific to various weed species in the classification process. SVM, random forests, and CNN are examples of machine learning models that are frequently used because of their capacity to identify intricate patterns in spectral data (Li et al., 2022).

Furthermore, the integration of ancillary data such as geographic information system (GIS) layers, soil properties, and weather conditions enhances the robustness of classification models. By harnessing the power of remote sensing and machine learning, farmers can swiftly identify prevalent weed species within their fields, facilitating targeted intervention strategies tailored to specific weed types (Mohammad, 2023).

Weed Density Mapping

Weed density mapping is a helpful tool for comprehending the quantity and geographical distribution of weeds in agricultural settings. Through the non-destructive and efficient assessment of weed density over large areas, remote sensing technologies assist farmers in more efficient resource allocation and management effort prioritization. Weed patches can be precisely recognized and measured by using high-resolution imagery and machine learning methods. The use of image segmentation algorithms allows for the detection and characterisation of weed-infested areas by dividing the landscape into homogeneous zones.

Moreover, as weed growth dynamics change during the growing season, the temporal dimension is critical to weed density mapping. Monitoring variations in weed density over time with time-series analysis of remote sensing data enables prompt intervention tactics and adaptive management techniques. Farmers can minimize the use of herbicides and slow the spread of herbicide-resistant weeds by employing targeted management strategies, such as spot spraying or manual removal, by creating accurate weed density maps (Lozano-Garzon et al., 2022).

Precision Herbicide Application

Precision herbicide application represents a paradigm shift in weed management, wherein herbicides are selectively applied to weed-infested areas while minimizing their impact on non-target vegetation. Remote sensing technologies, coupled with machine learning algorithms, play a pivotal role in enabling precise herbicide application through the delineation of weed-infested zones and the generation of prescription maps. By integrating spectral information derived from remote sensing data with spatially explicit weed maps, precision spraying systems can autonomously target weeds with a high degree of accuracy. Machine learning algorithms, trained on labelled data encompassing spectral signatures of both weeds and crops, facilitate real-time decision-making regarding herbicide application rates and target locations (Meroni et al., 2021).

Furthermore, advances in sensor technology, such as proximal sensors and UAVs, enable on-the-fly monitoring of weed distribution during spraying operations, ensuring precise and efficient herbicide delivery. Precision herbicide application not only reduces herbicide usage and associated environmental impacts but also enhances cost-effectiveness and operational efficiency. By minimizing herbicide drift and off-target effects, farmers can achieve effective weed control while preserving crop health and environmental quality (Alibabaei et al., 2022).

Economic Analysis and Decision Support Systems

Enhancing farm output, resource management, and economic returns can be greatly increased by integrating remote sensing technologies with economic frameworks and decision support systems. A few of the components that are examined in this section are economic effect assessment, cost-benefit analysis, and the creation of decision support systems specifically for the agriculture industry.

Cost-Benefit Analysis of Remote Sensing Applications

A key instrument for assessing the viability and effectiveness of using remote sensing technologies in agriculture is the cost-benefit analysis (CBA). It facilitates decision-making by quantifying the costs and benefits associated with such applications over a specified time horizon. NPV, a key metric in CBA, enables the comparison of the present value of costs against the present value of benefits, considering the time value of money (Kaur and Sivia, 2024). The formula for NPV can be expressed as:

$$NPV = \sum_{t=1}^{n} \frac{CF_t}{(1+r)^t} - Initial\ Investment \tag{1}$$

where:

- NPV = Net Present Value
- CF_t = Cash flow at time t
- r = Discount rate
- n = Number of periods
- "Initial Investment" refers to the initial outlay or investment required for the project.

Each cash flow (both costs and benefits) is discounted to its present value using the formula:

$$PV = \frac{CF_t}{(1+r)^t} \tag{2}$$

where:

- PV = Present value
- CF = Cash flow at time t
- r = Discount rate
- t = Time period

Then, the present values of all cash flows are summed up, and the initial investment is subtracted from the total to obtain the NPV.

Methodology:

- *Identification of Costs and Benefits*: The first stage is to list and classify all pertinent expenses and advantages related to the use of remote sensing applications in agriculture. Costs might include opportunity costs, operational costs (like maintenance and training), and initial investment costs (like software development and equipment acquisition). Conversely, advantages include increased crop output, more effective use of resources, lower input costs (fertilizers, water, etc.), and possible income increases.
- *Time Frame Selection*: The time horizon for the cost-benefit analysis should be carefully chosen to capture both the short-term and long-term effects of remote sensing technologies. Typically, a multi-year time frame is preferred to encompass the lifecycle of the technology and its impact on agricultural productivity and profitability.
- *Discount Rate Determination*: Because it takes the time worth of money into account and represents the opportunity cost of capital, choosing the right discount rate is essential. Several variables, including the rate of inflation, the risk profile, and the current interest rates in the market, might affect the discount rate.
- *Cash Flow Estimation*: Next, cash flows associated with costs and benefits are estimated for each period within the chosen time horizon. Costs are treated

as negative cash flows, while benefits are considered positive cash flows. It is imperative to account for uncertainties and risks associated with cash flow projections.

- *Net Present Value Calculation*: The NPV is computed by discounting the net cash flows (benefits minus costs) from each period to their present value using the chosen discount rate. The summation of discounted cash flows yields the NPV, indicating the net economic value generated by the remote sensing applications (Wongchai et al., 2022; Yoosefzadeh-Najafabadi et al., 2021).

Example: Suppose, for this analysis, that a farmer spends $50,000 on remote sensing equipment for crop management and monitoring. The benefits of higher crop output and lower input costs are projected to be worth $20,000 yearly, while the operational costs come in at $10,000. With a five-year timeframe and a 5% discount rate, we can compute the net present value (NPV) as follows:

- Year 1: Benefit = $20,000 – $10,000 (Cost) = $10,000
- Year 2: Benefit = $20,000 – $10,000 (Cost) = $10,000
- Year 3: Benefit = $20,000 – $10,000 (Cost) = $10,000
- Year 4: Benefit = $20,000 – $10,000 (Cost) = $10,000
- Year 5: Benefit = $20,000 – $10,000 (Cost) = $10,000

Calculating the present value of cash flows for each year and summing them up, we obtain:

$$NPV = \frac{10{,}000}{(1+0.05)^1} + \frac{10{,}000}{(1+0.05)^2} + \frac{10{,}000}{(1+0.05)^3} + \frac{10{,}000}{(1+0.05)^4} + \frac{10{,}000}{(1+0.05)^5} - 50{,}000 \tag{3}$$

$$NPV = \frac{10{,}000}{1.05} + \frac{10{,}000}{1.1025} + \frac{10{,}000}{1.1576} + \frac{10{,}000}{1.2155} + \frac{10{,}000}{1.2763} - 50{,}000 \tag{4}$$

$$NPV = 9523.81 + 9072.64 + 8638.55 + 8222.41 + 7823.48 - 50{,}000 \tag{5}$$

$NPV = \$12{,}280.89$

The calculated NPV of $12,280.89 suggests that the investment in remote sensing technology generates a positive net economic value over the specified time horizon. This indicates that the benefits accrued from improved crop yield and reduced input costs outweigh the initial investment and operational expenses, thereby justifying the adoption of remote sensing applications in agriculture.

Economic Impact Assessment

Beyond the immediate cost-benefit analysis, it is imperative to undertake a broader economic impact assessment to gauge the ramifications of remote sensing adoption on the agricultural sector and the wider economy. This assessment involves analyzing the ripple effects of enhanced agricultural productivity facilitated by remote sensing technologies. One aspect of economic impact assessment involves evaluating the income generation potential for farmers through increased yields and improved crop management practices. Additionally, the savings accrued from optimized resource utilization, reduced input costs, and more efficient pest control measures contribute to the overall economic benefit.

Furthermore, the adoption of remote sensing can stimulate ancillary industries, such as technology providers, data analytics firms, and agricultural consultancy services, thereby fostering economic growth and job creation within these sectors. By quantifying the multiplier effects of remote sensing adoption, policymakers and industry stakeholders can better understand its broader economic implications (Kumar et al., 2023; Zaman, 2023). However, it is also essential to consider any potential adverse effects or economic disruptions that may arise from the widespread adoption of remote sensing technologies. This includes the displacement of labour in traditional agricultural roles, market distortions due to unequal access to technology, and potential privacy or data security concerns. Economic impact assessment provides a holistic understanding of the macroeconomic implications of remote sensing adoption, guiding policymakers and stakeholders in formulating strategies to maximize its socio-economic benefits while mitigating potential drawbacks (Ennouri and Kallel, 2019).

Development of Decision Support Systems

With the incorporation of remote sensing data into decision support systems (DSS), agriculture management has undergone a paradigm shift, offering farmers timely insights and useful intelligence. This paragraph explains how to design dependable decision support systems that are tailored to the specific needs of agricultural stakeholders (Zhang et al., 2021). Since it provides a comprehensive picture of agricultural landscapes, the integration of data from multiple sources, including satellite imagery, aerial surveys, and ground-based sensors, is crucial to the development of DSSs. Algorithms for machine learning are essential for analyzing these enormous datasets, finding trends, and creating prediction models that help in decision-making.

The development process begins with the identification of key decision variables, such as planting schedules, irrigation strategies, and pest management protocols, based on which the DSS architecture is designed. Subsequently, historical and real-time remote sensing data are ingested into the system, undergoing preprocessing and feature extraction to facilitate algorithmic analysis. Predictive models are trained using machine learning techniques like supervised learning, unsupervised learning, and reinforcement learning to forecast agricultural outcomes based on input parameters. To improve the precision and dependability of these models, they are regularly improved and verified using ground truth data (Bal and Kayaalp, 2021).

The user interface of the decision support system is designed to be intuitive and user-friendly, providing farmers and agricultural practitioners with actionable insights presented in a visually comprehensible manner. This facilitates informed decision-making in areas such as crop selection, resource allocation, and risk management. Moreover, the adaptability and scalability of decision support systems enable customization to suit varying agricultural contexts and evolving environmental conditions (Tiwari et al., 2023). By empowering farmers with data-driven decision-making tools, remote sensing-enabled DSS fosters sustainable agricultural practices, optimizing yields, minimizing input costs, and mitigating environmental impact. The development of decision support systems represents a pivotal advancement in leveraging remote sensing technologies for sustainable agriculture. By harnessing the power of machine learning and real-time data analytics, these systems facilitate evidence-based decision-making, ultimately enhancing farm productivity, economic returns, and environmental stewardship (de Camargo et al., 2021).

Future Directions and Challenges

As we look towards the future of remote sensing for sustainable agriculture, it is evident that numerous opportunities and challenges lie ahead. When considering how remote sensing for sustainable agriculture will develop in the future and how it will be integrated with machine learning techniques, several important directions and obstacles become apparent that will influence the course of this subject.

Emerging Technologies in Remote Sensing

The rapid advancement of remote sensing technologies offers promising opportunities to improve agricultural practices. The widespread use of multispectral and hyperspectral imaging systems, which provide better spectral resolution and sensitivity, is one noteworthy development. With the use of these technologies, crop health, nutrient levels, and environmental variables may be monitored with greater accuracy (Ndikumana et al., 2018; Hassan-Esfahani et al., 2015). Additionally, the advent of UAVs and satellites equipped with advanced sensors has expanded the scope and accessibility of remote sensing applications. These platforms facilitate comprehensive and timely data collection over large agricultural areas, empowering farmers with actionable insights for informed decision-making (Ali et al., 2016).

Furthermore, the integration of LiDAR (Light Detection and Ranging) technology holds promise for enhanced three-dimensional mapping of agricultural landscapes. By capturing detailed elevation data and identifying terrain variations, LiDAR enables precise delineation of field boundaries and topographic features. This information is invaluable for optimizing irrigation, drainage, and land management strategies, ultimately contributing to resource conservation and productivity gains (Cai et al., 2018).

However, despite these advancements, challenges remain in harnessing the full potential of emerging remote sensing technologies. Issues such as data integration, processing complexity, and cost-effectiveness pose significant barriers to widespread adoption. Addressing these challenges will require interdisciplinary collaboration, innovative algorithm development, and investment in infrastructure and capacity building.

Integration of AI and Computer Vision

Applications for remote sensing in agriculture could be revolutionized by combining AI with computer vision. In particular, machine learning algorithms have proven to be exceptionally adept at analyzing vast amounts of remote sensing data and deriving useful insights with previously unheard-of efficiency and precision. For applications like crop classification, disease detection, and yield prediction, researchers can create reliable models by utilizing AI techniques like deep learning, CNNs, and recurrent neural networks (RNNs). These models provide comprehensive evaluation of crop health and climatic conditions by learning intricate patterns and connections from a variety of data sources, including as satellite photography, UAV footage, and ground-based sensors (Singh and Kaur, 2022).

Moreover, the integration of computer vision algorithms enables automated feature extraction and object detection from remote sensing imagery. This capability facilitates the identification of crop types, pest infestations, and other critical factors influencing agricultural productivity. Furthermore, advancements in edge computing and onboard processing enable real-time analysis of remote sensing data, empowering farmers with timely insights for proactive decision-making in the field (Xu et al., 2020). Nevertheless, the widespread adoption of AI and computer vision technologies in agriculture faces challenges related to data quality, model interpretability, and algorithm scalability. Addressing these challenges will require ongoing research efforts to improve algorithm robustness, enhance data integration pipelines, and ensure the ethical and responsible use of AI in agricultural decision support systems (Veerachamy and Ramar, 2022; Dhanaraju et al., 2022).

Addressing Scalability and Data Accessibility Issues

Scalability of data collection and analysis workflows is one of the main issues with remote sensing for sustainable agriculture. Scalable solutions for data collection, processing, and distribution are required as agricultural landscapes continue to grow and diversity and as there is a growing need for thorough and timely information. Researchers need to look for new ways to optimize resource allocation and streamline data collection workflows to overcome scaling issues. This could entail creating automated data gathering technologies that can effectively cover vast agricultural areas with little assistance from humans, like satellite constellations or UAV swarms. Furthermore, advancements in cloud computing and distributed processing architectures can facilitate parallelized analysis of large-scale remote sensing datasets, enabling rapid generation of actionable insights at scale (Chatterjee et al., 2020).

In addition to scalability concerns, ensuring data accessibility and interoperability is crucial for maximizing the impact of remote sensing technologies in agriculture. Open data initiatives, data sharing agreements, and standardized data formats play a pivotal role in promoting collaboration and knowledge exchange among stakeholders (Mishra, 2024). However, achieving widespread data accessibility and interoperability requires concerted efforts to overcome technical, institutional, and cultural barriers. This necessitates the development of data governance frameworks, data management protocols, and capacity-building initiatives to promote data literacy and empower stakeholders with the necessary skills and resources to harness the full potential of remote sensing technologies for sustainable agriculture (McLennon et al., 2021).

Conclusion

Remote sensing coupled with machine learning has emerged as a powerful toolkit for transforming agricultural practices towards a more sustainable and data-driven future. By leveraging diverse remote sensing platforms – from high-resolution satellites to manoeuvrable UAVs - farmers gain unprecedented access to real-time information about their fields. Pre-processed and analyzed through machine learning algorithms, this data empowers them to make informed decisions across various aspects of farm management. Crop monitoring and yield prediction become significantly more precise. Studies have shown that remote sensing-based yield prediction models can achieve accuracy rates exceeding 80%, significantly reducing uncertainties and allowing for proactive resource allocation. Water management, a critical challenge facing agriculture globally, is revolutionized with remote sensing data enabling estimation of soil moisture and water stress. This facilitates the adoption of precision irrigation practices, potentially saving billions of cubic meters of water annually. Furthermore, remote sensing empowers creation of detailed soil maps, aiding in targeted nutrient application and improved soil fertility. The economic benefits of remote sensing are undeniable. Studies reveal that for every dollar invested in precision agriculture using remote sensing, farmers can expect an average return of 4.5 dollars. Decision support systems built upon this data empower farmers to optimize resource use, minimize waste, and maximize economic returns. The future of remote sensing in agriculture is bright. Emerging technologies like hyperspectral imaging offer even more detailed data for comprehensive crop health analysis. Integration of artificial intelligence and computer vision will further enhance automation and real-time decision-making capabilities. However, challenges remain. Scalability and data accessibility issues need to be addressed to ensure the widespread adoption of these technologies, particularly by small-scale farmers. Nevertheless, with ongoing research and development, remote sensing holds immense potential to ensure agricultural sustainability, food security, and economic prosperity for future generations.

References

Abbas, F., Afzaal, H., Farooque, A.A. and Tang, S. 2020. Crop yield prediction through proximal sensing and machine learning algorithms. *Agronomy*, 10: 1046.

Abioye, E.A., Hensel, O., Esau, T.J., Elijah, O., Abidin, M.S.Z., Ayobami, A.S. and Nasirahmadi, A. 2022. Precision irrigation management using machine learning and digital farming solutions. *Agri. Engineering*, 4: 70–103.

Al-Adhaileh, M.H. and Aldhyani, T.H. 2022. Artificial intelligence framework for modeling and predicting crop yield to enhance food security in Saudi Arabia. *PeerJ Computer Science*, 8: e1104.

Ali, I., Cawkwell, F., Dwyer, E. and Green, S. 2016. Modeling managed grassland biomass estimation by using multitemporal remote sensing data—A machine learning approach. *IEEE Journal of Selected Topics in Applied Earth Observations and Remote Sensing*, 10: 3254–3264.

Alibabaei, K., Gaspar, P.D., Lima, T.M., Campos, R.M., Girão, I., Monteiro, J. and Lopes, C.M. 2022. A review of the challenges of using deep learning algorithms to support decision-making in agricultural activities. *Remote Sensing*, 14: 638.

Bal, F. and Kayaalp, F. 2021. Review of machine learning and deep learning models in agriculture. *International Advanced Researches and Engineering Journal*, 5: 309–323.

Balducci, F., Impedovo, D. and Pirlo, G. 2018. Machine learning applications on agricultural datasets for smart farm enhancement. *Machines*, 6: 38.

Bali, A., Chouhan, S.S., Kumar, G., Kumar, R. and Singh, U.P. 2023. Adaptive fault-tolerant control for pure-feedback stochastic nonlinear systems with sensor and actuator faults. *Circuits, Systems and Signal Processing*, 42: 5838–5867.

Benami, E., Jin, Z., Carter, M.R., Ghosh, A., Hijmans, R.J., Hobbs, A. and Lobell, D.B. 2021. Uniting remote sensing, crop modelling and economics for agricultural risk management. *Nature Reviews Earth & Environment*, 2: 140–159.

Benos, L., Tagarakis, A.C., Dolias, G., Berruto, R., Kateris, D. and Bochtis, D. 2021. Machine learning in agriculture: A comprehensive updated review. *Sensors*, 21: 3758.

Bharadiya, J.P., Tzenios, N.T. and Reddy, M. 2023. Forecasting of crop yield using remote sensing data, agrarian factors and machine learning approaches. *Journal of Engineering Research and Reports*, 24: 29–44.

Bian, C., Shi, H., Wu, S., Zhang, K., Wei, M., Zhao, Y. and Chen, S. 2022. Prediction of field-scale wheat yield using machine learning method and multi-spectral UAV data. *Remote Sensing*, 14: 1474.

Cai, Y., Guan, K., Lobell, D., Potgieter, A.B., Wang, S., Peng, J. and Peng, B. 2019. Integrating satellite and climate data to predict wheat yield in Australia using machine learning approaches. *Agricultural and Forest Meteorology*, 274: 144–159.

Cai, Y., Guan, K., Peng, J., Wang, S., Seifert, C., Wardlow, B. and Li, Z. 2018. A high-performance and in-season classification system of field-level crop types using time-series Landsat data and a machine learning approach. *Remote Sensing of Environment*, 210: 35–47.

Canata, T.F., Wei, M.C.F., Maldaner, L.F. and Molin, J.P. 2021. Sugarcane yield mapping using high-resolution imagery data and machine learning technique. *Remote Sensing*, 13: 232.

Cao, J., Zhang, Z., Tao, F., Zhang, L., Luo, Y., Zhang, J. and Xie, J. 2021. Integrating multi-source data for rice yield prediction across China using machine learning and deep learning approaches. *Agricultural and Forest Meteorology*, 297: 108275.

Chatterjee, S., Dey, N. and Sen, S. 2020. Soil moisture quantity prediction using optimized neural supported model for sustainable agricultural applications. *Sustainable Computing: Informatics and Systems*, 28: 100279.

Chlingaryan, A., Sukkarieh, S. and Whelan, B. 2018. Machine learning approaches for crop yield prediction and nitrogen status estimation in precision agriculture: A review. *Computers and Electronics in Agriculture*, 151: 61–69.

Darwin, B., Dharmaraj, P., Prince, S., Popescu, D.E. and Hemanth, D.J. 2021. Recognition of bloom/yield in crop images using deep learning models for smart agriculture: A review. *Agronomy*, 11: 646.

de Camargo, T., Schirrmann, M., Landwehr, N., Dammer, K.H. and Pflanz, M. 2021. Optimized deep learning model as a basis for fast UAV mapping of weed species in winter wheat crops. *Remote Sensing*, 13: 1704.

De Lara, A., Mieno, T., Luck, J.D. and Puntel, L.A. 2023. Predicting site-specific economic optimal nitrogen rate using machine learning methods and on-farm precision experimentation. *Precision Agriculture*, 24: 1792–1812.

De Rosa, D., Basso, B., Fasiolo, M., Friedl, J., Fulkerson, B., Grace, P.R. and Rowlings, D.W. 2021. Predicting pasture biomass using a statistical model and machine learning algorithm implemented with remotely sensed imagery. *Computers and Electronics in Agriculture*, 180: 105880.

Delgado, J.A., Short Jr, N.M., Roberts, D.P. and Vandenberg, B. 2019. Big data analysis for sustainable agriculture on a geospatial cloud framework. *Frontiers in Sustainable Food Systems*, 3: 54.

Dhanaraju, M., Chenniappan, P., Ramalingam, K., Pazhanivelan, S. and Kaliaperumal, R. 2022. Smart farming: Internet of Things (IoT)-based sustainable agriculture. *Agriculture*, 12: 1745.

Di, Y., Gao, M., Feng, F., Li, Q. and Zhang, H. 2022. A new framework for winter wheat yield prediction integrating deep learning and Bayesian optimization. *Agronomy*, 12: 3194.

Diaz-Gonzalez, F.A., Vuelvas, J., Correa, C.A., Vallejo, V.E. and Patino, D. 2022. Machine learning and remote sensing techniques applied to estimate soil indicators–review. *Ecological Indicators*, 135: 108517.

Elavarasan, D. and Vincent, P.D. 2020. Crop yield prediction using deep reinforcement learning model for sustainable agrarian applications. *IEEE Access*, 8: 86886–86901.

Ennouri, K. and Kallel, A. 2019. Remote sensing: an advanced technique for crop condition assessment. *Mathematical Problems in Engineering*, 2019(1): 1–8.

Escalante, H.J., Rodríguez-Sánchez, S., Jiménez-Lizárraga, M., Morales-Reyes, A., De La Calleja, J. and Vazquez, R. 2019. Barley yield and fertilization analysis from UAV imagery: A deep learning approach. *International Journal of Remote Sensing*, 40: 2493–2516.

Eugenio, F.C., Grohs, M., Venancio, L.P., Schuh, M., Bottega, E.L., Ruoso, R. and Fernandes, P. 2020. Estimation of soybean yield from machine learning techniques and multispectral RPAS imagery. *Remote Sensing Applications: Society and Environment*, 20: 100397.

Gómez, D., Salvador, P., Sanz, J. and Casanova, J.L. 2019. Potato yield prediction using machine learning techniques and sentinel data. *Remote Sensing*, 11: 1745.

Gupta, S., Geetha, A., Sankaran, K.S., Zamani, A.S., Ritonga, M., Raj, R. and Mohammed, H.S. 2022. Machine learning-and feature selection-enabled framework for accurate crop yield prediction. *Journal of Food Quality*, 1–7.

Han, J., Zhang, Z., Cao, J., Luo, Y., Zhang, L., Li, Z. and Zhang, J. 2020. Prediction of winter wheat yield based on multi-source data and machine learning in China. *Remote Sensing*, 12: 236.

Hara, P., Piekutowska, M. and Niedbała, G. 2021. Selection of independent variables for crop yield prediction using artificial neural network models with remote sensing data. *Land*, 10: 609.

Hassan-Esfahani, L., Torres-Rua, A. and McKee, M. 2015. Assessment of optimal irrigation water allocation for pressurized irrigation system using water balance approach, learning machines and remotely sensed data. *Agricultural Water Management*, 153: 42–50.

Johnson, M.D., Hsieh, W.W., Cannon, A.J., Davidson, A. and Bédard, F. 2016. Crop yield forecasting on the Canadian Prairies by remotely sensed vegetation indices and machine learning methods. *Agricultural and Forest Meteorology*, 218: 74–84.

Ju, S., Lim, H., Ma, J.W., Kim, S., Lee, K., Zhao, S. and Heo, J. 2021. Optimal county-level crop yield prediction using MODIS-based variables and weather data: A comparative study on machine learning models. *Agricultural and Forest Meteorology*, 307: 108530.

Kaur, G. and Sivia, J.S. 2024. Integrating data envelopment analysis and machine learning approaches for energy optimization, decreased carbon footprints and wheat yield prediction across North-Western India. *Journal of Soil Science and Plant Nutrition*, 1–24.

Kayad, A., Sozzi, M., Gatto, S., Marinello, F. and Pirotti, F. 2019. Monitoring within-field variability of corn yield using Sentinel-2 and machine learning techniques. *Remote Sensing*, 11: 2873.

Khan, A., Vibhute, A.D., Mali, S. and Patil, C.H. 2022. A systematic review on hyperspectral imaging technology with a machine and deep learning methodology for agricultural applications. *Ecological Informatics*, 69: 101678.

Khanal, S., Fulton, J., Klopfenstein, A., Douridas, N. and Shearer, S. 2018. Integration of high resolution remotely sensed data and machine learning techniques for spatial prediction of soil properties and corn yield. *Computers and Electronics in Agriculture*, 153: 213–225.

Khanal, S., Kc, K., Fulton, J.P., Shearer, S. and Ozkan, E. 2020. Remote sensing in agriculture—Accomplishments, limitations and opportunities. *Remote Sensing*, 12: 3783.

Kumar, N., Kushwaha, R.R., Meena, N.R., Mishra, H. and Yadav, A.P.S. 2023. A study on costs and returns of paddy cultivation in Ambedkar Nagar district of Uttar Pradesh. *International Journal of Statistics and Applied Mathematics*, 8: 107–111.

Kumar, R., Singh, M.P., Kumar, P. and Singh, J.P. 2015. Crop Selection Method to maximize crop yield rate using machine learning technique. In *2015 International Conference on Smart Technologies and Management for Computing, Communication, Controls, Energy and Materials (ICSTM)* (pp. 138–145). IEEE.

Kumar, S., Meena, R.S., Sheoran, S., Jangir, C.K., Jhariya, M.K., Banerjee, A. and Raj, A. 2022. Remote sensing for agriculture and resource management. In *Natural Resources Conservation and Advances for Sustainability* (pp. 91–135). Elsevier.

Li, K.Y., Sampaio de Lima, R., Burnside, N.G., Vahtmäe, E., Kutser, T., Sepp, K. and Sepp, K. 2022. Toward automated machine learning-based hyperspectral image analysis in crop yield and biomass estimation. *Remote Sensing*, 14: 1114.

Li, L., Wang, B., Feng, P., Li Liu, D., He, Q., Zhang, Y. and Yu, Q. 2022. Developing machine learning models with multi-source environmental data to predict wheat yield in China. *Computers and Electronics in Agriculture*, 194: 106790.

Lin, F., Weng, Y., Chen, H. and Zhuang, P. 2021. Intelligent greenhouse system based on remote sensing images and machine learning promotes the efficiency of agricultural economic growth. *Environmental Technology & Innovation*, 24: 101758.

Lozano-Garzon, C., Bravo-Córdoba, G., Castro, H., González-Rodríguez, G., Niño, D., Nuñez, H. and Suárez, L.I. 2022. Remote sensing and machine learning modeling to support the identification of sugarcane crops. *IEEE Access*, 10: 17542–17555.

Machwitz, M., Pieruschka, R., Berger, K., Schlerf, M., Aasen, H., Fahrner, S. and Rascher, U. 2021. Bridging the gap between remote sensing and plant phenotyping—Challenges and opportunities for the next generation of sustainable agriculture. *Frontiers in Plant Science*, 12: 749374.

Martos, V., Ahmad, A., Cartujo, P. and Ordoñez, J. 2021. Ensuring agricultural sustainability through remote sensing in the era of agriculture 5.0. *Applied Sciences*, 11: 5911.

McLennon, E., Dari, B., Jha, G., Sihi, D. and Kankarla, V. 2021. Regenerative agriculture and integrative permaculture for sustainable and technology driven global food production and security. *Agronomy Journal*, 113: 4541–4559.

Mekonnen, Y., Namuduri, S., Burton, L., Sarwat, A. and Bhansali, S. 2019. Machine learning techniques in wireless sensor network based precision agriculture. *Journal of the Electrochemical Society*, 167: 037522.

Meraj, G., Kanga, S., Ambadkar, A., Kumar, P., Singh, S.K., Farooq, M. and Sahu, N. 2022. Assessing the yield of wheat using satellite remote sensing-based machine learning algorithms and simulation modeling. *Remote Sensing*, 14: 3005.

Meroni, M., Waldner, F., Seguini, L., Kerdiles, H. and Rembold, F. 2021. Yield forecasting with machine learning and small data: What gains for grains? *Agricultural and Forest Meteorology*, 308: 108555.

Mhango, J.K., Harris, E.W., Green, R. and Monaghan, J.M. 2021. Mapping potato plant density variation using aerial imagery and deep learning techniques for precision agriculture. *Remote Sensing*, 13: 2705.

Mishra, H. 2024. Counting the costs: Understanding agricultural economics and the concept of costs. pp. 198–210. *In*: Krashankant Sharma, Pranjal Singh, Shreya Singh, Ashwani Yadav and Vedangi Awasthi (eds.). *Text Book on Practices for Agricultural Sustainability*. Golden Leaf Publishers.

Mishra, H. 2024. Molecular technology in agriculture. pp. 204–226. *In*: Deepak Kumar Patel, Shripati Dwivedi, Anupam Dubey, Abhishek Kumar and Waris Ali (eds.). *Agricultural Transformation in the Modern Era*. Elite Publishing House.

Mishra, H. and Mishra, D. 2023. Artificial intelligence and machine learning in agriculture: Transforming farming systems. *Res. Trends Agric. Sci.*, 1: 1–16.

Mishra, H. and Mishra, D. 2024. Economic evaluation of UAV-based soil sampling approaches. In *Applications of Computer Vision and Drone Technology in Agriculture 4.0* (pp. 271–291). Singapore: Springer Nature Singapore.

Mishra, H., Tiwari, A.K. and Nishad, D.C. 2024. Economic viability of sustainable agriculture practices in modern farming. *Advances in Agriculture Sciences Volume II*, 24: 105.

Mishra, S., Mishra, D. and Santra, G.H. 2016. Applications of machine learning techniques in agricultural crop production: A review paper. *Indian Journal of Science and Technology*, 9(38): 1–14.

Mohamed, M. 2023. Agricultural sustainability in the age of deep learning: Current trends, challenges and future trajectories. *Sustainable Machine Intelligence Journal*, 4: 2–1.

Muniasamy, A. 2020. Machine learning for smart farming: a focus on desert agriculture. In *2020 International Conference on Computing and Information Technology (ICCIT-1441)* (pp. 1–5). IEEE.

Nay, J., Burchfield, E. and Gilligan, J. 2018. A machine-learning approach to forecasting remotely sensed vegetation health. *International Journal of Remote Sensing*, 39: 1800–1816.

Ndikumana, E., Ho Tong Minh, D., Baghdadi, N., Courault, D. and Hossard, L. 2018. Deep recurrent neural network for agricultural classification using multitemporal SAR Sentinel-1 for Camargue, France. *Remote Sensing*, 10: 1217.

Nguyen, H.A.T., Sophea, T., Gheewala, S.H., Rattanakom, R., Areerob, T. and Prueksakorn, K. 2021. Integrating remote sensing and machine learning into environmental monitoring and assessment of land use change. *Sustainable Production and Consumption*, 27: 1239–1254.

Nishad, D.C., Mishra, H., Tiwari, A.K. and Mishra, D. 2024. Post-harvest management: Enhancing food security and sustainability. *Advances in Agriculture Sciences Volume II*, 24: 136.

Paudel, D., Boogaard, H., de Wit, A., Janssen, S., Osinga, S., Pylianidis, C. and Athanasiadis, I.N. 2021. Machine learning for large-scale crop yield forecasting. *Agricultural Systems*, 187: 103016.

Prodhan, F.A., Zhang, J., Yao, F., Shi, L., Pangali Sharma, T.P., Zhang, D. and Mohana, H.P. 2021. Deep learning for monitoring agricultural drought in South Asia using remote sensing data. *Remote Sensin.*, 13: 1715.

Radočaj, D., Jurišić, M. and Gašparović, M. 2022. The role of remote sensing data and methods in a modern approach to fertilization in precision agriculture. *Remote Sensing*, 14: 778.

Rashid, M., Bari, B.S., Yusup, Y., Kamaruddin, M.A. and Khan, N. 2021. A comprehensive review of crop yield prediction using machine learning approaches with special emphasis on palm oil yield prediction. *IEEE Access*, 9: 63406–63439.

Ruan, G., Li, X., Yuan, F., Cammarano, D., Ata-UI-Karim, S.T., Liu, X. and Cao, Q. 2022. Improving wheat yield prediction integrating proximal sensing and weather data with machine learning. *Computers and Electronics in Agriculture*, 195: 106852.

Saikai, Y., Patel, V. and Mitchell, P.D. 2020. Machine learning for optimizing complex site-specific management. *Computers and Electronics in Agriculture*, 174: 105381.

Saikanth, D.R.K., Ragini, M., Tripathi, G., Rajesh, K., Giri, A., Pandey, S.K. and Verma, L.K. 2024. The impact of emerging technologies on sustainable agriculture and rural development. *International Journal of Environment and Climate Change*, 14: 253–263.

Salaria, M. and Suresh, A. 1990. Remote sensing in agriculture. *Smart and Sustainable Agricultural Technology*, 408.

Saxena, A., Chouhan, S.S., Aziz, R.M. and Agarwal, V. 2024. A comprehensive evaluation of Marine predator chaotic algorithm for feature selection of COVID-19. *Evolving Systems*, 1–14.

Schwalbert, R.A., Amado, T., Corassa, G., Pott, L.P., Prasad, P.V. and Ciampitti, I.A. 2020. Satellite-based soybean yield forecast: Integrating machine learning and weather data for improving crop yield prediction in southern Brazil. *Agricultural and Forest Meteorology*, 284: 107886.

Shaikh, T.A., Rasool, T. and Lone, F.R. 2022. Towards leveraging the role of machine learning and artificial intelligence in precision agriculture and smart farming. *Computers and Electronics in Agriculture*, 198: 107119.

Shankar, P., Werner, N., Selinger, S. and Janssen, O. 2020. Artificial intelligence driven crop protection optimization for sustainable agriculture. In *2020 IEEE/ITU International Conference on Artificial Intelligence for Good (AI4G)* (pp. 1–6). IEEE.

Sharifi, A. 2021. Yield prediction with machine learning algorithms and satellite images. *Journal of the Science of Food and Agriculture*, 101: 891–896.3.

Sharma, A., Jain, A., Gupta, P. and Chowdary, V. 2020. Machine learning applications for precision agriculture: A comprehensive review. *IEEE Access*, 9: 4843–4873.

Sharma, R., Kamble, S.S., Gunasekaran, A., Kumar, V. and Kumar, A. 2020. A systematic literature review on machine learning applications for sustainable agriculture supply chain performance. *Computers & Operations Research*, 119: 104926.

Shendryk, Y., Davy, R. and Thorburn, P. 2021. Integrating satellite imagery and environmental data to predict field-level cane and sugar yields in Australia using machine learning. *Field Crops Research*, 260: 107984.

Singh, P. and Kaur, A. 2022. A systematic review of artificial intelligence in agriculture. *Deep Learning for Sustainable Agriculture*, 57–80.

Sishodia, R.P., Ray, R.L. and Singh, S.K. 2020. Applications of remote sensing in precision agriculture: A review. *Remote Sensing*, 12: 3136.

Solanki, S., Singh, U.P. and Chouhan, S.S. 2023. Brain tumor classification using ML and DL approaches. In *2023 IEEE 5th International Conference on Cybernetics, Cognition and Machine Learning Applications (ICCCMLA)* (pp. 204–208). IEEE.

Solanki, S., Singh, U.P., Chouhan, S.S. and Jain, S. 2023. Brain tumour detection and classification by using deep learning classifier. *Int. J. Intell. Syst. Appl. Eng.*, 11: 279.

Solanki, S., Singh, U.P., Chouhan, S.S. and Jain, S. 2024. A systematic analysis of magnetic resonance images and deep learning methods used for diagnosis of brain tumor. *Multimedia Tools and Applications*, 83: 23929–23966.

Teodoro, P.E., Teodoro, L.P.R., Baio, F.H.R., da Silva Junior, C.A., dos Santos, R.G., Ramos, A.P.M. and Shiratsuchi, L.S. 2021. Predicting days to maturity, plant height and grain yield in soybean: A machine and deep learning approach using multispectral data. *Remote Sensing*, 13: 4632.

Tian, H., Wang, P., Tansey, K., Han, D., Zhang, J., Zhang, S. and Li, H. 2021. A deep learning framework under attention mechanism for wheat yield estimation using remotely sensed indices in the Guanzhong Plain, PR China. *International Journal of Applied Earth Observation and Geoinformation*, 102: 102375.

Tiwari, A.K. and Mishra, H. 2024. Sustainable agriculture: The Future of World. pp. 167–192. *In*: Deepak Kumar Patel, Shripati Dwivedi, Anupam Dubey, Abhishek Kumar and Waris Ali (eds.). *Agricultural Transformation in the Modern Era*. Elite Publishing House.

Tiwari, A.K., Mishra, H. and Nishad, D.C. 2024. Market dynamics and consumer perceptions of organic produce in contemporary agriculture. *Advances in Agriculture Sciences Volume II*, 24: 120.

Tiwari, A.K., Mishra, H., Nishad, D.C. and Pandey, A. 2023. Sustainable water management in agriculture: irrigation techniques and water conservation. *Dr. Ajay B. Jadhao*, 53.

Tiwari, P., Poudel, K.P., Yang, J., Silva, B., Yang, Y. and McConnell, M. 2023. Marginal agricultural land identification in the Lower Mississippi Alluvial Valley based on remote sensing and machine learning model. *International Journal of Applied Earth Observation and Geoinformation*, 125: 103568.

Tripathi, A., Tiwari, R.K. and Tiwari, S.P. 2022. A deep learning multi-layer perceptron and remote sensing approach for soil health based crop yield estimation. *International Journal of Applied Earth Observation and Geoinformation*, 113: 102959.

Veerachamy, R. and Ramar, R. 2022. Agricultural Irrigation Recommendation and Alert (AIRA) system using optimization and machine learning in Hadoop for sustainable agriculture. *Environmental Science and Pollution Research*, 29: 19955–19974.

Virnodkar, S.S., Pachghare, V.K., Patil, V.C. and Jha, S.K. 2020. Remote sensing and machine learning for crop water stress determination in various crops: A critical review. *Precision Agriculture*, 21: 1121–1155.

Wang, D., Cao, W., Zhang, F., Li, Z., Xu, S. and Wu, X. 2022. A review of deep learning in multiscale agricultural sensing. *Remote Sensing*, 14: 559.

Wei, M.C.F., Maldaner, L.F., Ottoni, P.M.N. and Molin, J.P. 2020. Carrot yield mapping: A precision agriculture approach based on machine learning. *AI*, 1: 229–241.

Weiss, M., Jacob, F. and Duveiller, G. 2020. Remote sensing for agricultural applications: A meta-review. *Remote Sensing of Environment*, 236: 111402.

Wongchai, A., Shukla, S.K., Ahmed, M.A., Sakthi, U. and Jagdish, M. 2022. Artificial intelligence-enabled soft sensor and internet of things for sustainable agriculture using ensemble deep learning architecture. *Computers and Electrical Engineering*, 102: 108128.

Xu, J.X., Ma, J., Tang, Y.N., Wu, W.X., Shao, J.H., Wu, W.B. and Guo, H.Q. 2020. Estimation of sugarcane yield using a machine learning approach based on UAV-lidar data. *Remote Sensing*, 12: 2823.

Xu, W., Chen, P., Zhan, Y., Chen, S., Zhang, L. and Lan, Y. 2021. Cotton yield estimation model based on machine learning using time series UAV remote sensing data. *International Journal of Applied Earth Observation and Geoinformation*, 104: 102511.

Yoosefzadeh-Najafabadi, M., Tulpan, D. and Eskandari, M. 2021. Using hybrid artificial intelligence and evolutionary optimization algorithms for estimating soybean yield and fresh biomass using hyperspectral vegetation indices. *Remote Sensing*, 13: 2555.

Zaman, Q.U. 2023. Precision agriculture technology: A pathway toward sustainable agriculture. In *Precision Agriculture* (pp. 1–17). Academic Press.

Zhang, J., Liu, J., Chen, Y., Feng, X. and Sun, Z. 2021. Knowledge mapping of machine learning approaches applied in agricultural management—A scientometric review with citespace. *Sustainability*, 13: 7662.

Zhang, L., Zhang, Z., Luo, Y., Cao, J. and Tao, F. 2019. Combining optical, fluorescence, thermal satellite and environmental data to predict county-level maize yield in China using machine learning approaches. *Remote Sensing*, 12: 21.

Zhang, L., Zhang, Z., Luo, Y., Cao, J., Xie, R. and Li, S. 2021. Integrating satellite-derived climatic and vegetation indices to predict smallholder maize yield using deep learning. *Agricultural and Forest Meteorology*, 311: 108666.

Chapter 8

Leveraging AI and CV Technologies to Advance Water Quality Assessment in Ecological Informatics

Karan Gehlot,[1] *Anil Dhanda*,[1] *Prawar*[1] and *Chintan Singh*[2,*]

Introduction

Overview of Water Quality Assessment

A complex process known as "water quality assessment" is used to determine if a body of water is safe to use for drinking, recreational activities, farming, or industrial processes. The measurement of physical, chemical, and biological parameters is included in this assessment. Physical characteristics that include temperature, turbidity, color, taste, and odour serve as fundamental water quality indicators (Neama et al., 2022). The solubility of gases and the metabolic rates of aquatic organisms are influenced by temperature, whereas light penetration and photosynthetic aquatic plants and creatures are affected by turbidity, which is the result of suspended particles (Petropoulou et al., 2023) (de Mello et al., 2020). Chemical parameters are used to quantify the concentrations of various substances, such as dissolved oxygen (DO), pH, nutrients, heavy metals, and organic pollutants. For example, phosphates and nitrates are important for the respiration of aquatic organisms; heavy metals, like mercury

[1] Amity Institute of Applied Sciences, Amity University Noida, Uttar Pradesh, 201301, India.
[2] Assistant Professor, Amity Institute of Forensic Sciences, Amity University, Sector 125, Noida, U.P., 201301, India.
* Corresponding author: chntnsngh@gmail.com, csingh5@amity.edu
ORCID: https://orcid.org/0000-0002-1096-2196

and lead, can be toxic even at low concentrations; and pesticides and herbicides are examples of organic pollutants that can be harmful to ecosystems and human health. Indicators of water quality include Biological Oxygen Demand (BOD), which reveals levels of organic pollution, and microbial indicators, which show pathogen contamination. Biological parameters evaluate the existence and health of aquatic species. The diversity and quantity of fish populations indicate environmental health. *In situ* sensors for ongoing monitoring, remote sensing for extensive observation, and field sampling for laboratory analysis are some of the techniques used in the assessment of water quality (Uddin and Nashi, 2021) (Krishnan et al., 2021). To model water quality and forecast future conditions, the gathered data is analyzed to find trends, pollution sources, and possible hazards. This is done using statistical and computational tools. For the sake of maintaining ecosystems, safeguarding public health, and assuring sustainable water management, this thorough examination is crucial. It offers the scientific foundation for creating laws and rules that manage resources, reduce pollution, and lessen the negative effects of climate change and human activity on aquatic habitats. This helps stakeholders make decisions that will preserve and enhance the quality of the water they work with (S. A., 2021).

Importance of Ecological Informatics in Environmental Monitoring

By combining data management, analysis, and computational modeling, ecological informatics contributes significantly to environmental monitoring and advances our knowledge of ecological systems. Ecological informatics provide the methods and tools required to effectively handle massive, complicated datasets in the context of water quality monitoring (Gupta, 2021). Conventional monitoring methods are frequently labor- and time-intensive since they mainly rely on manual sampling and laboratory analysis. To overcome these constraints, ecological informatics uses cutting-edge data gathering methods including automated sensors, remote sensing, and geographic information systems (GIS) to quickly and thoroughly collect data. This makes it possible to continuously monitor water bodies, which makes it easier to identify pollution incidents early on and put mitigation plans in place promptly.

Furthermore, by using advanced statistical and computational methods, ecological informatics improves data analysis. For example, machine learning algorithms can spot patterns and trends in water quality data that traditional approaches would miss. Based on past data, these algorithms can forecast future conditions, offering important insights into possible hazards and assisting in the direction of resource management choices. Furthermore, ecological informatics facilitates the creation of ecological models that replicate the interplay among diverse environmental elements (Yuh et al., 2023) (Yang, 2023). These models can forecast how various management approaches will affect the quality of the water, which can help with the creation of sensible laws and rules. Ecological informatics enables scientists, decision-makers, and stakeholders to make well-informed decisions about maintaining and enhancing water quality by converting unprocessed data into useful knowledge.

Additionally, the incorporation of ecological informatics into environmental monitoring fosters cross-disciplinary cooperation and data exchange. Standardized procedures and open data platforms make it easier for information to be shared, allowing for a more comprehensive understanding of environmental challenges. When dealing with intricate, multifaceted issues like water contamination, which frequently have multiple sources and channels, a coordinated approach is crucial (L. G. Jr et al., 2021). Ecological informatics makes it possible for us to work together across disciplines and break down silos to better monitor, manage, and safeguard our water resources. To summarize, the significance of ecological informatics in environmental monitoring is immense; it offers the technological framework and analytical instruments required to progress our comprehension and management of aquatic environments.

Role of AI and Computer Vision (CV) Technologies

Water quality evaluation is being revolutionized by Artificial Intelligence (AI) and Computer Vision (CV) technologies, which offer novel approaches for accurate analysis and real-time monitoring. With unmatched accuracy and speed, AI can handle and analyze enormous volumes of data about water quality through its many branches, including machine learning and deep learning (Khang et al., 2024). Traditional monitoring methods may miss minor patterns and anomalies that point to pollution or changes in water quality, but these technologies can identify them. Predictive analytics is made possible by the ability of machine learning algorithms, for instance, to identify trends in data sets about chemical concentrations, temperature variations, or biological markers (Ward et al., 2021). This enables the prediction of possible environmental problems before they become serious, an approach towards the application of AI is illustrated in Fig. 1.

Environmental monitoring greatly benefits from the application of computer vision, a branch of AI that focuses on giving computers the ability to read and comprehend visual data. With the use of CV technology, water quality factors including turbidity, color changes, and the presence of floating trash or algae blooms can be evaluated by analyzing photos and videos taken by satellites, drones, and underwater cameras (Ma et al., 2023) (O'Mahony et al., 2020). For large-scale and remote water bodies where traditional sampling is logistically difficult, this capability is vital. For example, CV-analyzed satellite images can track the evolution of algal blooms over time, offering vital information for water resource management and eutrophication

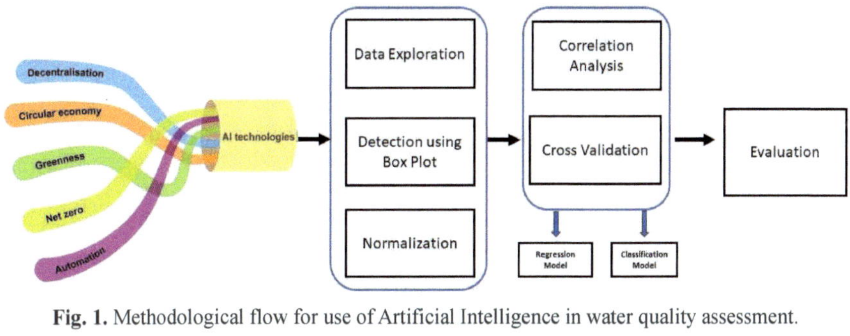

Fig. 1. Methodological flow for use of Artificial Intelligence in water quality assessment.

mitigation. CV decreases the need for manual inspection and expedites response times by automating the detection and measurement of these visual cues.

The creation of automated monitoring systems that continuously evaluate water quality in real-time is also made easier by the combination of AI and CV technologies. These systems can be installed in a variety of settings, including distant natural lakes and rivers and metropolitan water delivery networks. These devices, which are outfitted with cameras and sensors, gather continuous streams of data that are instantly analyzed by AI algorithms. Through real-time analysis, water quality problems, such as abrupt declines in dissolved oxygen levels or spikes in pollutant concentrations, can be identified and addressed quickly (Khan et al., 2021). This results in alarms being sent out. These automated solutions drastically lower operating costs and resource needs while simultaneously improving the efficacy and accuracy of water quality monitoring.

Furthermore, the use of AI and CV technologies in evaluating water quality encourages the development of extensive and cohesive data platforms. These platforms provide a comprehensive picture of water quality over time and place by aggregating data from several sources, such as historical records, satellite imaging, and field sensors (Khang et al., 2024) (Chai et al., 2021). AI-powered advanced data analytics and visualisation tools can promote evidence-based decision-making by assisting stakeholders in understanding intricate relationships and trends. Predictive models, for example, can mimic the results of different pollution control strategies, assisting legislators in creating environmental legislation that are more successful. In conclusion, by improving data collection, analysis, and administration, AI and CV technologies are revolutionizing the assessment of water quality and eventually paving the way for more proactive and knowledgeable management of water resources.

Fundamentals of Water Quality Assessment

Parameters and Indicators of Water Quality

To determine the health and fitness of water bodies for various uses, a range of physical, chemical, and biological factors are measured and analyzed. These processes provide the basis of the water quality assessment process. Physical characteristics of water, including temperature, turbidity, color, taste, and odor, offer fundamental insights into its quality. Temperature impacts the solubility of gases such as oxygen and the metabolic rates of aquatic organisms, whereas turbidity—a result of suspended particles—affects light penetration and, in turn, aquatic plant photosynthesis. Variations in color, flavor, and odor can be indicators of possible pollution and ecological disruptions caused by organic waste, algae, and industrial pollutants (Kothari et al., 2021) (Aydin et al., 2021). Figure 2 illustrates such parameters required for water quality assessment.

Chemical parameters have a substantial impact on aquatic ecosystems and human health by providing information about the concentration of different compounds in water. pH, dissolved oxygen (DO), nutrients, heavy metals, and organic contaminants are important chemical indicators. Extreme pH levels can be hazardous

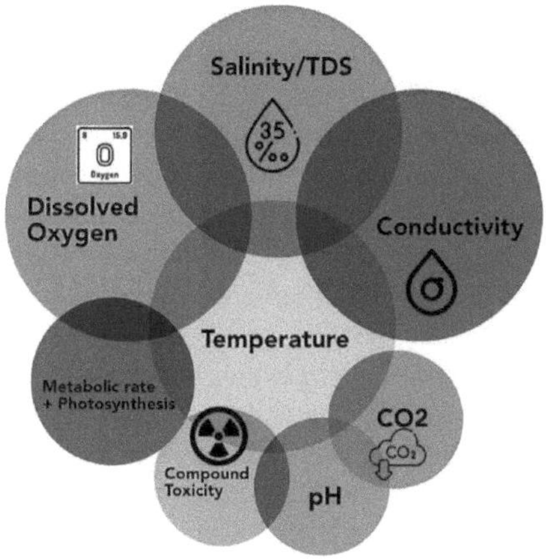

Fig. 2. Illustration depicting parameters for water quality assessment.

to aquatic life and have an impact on the toxicity of chemicals. pH levels also alter the solubility and availability of nutrients and metals. Fish and other aquatic life depend on DO for respiration, and low DO levels are frequently indicative of organic contamination. While excess of certain nutrients, such as phosphates and nitrates, can cause eutrophication, which can result in toxic algal blooms and oxygen depletion, is necessary for plant growth. Heavy metals that come from air deposition and industrial discharges, such as lead, mercury, and cadmium, are hazardous even in small amounts. Because of their toxicity and endurance, organic pollutants, such as pesticides and herbicides, pose long-term dangers to human health as well as ecosystems (M. V.-E., 2020).

By evaluating aquatic species' existence and condition, biological factors serve as clear markers of the quality of the water. High (BOD) levels indicate considerable organic pollution and possible oxygen depletion (Wen et al., 2020). BOD measures the quantity of oxygen required by microorganisms to breakdown organic materials. Microbial markers, such as coliform bacteria, indicate pathogen contamination, frequently from sewage and other sources of faecal pollution. Fish population diversity and abundance, as well as variations in species composition, are important markers of the general health of aquatic ecosystems. These indicators can also reveal ecological disruptions and the effects of pollution. To sum up, a thorough evaluation of the physical, chemical, and biological aspects of water bodies is necessary to give a thorough grasp of their appropriateness and overall health. These indicators aid in locating pollution sources, evaluating the effects on the environment, and directing the creation of management and conservation plans for water resources. Water quality assessments can guarantee the sustainability and safety of aquatic habitats for a range of uses by integrating these characteristics.

Traditional Methods vs. Modern Approaches

Over time, there has been a major evolution in the assessment of water quality. Traditional methodologies have given way to the integration of contemporary technological technologies. Conventional techniques mostly rely on laboratory analysis and field sampling, which can be labor- and time-intensive despite being extremely precise and particular. The process of field sampling involves gathering water samples by hand from different depths and places. These samples are then subjected to laboratory analysis utilizing methods like spectrophotometry, chromatography, and titration. These techniques offer accurate measurements of water quality indicators such as microbiological content and chemical concentrations, but they are generally restricted in terms of their temporal and spatial coverage, necessitating a large investment of both financial and human resources.

On the other hand, contemporary methods make use of cutting-edge technologies to improve the effectiveness, scope, and real-time capabilities of water quality evaluation. For example, automated *in situ* sensors are placed in water bodies to continuously check for elements including temperature, turbidity, dissolved oxygen, and pH. Because these sensors give real-time data, fewer human samplings are required, and problems with water quality can be identified right away. Observing long-term, large-scale changes in water quality through satellite imaging and aerial photography is another way that remote sensing and Geographic Information Systems (GIS) broaden the scope of monitoring. The integration of spatial and temporal data made possible by GIS technology enables thorough study and visualization of pollution sources and trends over large geographic areas.

Another advancement is the incorporation of machine learning (ML) and AI into the evaluation of water quality. Large datasets can be analyzed by AI algorithms, which can then be used more effectively and accurately than traditional statistical approaches to find patterns, forecast trends, and spot abnormalities. Trained on previous data, machine learning models are able to predict future water quality conditions, offering useful insights for proactive management.

Water quality testing is made more thorough and efficient by combining old techniques with new technology. For specific pollutants and biological assessments in particular, traditional approaches continue to be crucial for getting accurate and comprehensive results. Modern methods, on the other hand, provide larger temporal and spatial coverage, allowing for large-scale, ongoing monitoring. Together, these techniques provide a more comprehensive understanding of water quality, facilitating improved decision-making and water resource management. In addition to improving assessment efficiency and accuracy, this integrated approach aids in the creation of more effective policies and interventions that preserve and protect aquatic habitats.

Introduction to Artificial Intelligence

Basics of AI and Machine Learning

AI is the term used to describe how technology, especially computer systems, can simulate human intelligence processes. Learning, reasoning, problem-solving, perception, and language comprehension are some of these processes. AI can be divided into two categories: broad AI, which can carry out any intellectual work that a human can, and narrow AI, which is made to accomplish a specific task (like facial recognition or internet searches). Machine learning (ML) is a subfield of AI that focuses on creating statistical models and algorithms that let computers analyze, forecast, and make decisions based on data (Singh et al., 2024).

The ability of models to get better at a task via practice, or by processing more data, is what defines machine learning. Three primary categories of machine learning algorithms exist: reinforcement learning, unsupervised learning, and supervised learning. A labeled dataset is used to train the algorithm in supervised learning, where each training example has an output label associated with it. The programme is supposed to be able to anticipate new, unobserved data with accuracy (Singh et al., 2023). The objective of unsupervised learning is to find underlying structures or hidden patterns in the input data by training on data without labeled replies. By rewarding good behaviors and penalizing undesired ones, reinforcement learning allows an algorithm to learn the best course of action through trial and error.

A kind of machine learning known as "deep learning" has become a potent tool in the field of artificial intelligence. It uses multi-layered neural networks, hence the name "deep" learning. Because these neural networks are made to resemble the structure and operation of the human brain, they can analyze and learn in complicated ways from vast amounts of data. Deep learning is especially helpful for applications like picture recognition, natural language processing, and speech recognition because it performs well on tasks involving unstructured data, such as text, audio, and photos (Singh et al., 2024). While recurrent neural networks (RNNs) are better suited for sequential data analysis, such time series or language modeling, convolutional neural networks (CNNs), a sort of deep learning model, are especially good at image-related tasks. Deep learning's progress has greatly improved AI systems' capabilities, enabling them to achieve previously unheard-of levels of accuracy and efficiency across a wide range of applications. Deep learning and AI have the potential to revolutionize many industries, including environmental sciences, as computing power and algorithmic complexity increase.

Applications of AI in Environmental Sciences

There are now more options for tracking, controlling, and reducing environmental problems thanks to the use of AI in environmental sciences. AI finds widespread use in environmental monitoring, where algorithms analyze massive amounts of data from several sources to spot patterns and abnormalities. AI is capable of analyzing remote sensing data and satellite pictures, for instance, to track the health of coral reefs, monitor deforestation, and identify illicit fishing activities, some of the applications are depicted in Fig. 3. By using past pollution data and climatic variables, machine learning algorithms can forecast air quality levels and issue early warnings for days with dangerously low air quality (Zhong et al., 2021).

Artificial Intelligence is revolutionary in the field of water quality evaluation. AI systems can continuously monitor water quality indicators including pH, turbidity, dissolved oxygen, and the presence of pollutants by analyzing data from *in situ* sensors, satellite photos, and historical records (Gupta et al., 2021). Proprietary analysis techniques may overlook some patterns and correlations in the data, but machine learning algorithms can, making it possible to anticipate changes in water quality more precisely. For example, by examining environmental variables like

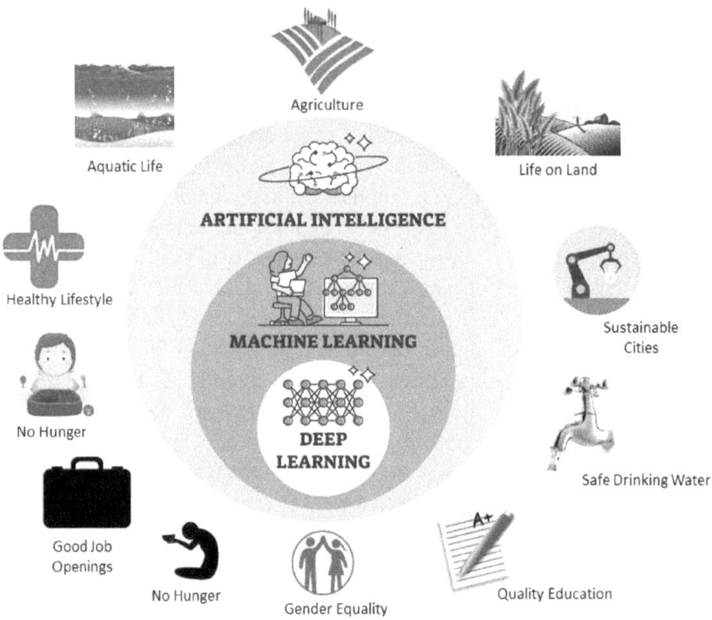

Fig. 3. Illustration for various applications of AI in environment sustainability.

temperature, light availability, and nutrient levels, AI models can predict dangerous algal blooms. This makes it possible to protect aquatic ecosystems and public health on time.

Resource management optimization is a key area in which artificial intelligence is being used in environmental sciences. By evaluating crop requirements, soil moisture content, and weather predictions, AI systems can optimize the use of water resources in agriculture by suggesting effective irrigation schedules. This increases agricultural yields while simultaneously conserving water. AI is used in wildlife conservation to examine data from acoustic sensors, drones, and video traps to track the movements of endangered species and keep an eye on animal populations (Zhong et al., 2021) (Sanchez et al., 2022). AI facilitates the creation of efficient conservation methods and mitigates the loss of biodiversity by processing massive datasets efficiently and accurately.

To sum up, artificial intelligence and machine learning have emerged as essential instruments in the field of environmental sciences, providing robust approaches to the observation, evaluation, and administration of natural resources. AI has the potential to handle enormous volumes of data, identify patterns, and make predictions. These capabilities will help us better understand complex environmental systems and respond to environmental concerns. The application of AI technology in environmental sciences is anticipated to grow as it develops, making a greater contribution to the sustainability and preservation of our natural world (Jain et al., 2020).

Computer Vision (CV) Technologies in Environmental Monitoring

Understanding Computer Vision and Image Processing

The artificial intelligence discipline of CV makes it possible for computers to analyze and comprehend visual data from their environment. Fundamentally, CV entails the retrieval of significant data from pictures and videos to carry out operations that would typically need human vision. A key element of CV is image processing, which includes a range of methods for improving, interpreting, and modifying images (Liu et al., 2024). Filtering, segmentation, feature extraction, and pattern recognition are some of these approaches that together enable the conversion of unprocessed visual data into meaningful insights.

Image processing begins with the acquisition of images through cameras, satellites, drones, or other sensors. These images often undergo preprocessing steps, such as noise reduction, contrast enhancement, and normalization, to improve their quality and suitability for further analysis. Segmentation divides an image into meaningful regions or objects, making it easier to analyze specific features (Valipoor and de Antonio, 2023). Feature extraction identifies relevant attributes, such as edges, textures, or shapes, that can be used for classification or recognition tasks. Pattern recognition, often powered by machine learning algorithms, enables the identification and categorization of objects or phenomena within the images.

Deep learning has transformed CV by dramatically increasing the precision and effectiveness of image analysis, especially when convolutional neural networks (CNNs) are employed. CNNs are very good at tasks like object identification, picture classification, and semantic segmentation because of their ability to automatically and adaptively learn spatial hierarchies of characteristics from images. These abilities have wide-ranging applications in many fields, including as autonomous driving, environmental monitoring, and medical imaging (Liu et al., 2024) (Liu et al., 2023).

Role of CV in Water Quality Assessment

The use of deep learning has transformed CV by dramatically increasing the precision and effectiveness of image analysis, especially when convolutional neural networks (CNNs) are employed. CNNs are very good at tasks like object identification, picture classification, and semantic segmentation because of their ability to automatically and adaptively learn spatial hierarchies of characteristics from images. These abilities have wide-ranging applications in several fields, including as autonomous driving, environmental monitoring, and medical imaging (Rawat et al., 2023).

The identification and tracking of turbidity is one of CV's main functions in the evaluation of water quality. Elevated levels of turbidity, frequently resulting from suspended particulates including silt, algae, and microbes, can negatively impact aquatic ecosystems by diminishing light penetration and interfering with photosynthesis. To precisely measure turbidity levels, CV algorithms examine how light is scattered and absorbed in photographs of water (Adelani et al., 2024). Compared to conventional manual sample techniques, this automated analysis is more

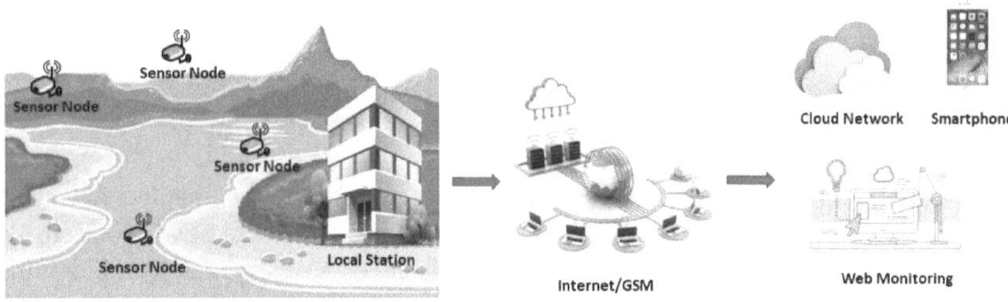

Fig. 4. An illustrated computer vision-based framework for water quality monitoring.

effective and covers a wider region, giving real-time insights into the clarity of the water and possible pollution sources.

Additionally, CV is essential for monitoring algal blooms, which can negatively impact aquatic life and water quality. Toxins produced by harmful algal blooms (HABs), which are frequently caused by nutrient pollution, pose a risk to both human health and animals. CV systems can identify and monitor the degree and evolution of algal blooms throughout time through the use of satellite imagery and aerial photography. These systems analyze concentrations of several forms of algae and distinguish between them using machine learning and pattern recognition techniques, one of the frameworks has been illustrated in Fig. 4. Prompt management efforts to lessen the effects of HABs are made possible through early discovery and ongoing monitoring (Tang et al., 2022).

Identifying contaminants and other items in water bodies is also made easier by CV technologies. Utilizing image processing techniques, for example, plastic garbage and other detritus can be visually identified and categorized. By identifying and measuring the amount of contaminants present in images captured by surface and underwater cameras, CV systems can provide important information for strategies aimed at preventing pollution and organizing cleanup operations (Punithavathi et al., 2022). CV can also assist in a thorough evaluation of the water quality by keeping an eye on variations in the color of the water, which may be a sign of dissolved materials or chemical contaminants. In conclusion, computer vision technologies are now essential to assessing the quality of modern water since they provide strong instruments for ongoing, extensive, and real-time monitoring. Through the use of sophisticated image processing and deep learning techniques, CV improves our capacity to identify, evaluate, and address a range of water quality concerns, ultimately leading to improved aquatic environment management and protection.

Integrating AI and CV for Water Quality Assessment

Data Collection and Preprocessing

Robust data collection and preprocessing are essential for the accuracy and dependability of subsequent studies, and they are the first steps in an effective evaluation of water quality. Information is gathered from a variety of sources, such as underwater cameras, satellite pictures, *in situ* sensors, remote sensing technology, and aerial photography, during the data collection process (Nti et al., 2023). Temperature, turbidity, dissolved oxygen, pH, and other physical and chemical characteristics can all be continuously and real-time measured with *in situ* sensors. Large-scale data are available through satellite imaging and remote sensing, which can record changes in water quality across wide areas both spatially and temporally.

To guarantee the acquired data's quality and usefulness for analysis, preprocessing is a must. Preprocessing procedures for sensor data could involve calibration, noise reduction, and handling of missing readings. The accuracy and consistency of the sensor readings with accepted criteria are guaranteed by calibration (Taşan, 2023). Random oscillations and outliers in the data can be eliminated with the use of noise reduction techniques like filtering. With a view to ensure that the dataset is full for analysis, handling missing values requires the use of interpolation or imputation techniques to estimate the missing data.

Preprocessing for image-based data from drones and satellites entails multiple processes. These include normalization to standardize the data range, radiometric correction to correct the pixel values for atmospheric effects and sensor noise, and geometric correction to correct distortions caused by the perspective of the sensor (Al-Janabi and Al-Barmani, 2023). Techniques for image segmentation and enhancement, like contrast adjustment and edge recognition, are used to draw attention to essential characteristics and make key details easier to see. Preprocessing checks that the data is correct, clean, and prepared for AI and Computer Vision (CV) model analysis.

AI Models for Water Quality Prediction

Predicting water quality is becoming more and more common with AI models, especially those that make use of machine learning and deep learning techniques. These models are capable of pattern recognition, in-depth dataset analysis, and future water quality forecasting. In water quality evaluation, supervised learning algorithms including random forests, decision trees, and support vector machines (SVM) are frequently employed for regression and classification tasks. Through training on historical data, these models discover the correlations between water quality indicators (such as pollutant concentrations and algal bloom occurrences) and input factors (such as temperature, pH, and nutrient levels) (Chhipi-Shrestha et al., 2023).

Convolutional neural networks (CNNs) and recurrent neural networks (RNNs) are two examples of deep learning models that provide sophisticated processing capabilities for handling big, unstructured datasets. CNNs are especially good in identifying and categorizing characteristics linked to water quality, like turbidity levels or algal blooms, by analyzing spatial data, such as satellite photography (Iqbal et al., 2023). Because RNNs can handle sequential data, they are a good fit for time-series predictions. This means that water quality trends can be predicted using sensor data from the past. Techniques like feature selection and dimensionality reduction, which aid in locating the most pertinent variables and simplifying the dataset, are advantageous to AI models, such depiction of machine learning based framework has been illustrated in Fig. 5. This enhances the interpretability and performance of the model. Furthermore, a lot of the time, ensemble learning techniques are used, which integrate several models to improve prediction accuracy. These AI models offer insightful information about the dynamics of water quality, facilitating proactive management and risk-averse decision-making (Irwan et al., 2023).

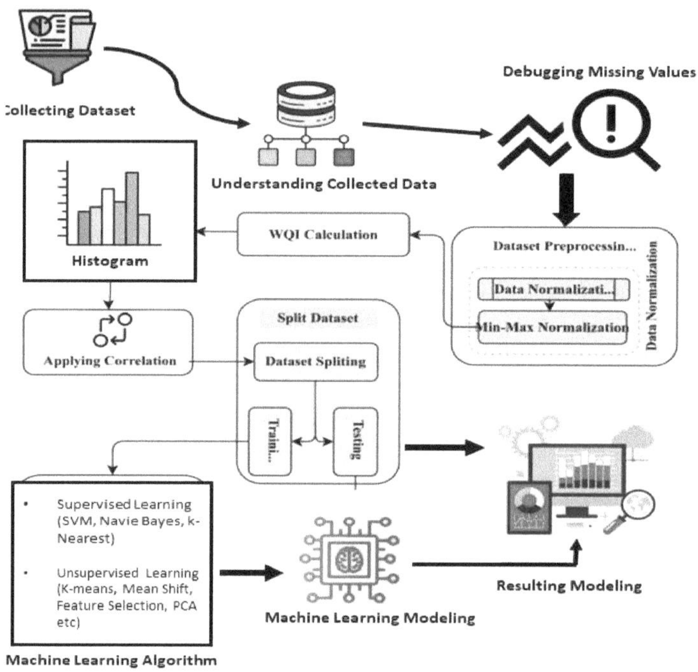

Fig. 5. Feature selection and dimensionality reduction based AI framework for water quality assessment.

CV Techniques for Image-Based Analysis

For the purpose to analyze image-based data for the purpose to assess water quality, CV techniques are essential. With the aim to extract valuable information from visual data, such as satellite imagery, drone footage, and underwater camera feeds, CV uses cutting edge image processing and deep learning techniques. These methods provide very accurate and efficient methods for the detection, categorization, and monitoring of several indicators of water quality. The analysis of image-based data in the assessment of water quality depends heavily on CV algorithms. Advanced image processing and deep learning techniques are utilized by CV to derive significant information from visual data, including underwater video feeds, drone footage, and satellite imagery. These methods allow for the very accurate and efficient detection, categorization, and monitoring of a variety of water quality indicators.

To detect certain pollutants and occurrences in water bodies, object detection and classification are essential. Using labelled datasets, convolutional neural networks (CNNs) and other deep learning models are trained to identify patterns and categorise things, such as invasive aquatic species, plastic garbage, and oil spills. With its ability to handle massive amounts of picture data quickly, these models can offer real-time insights on the distribution and presence of pollutants (Chhipi-Shrestha et al., 2023; Jadhav et al., 2023; Taşan, 2023; Chouhan et al., 2024; Chouhan et al., 2024; Kumar et al., 2024).

Additionally, CV approaches make it possible to track how factors related to water quality vary over time. Through the examination of image sequences, CV systems can identify changes in features such as color, turbidity, and the size of algal blooms over time. Understanding the evolution of water quality problems and assessing the success of management actions require this temporal study. Thus, the use of AI and CV for water quality evaluation improves monitoring efforts' precision, scalability, and promptness, leading to better-informed and successful environmental management plans.

Case Studies and Applications

Bearing an Eye on Coastal Water Algal Blooms

The monitoring of harmful algal blooms (HABs) in coastal waters is one of the most important uses of AI and computer vision (CV) technology in the assessment of water quality. Because HABs can create poisons that contaminate seafood and drinking water, they pose serious dangers to human health, the local economy, and marine ecosystems. Using AI models, *in situ* sensor data and satellite pictures were combined in a case study throughout the Gulf of Mexico to identify and forecast HAB occurrences. High-resolution satellite photos were analyzed using deep learning methods, namely convolutional neural networks (CNNs), to look for indicators of algal blooms, such as changes in water color and chlorophyll concentrations. The AI models were able to recognize the temporal and spatial patterns connected to HAB occurrences since they had been trained on historical bloom data. Early warning capabilities offered by this strategy made it possible to promptly issue public health alerts and take mitigation measures. The accomplishment of this case study demonstrates how AI and CV technologies can improve coastal water quality monitoring and management, protecting both human communities and marine life.

Machine Learning for River Pollution Prediction

Machine learning models were created to forecast pollution levels based on different environmental and anthropogenic parameters in a case study that focused on the Ganges River in India. One of the world's most polluted rivers, the Ganges, is hampered by home sewage, agricultural runoff, and industrial discharge. Machine learning models were trained using data from many sources, such as weather stations, remote sensing, and manual water sampling. The data was analyzed using supervised learning techniques, such as support vector machines (SVM) and random forests, to forecast the quantities of important contaminants, such as heavy metals, nitrates, and biological oxygen demand (BOD). Significant indicators of pollution levels, including population density, industrial activity, and rainfall patterns, were identified by the models. The identification of important pollution hotspots and the streamlining of remediation activities were made possible by these predictive capabilities, which also made it possible to implement more effective pollution management techniques. This case study's incorporation of machine learning showed how it may improve the efficiency and predicted accuracy of water quality tests in vast, intricate river systems.

Monitoring Urban Water Bodies in Real-Time with CV

A case study was carried out in an urban environment to use Computer Vision (CV) technology to monitor the water quality of lakes and reservoirs. Pollution from stormwater runoff, industrial discharges, and recreational activities is a common problem for urban aquatic bodies. Throughout many urban lakes, a network of smart cameras and drones with high-resolution imaging capabilities was set up for this study. To process the visual data that these devices collected, CV algorithms were created with an emphasis on surface litter, turbidity, and the presence of algae. Specific contaminants, such as plastic garbage and oil spills, were recognized by object identification models, while areas of interest were isolated using image segmentation techniques. Due to the real-time analysis that CV systems provide, local officials can now continuously monitor water quality and react quickly to pollution situations. Furthermore, the information gathered helped create long-term strategies for improving the quality of the water, proving the usefulness of CV in managing the water quality of cities.

Improving Agricultural Areas' Water Quality

A significant contributor to water pollution is agricultural runoff, which enters neighboring water bodies and introduces sediments, fertilizers, and pesticides. A case study examined the use of AI and CV technologies to monitor and manage water quality in streams and rivers affected by farming activities in a Midwest agricultural region of the United States. A comprehensive monitoring system was created by combining data from ground-based sensors with remote sensing data from drones and satellites. To forecast nutrient levels (such as nitrogen and phosphorus) in water bodies based on meteorological conditions, soil types, and agricultural activities, artificial intelligence models—such as neural networks and regression algorithms—were employed. Drone imagery was analyzed using CV techniques to determine the amount of stream silt deposition and erosion. By using an integrated approach, farmers and water managers were able to reduce nitrogen runoff and enhance water quality by using targeted best management practices (BMPs) including buffer strips and cover crops. This case study demonstrated how AI and CV may support environmentally friendly farming methods and safeguard rural communities' water supplies.

Automated Ocean Microplastic Detection

Microplastics represent a growing threat to human health and aquatic life as they contribute to marine contamination. A case study that used CV technology to automatically detect and quantify microplastics in ocean waters was presented. The manual collection and microscopic inspection of samples are involved in the labor-intensive and time-consuming traditional methods of microplastic analysis. Underwater robots fitted with cutting-edge imaging equipment were used in this investigation to gather visual data at different depths. The size, shape, and color of microplastic particles were used to identify and categorize them using CV algorithms, which are specifically made for particle detection. Researchers were able to pinpoint pollution hotspots and comprehend the patterns of microplastic distribution in the water thanks to the automated system's real-time data on microplastic concentrations. The case study supported attempts to lessen the effects of microplastic contamination on marine ecosystems by demonstrating the viability and effectiveness of employing CV for extensive monitoring of pollution.

These case studies demonstrate how combining AI and CV technologies in water quality evaluation across a range of situations can have revolutionary effects. These cutting-edge instruments offer precise, thorough, and real-time information on the quality of water in a variety of settings, including urban lakes, agricultural areas, and the open ocean. AI and CV technologies are critical to the preservation of aquatic ecosystems and the sustainability of water resources because they improve our capacity to observe, anticipate, and handle problems related to water quality.

Challenges and Future Directions

The availability and quality of data is a major obstacle to the integration of AI and CV technologies for the assessment of water quality. Resolved, real-time data with high resolution from several sources are necessary for precise analysis and forecasting. However, a lack of thorough monitoring systems in many areas results in inconsistent and gapped data. Strong preprocessing and validation techniques are also required since data quality can be harmed by environmental interferences, calibration problems, and sensor malfunctions.

Complexity of Water Quality Parameters: A wide range of physical, chemical, and biological factors can interact in intricate and non-linear ways to affect water quality. Accurately representing these interactions in AI models necessitates deep subject expertise and advanced mode ling methods. The assessment method is further complicated by the fluctuation of these factors over various temporal and geographical ranges.

Algorithmic Interpretability and Transparency: It can be difficult to comprehend how predictions are formed in many AI models because of their "black-box" nature, especially deep learning algorithms. The public, lawmakers, and environmental scientists are just a few of the stakeholders that may find it difficult to trust and accept this lack of transparency. Creating interpretable models that provide light on the decision-making process is essential to the general use of AI in the evaluation of water quality.

Integration of Multisource Data: Assessing the quality of water frequently entails combining data from a variety of sources, including in-situ sensors, satellite images, and laboratory investigations. The difficulties in achieving a smooth integration and harmonization of these disparate datasets stem from variations in data formats, resolutions, and temporal frequencies. Coherent and extensive datasets for analysis require the use of effective data fusion techniques.

Scalability and Real-Time Processing: The scalability of AI and CV solutions becomes a crucial concern as the amount of environmental data continues to expand. Real-time processing and analysis of big datasets necessitate the use of powerful computers and effective algorithms. Effective monitoring of water quality depends on AI systems being able to process large amounts of data and deliver timely insights without sacrificing accuracy.

Future Aspects

Enhanced Sensor Networks and IoT Integration: Using Internet of Things (IoT) devices and advanced sensor networks can greatly enhance data collection efforts. These sensors can monitor water quality indicators in real time and at high resolution, providing AI models with continuous data streams. Water quality assessment systems will become more dynamic and responsive with the combination of IoT, AI, and CV technologies.

Fabrication of Hybrid Models: Predicting water quality can be made more accurate and reliable by fusing data-driven AI techniques with conventional mechanistic models. Hybrid models offer a more thorough knowledge of intricate environmental processes by combining the advantages of both approaches. This method can make models easier to understand and make it easier to incorporate domain knowledge into AI systems.

Developments in Edge Computing: By processing data closer to the source instead of depending on centralized cloud servers, edge computing can improve the effectiveness of water quality monitoring systems. AI algorithms can be deployed on edge devices, including sensors and drones, to do real-time analysis locally, which lowers latency and bandwidth needs. This will make it possible to make decisions more quickly and to act more quickly.

Including Citizen Science: Using citizen science programmes to involve the public in water quality monitoring can increase data collection and increase public knowledge of environmental issues. Communities can provide useful data by using inexpensive sensors and mobile apps, and this data can then be included in AI models. In addition to improving data coverage, this participatory approach can instill a sense of ownership in the local populace.

Support from Policies and Regulations: Adequate policies and regulations are necessary for the effective application of AI and CV technologies in the evaluation of water quality. To guarantee data sharing among stakeholders, standardize data gathering techniques, and encourage the uptake of cutting-edge technologies, governments, and regulatory organizations can play a critical role. Effective policy frameworks that encourage creativity and teamwork will be essential for tackling problems with water quality.

Ethical and Social Issues: Taking ethical and social issues into account is crucial to the correct application of AI in the evaluation of water quality. Building trust and acceptance requires protecting data privacy, avoiding biases in AI models, and encouraging openness in decision-making procedures. Furthermore, it is imperative to guarantee that technological innovations yield benefits for all communities, particularly those that are marginalized and neglected.

Conclusion

The introduction of cutting-edge technology such as computer vision (CV) and artificial intelligence (AI) into ecological informatics has dramatically revolutionized water quality assessment and tracking. It is clear from the thorough overview given in the introduction and the in-depth discussion of the parameters for the water quality assessment in Section 1.1 that state-of-the-art technological solutions are being added to and improving traditional ways of monitoring water bodies. These developments are essential for tackling the escalating problems brought on by pollution, climate change, and the growing need for clean water. We stated in the introduction how vital it is to monitor water quality to maintain ecosystems, human populations, and economic activity. Applying ecological informatics, which includes gathering, analyzing, and interpreting environmental data, provides an effective method for comprehending and managing water resources. The capacity of ecological informatics to present a full picture of water quality by combining data from many sources is what makes it so important for environmental monitoring. AI algorithms can evaluate big and complicated datasets to forecast trends in water quality and pinpoint the sources of contamination. This is especially true for models that make use of machine learning and deep learning. As new data becomes available, these models gain from ongoing learning and development, which eventually increases their predicted accuracy. Conversely, CV technologies offer strong capabilities for image-based analysis, making it possible to monitor and evaluate visual indications including turbidity, surface detritus, and algal blooms in real time. By automating water quality monitoring and minimizing the need for human involvement, AI and CV work together to enable quicker reactions to new problems as they arise. Even with these tremendous advances, there are still many difficulties. These include overcoming the opacity of AI models, addressing the complexity of water quality parameters, and guaranteeing data availability and quality. Furthermore, academics and practitioners must continue to innovate and work together to scale AI and CV solutions for real-time processing, as well as integrate multisource data. To fully realize the promise of AI and CV in water quality monitoring, these issues must be resolved. In terms of the future, the development and integration of cutting-edge technologies will be crucial to the assessment of water quality. Edge computing, IoT integration, hybrid modeling techniques, and improved sensor networks are all set to advance the capabilities of data processing, analysis, and gathering. Water quality monitoring efforts will also be greatly aided by encouraging favorable legislative and regulatory frameworks and involving the public through citizen science projects. Last but not least, the field of ecological informatics, strengthened by AI and CV technologies, offers transformative potential for the assessment of water quality. By utilizing these developments, we can accomplish more precise, effective, and thorough monitoring of water resources, ultimately contributing to the sustainability and health of our aquatic ecosystems. More research, creativity, and collaboration will be required to overcome current difficulties and maximize the promise of these technologies.

Declaration

Ethical Responsibilities of Authors

"All authors have read, understood, and have complied as applicable with the statement on "Ethical responsibilities of Authors" as found in the Instructions for Authors".

Ethical Approval

Not applicable.

Use of AI tool

Example 1: The corresponding author declares that no AI tool has been used to prepare this manuscript.

Example 2: The corresponding author declares that ***Gemini and Chat-Gpt*** *has been used to prepare this manuscript.*

Conflict of Interest

On behalf of all the authors the corresponding author states that there is no conflict of interest to disclose.

Competing Interests

Not applicable.

Authors' Contributions

Karan[1]: Conceptualization, Methodology, Writing - Original Draft, Writing - Review & Editing, **Anil Dhanda**[2]: Investigation, Formal analysis, Visualization, Software, Data Curation, Supervision, **Prawar**[3]: Writing - Original Draft, Formal analysis, Methodology, **Chintan**[4]: Formal analysis, Visualization, Supervision, Formal analysis.

Funding

No funding has been received.

References

Adeshina Adelani, F., Stefano Okafor, E., Sonimiteim Jacks, B., Abayomi Ajala, O. and Author, C. 2024. Theoretical frameworks for the role of ai and machine learning in water cybersecurity: insights from African and US applications. *fepbl.com*, 5(3): 681–692, doi: 10.51594/csitrj.v5i3.928.

Al-Janabi, S. and Al-Barmani, Z. 2023. Intelligent multi-level analytics of soft computing approach to predict water quality index (IM12CP-WQI). *Soft Comput.*, 27(12): 7831–7861, doi: 10.1007/S00500-023-07953-Z.

Aydin, H., Ustaoğlu, F., Tepe, Y., E, S.-E. 2021. forensics, and undefined 2021. Assessment of water quality of streams in northeast Turkey by water quality index and multiple statistical methods. *Taylor & Francis*, 22(1-2): 270–287, doi: 10.1080/15275922.2020.1836074.

Chai, J., Zeng, H., Li, A., E. N.-M. L. 2021. with Applications, and undefined 2021. Deep learning in computer vision: A critical review of emerging techniques and application scenarios. Elsevier, Accessed: Jun. 01, 2024. [Online]. Available: https://www.sciencedirect.com/science/article/pii/S2666827021000670.

Chhipi-Shrestha, G., Mian, H.R., Mohammadiun, S., Rodriguez, M., Hewage, K. and Sadiq, R. 2023. Digital water: Artificial intelligence and soft computing applications for drinking water quality assessment. *Clean Technol. Environ. Policy*, 25(5): 1409–1438, doi: 10.1007/S10098-023-02477-4.

Chhipi-Shrestha, G., Mian, H.R., Mohammadiun, S., Rodriguez, M., Hewage, K. and Sadiq, R. 2023. Digital water: Artificial intelligence and soft computing applications for drinking water quality assessment. *Clean Technologies and Environmental Policy*, 25(5): 1409–1438. Available: https://link.springer.com/article/10.1007/s10098023-02477-4.

Chouhan, S.S., Singh, U.P. and Jain, S. 2024. Introduction to computer vision and drone technology. *In*: Chouhan, S.S., Singh, U.P. and Jain, S. (eds.). *Applications of Computer Vision and Drone Technology in Agriculture 4.0.* Springer, Singapore. https://doi.org/10.1007/978-981-99-8684-2_1.

Chouhan, S.S., Singh, U.P., Sharma, U. and Jain, S. 2021. Leaf disease segmentation and classification of *Jatropha Curcas* L. and *Pongamia Pinnata* L. biofuel plants using computer vision based approaches. *Measurement*, 171: 108796, ISSN 0263-2241, https://doi.org/10.1016/j.measurement.2020.108796.

Chouhan, S.S., Singh, U.P., Sharma, U. et al. 2024. Classification of different plant species using Deep Learning and Machine Learning algorithms. *Wireless Pers. Commun.*, 136: 2275–2298. https://doi.org/10.1007/s11277-024-11374-y.

de Mello, K., Taniwaki, R.H., de Paula, F.R., Valente, R.A., Randhir, T.O., Macedo, D.R., Leal, C.G., Rodrigues, C.B. and Hughes, R.M. 2020. Multiscale land use impacts on water quality: Assessment, planning, and future perspectives in Brazil. *Journal of Environmental Management*, 270: 110879. Available: https://www.sciencedirect.com/science/article/pii/S0301479720308094.

Gadelha Jr, L.M., de Siracusa, P.C., Dalcin, E.C., da Silva, L.A.E., Augusto, D.A., Krempser, E., Affe, H.M., Costa, R.L., Mondelli, M.L., Meirelles, P.M. and Thompson, F. 2021. A survey of biodiversity informatics: Concepts, practices, and challenges. *Wiley Interdisciplinary Reviews: Data Mining and Knowledge Discovery*, 11(1): e1394. Available: https://wires.onlinelibrary.wiley.com/doi/abs/10.1002/widm.1394.

Gupta, S., Aga, D., Pruden, A., … L. Z.-E. S., 2021. and undefined 2021. Data analytics for environmental science and engineering research. *ACS Publications*, 55(16): 10895–10907, doi: 10.1021/acs.est.1c01026.

Gupta, S., S. 2021. G.-E. Informatics, and undefined 2021. A critical review on water quality index tool: Genesis, evolution and future directions. Elsevier, Accessed: Jun. 01, 2024. [Online]. Available: https://www.sciencedirect.com/science/article/pii/S157495412100090X?casa_token=AvPFU9gQpt8AAAAA:xAItMoLjPam9I5CUsOaHmpFca7RJheTrHs-oAeiIAxHLLI6g_wFydqV1Tq5227QdGHBgU_Ssy5fx.

Iqbal, U., Bin Riaz, M.Z., Barthelemy, J., Perez, P. and Idrees, M.B. 2023. The last two decades of computer vision technologies in water resource management: A bibliometric analysis. *Water and Environment Journal*, 37(3): 373–389, doi: 10.1111/WEJ.12845.

Irwan, D. et al. 2023. Predicting water quality with Artificial Intelligence: A review of methods and applications. *Archives of Computational Methods in Engineering*, 30(8): 4633–4652, doi: 10.1007/S11831-023-09947-4.

Jadhav, A.R., Pathak, P.D. and Raut, R.Y. 2023. Water and wastewater quality prediction: Current trends and challenges in the implementation of artificial neural network. *Environ. Monit. Assess.*, 195(2), doi: 10.1007/S10661-022-10904-0.

Jain, P. et al. 2020. A review of machine learning applications in wildfire science and management. *cdnsciencepub.com*, 28(4): 478–505, doi: 10.1139/er-2020-0019.

Khan, A., Laghari, A. and S. A. 2021.-E. E. T. on, and undefined 2021. Machine learning in computer vision: A review. publications.eai.eu, 2021, doi: 10.4108/eai.21-4-2021.169418.

Khang, A., Abdullayev, V., … E. L.-C. V. 2024. and undefined 2024. Application of Computer Vision (CV) in the Healthcare Ecosystem. taylorfrancis.com, Accessed: Jun. 01, 2024. [Online]. Available: https://www.taylorfrancis.com/chapters/edit/10.1201/9781003429609-1/application-computer-vision-cv-healthcare-ecosystem-alex-khang-vugar-abdullayev-eugenia-litvinova-svetlana-chumachenko-abuzarova-vusala-alyar-anh.

Khang, A., Abdullayev, V., Litvinova, E., Chumachenko, S., Alyar, A.V. and Anh, P.T.N. 2024. Application of Computer Vision (CV) in the healthcare ecosystem. *Computer Vision and AI-Integrated IoT Technologies in the Medical Ecosystem*, pp. 1–16, Jan. 2024, doi: 10.1201/9781003429609-1/Application-Computer-Vision-Cv-Healthcare-Ecosystem-Alex-Khang-Vugar-Abdullayev-Eugenia-Litvinova-Svetlana-Chumachenko-Abuzarova-Vusala-Alyar-Anh.

Kothari, V., Vij, S. and Sharma, S. 2021.N. G. S. Indicators, and undefined 2021. Correlation of various water quality parameters and water quality index of districts of Uttarakhand. Elsevier, Accessed: Jun. 01, 2024. [Online]. Available: https://www.sciencedirect.com/science/article/pii/S2665972720300775.

Krishnan, A. et al. 2021. Monitoring of heavy metal contamination in Netravati river basin: Overview of pollution indices and risk assessment. *Springer*, 7(2): 20, doi: 10.1007/s40899-021-00502-2.

Kumar, R. et al. 2024. Hybrid approach of cotton disease detection for enhanced crop health and yield. In *IEEE Access*, https://ieeexplore.ieee.org/document/10589357.

Liu, B. et al. 2024. Integration and performance analysis of Artificial Intelligence and computer vision based on Deep Learning algorithms. arxiv.org, Accessed: Jun. 01, 2024. [Online]. Available: https://arxiv.org/abs/2312.12872.

Liu, B., Yu, L., Che, C., Lin, Q., Hu, H. and Zhao, X. 2024. Integration and performance analysis of Artificial Intelligence and computer vision based on Deep Learning algorithms, Accessed: Jun. 01, 2024. [Online]. Available: http://arxiv.org/abs/2312.12872.

M. V.-E. 2020. Pollution and undefined 2020. Use of water quality index and multivariate statistical methods for the evaluation of water quality of a stream affected by multiple stressors: A case study. Elsevier, Accessed: Jun. 01, 2024. [Online]. Available: https://www.sciencedirect.com/science/article/pii/S0269749120361054.

Ma, D., Dang, B., Li, S. and Zang, H. 2023. … X. D.-J. of C. S. and, and undefined 2023. Implementation of computer vision technology based on artificial intelligence for medical image analysis. wepub.org, Accessed: Jun. 01, 2024. [Online]. Available: http://wepub.org/index.php/IJCSIT/article/view/529.

Neama, M., Attia, M., Negm, A. 2022. M. N.-W. Quality, A. and, and undefined 2022. Overview of water resources, quality, and management in India. Springer, Accessed: Jun. 01, 2024. [Online]. Available: https://link.springer.com/chapter/10.1007/978-3-030-95687-5_1.

Nti, E.K. et al. 2023. Water pollution control and revitalization using advanced technologies: Uncovering artificial intelligence options towards environmental health protection. *cell.com*, 9: e18170, doi: 10.1016/j.heliyon.2023.e18170.

O'Mahony, N. et al. 2020. Deep learning vs. traditional computer vision. *Advances in Intelligent Systems and Computing*, 943: 128–144, doi: 10.1007/978-3-030-17795-9_10.

Petropoulou, I., Frousiou, M. and Vasileiou, E. 2023. An overview of quality assessment methods for water and soil in mining regions. *Mater. Proc.*, 15: 31, Accessed: Jun. 01, 2024. [Online]. Available: https://www.academia.edu/download/108156195/pdf.pdf.

Punithavathi, R. et al. 2022. Computer vision and deep learning-enabled weed detection model for precision agriculture. researchgate.net, doi: 10.32604/csse.2023.027647.

Rawat, P., Bajaj, M., Vats, S. and Sharma, V. 2023, March. An analysis of crop recommendation systems employing diverse machine learning methodologies. In *2023 International Conference on Device Intelligence, Computing and Communication Technologies, (DICCT)* (pp. 619–624). IEEE. Available: https://ieeexplore.ieee.org/abstract/document/10110085/.

S. A.-H. of W. P. and Quality and undefined 2021. Overview: Modern water purity and quality. Elsevier, Accessed: Jun. 01, 2024. [Online]. Available: https://www.sciencedirect.com/science/article/pii/B9780128210574000148.

Sanchez, T.W., Shumway, H., Gordner, T. and Lim, T 2023. The prospects of artificial intelligence in urban planning. *Taylor & Francis*, 41(4): 382–395, doi: 10.1080/12265934.2022.2102538.

Singh, C., Khajuria, H. and Nayak, B.P. 2023. A study of implementing a Blockchain-Based Forensic Model Integration (BBFMI) for IoT devices in digital forensics. pp. 318–327. *In*: Hu, Z., Dychka, I. and He, M. (eds.). *Advances in Computer Science for Engineering and Education VI*, Cham: Springer Nature Switzerland.

Singh, C., Khajuria, H. and Nayak, B.P. 2024. Machine Learning driven strategies for safeguarding India's digital environment against cyberstalking. Feb. 2024, doi: 10.3233/ATDE231337.

Singh, C., Khajuria, H. and Nayak, B.P. 2024. Rapid advancement and trends of big data analytics and cyber-physical system embedded in healthcare and Industry 4.0. *Intelligent Security Solutions for Cyber-Physical Systems*, pp. 216–233, doi: 10.1201/9781003406105-19.

Tang, B., Chen, L. and Sun, W. 2023. Z. L.-I. I. Processing, and undefined 2023. Review of surface defect detection of steel products based on machine vision. *Wiley Online Library*, 17(2): 303–322, Feb. 2022, doi: 10.1049/ipr2.12647.

Taşan, S. 2023. Estimation of groundwater quality using an integration of water quality index, Artificial Intelligence methods and GIS: Case study, Central Mediterranean Region of Turkey. *Appl. Water Sci.*, 13(1), doi: 10.1007/S13201-022-01810-4.

Taşan, S. 2023. Science and undefined 2023. Estimation of groundwater quality using an integration of water quality index, artificial intelligence methods and GIS: Case study, Central Mediterranean. Springer, Accessed: Jun. 01, 2024. [Online]. Available: https://link.springer.com/article/10.1007/s13201-022-01810-4.

Uddin, M. and Nash, S. 2021. A. O.-E. Indicators, and undefined 2021. A review of water quality index models and their use for assessing surface water quality. Elsevier, Accessed: Jun. 01, 2024. [Online]. Available: https://www.sciencedirect.com/science/article/pii/S1470160X20311572.

Valipoor, M.M. and de Antonio, A. 2023. Recent trends in computer vision-driven scene understanding for VI/blind users: A systematic mapping. *Univers. Access Inf. Soc.*, 22(3): 983–1005, Aug. 2023, doi: 10.1007/S10209-022-00868-W.

Ward, T., Mascagni, P., Ban, Y. and Rosman, G. 2021. N. P.- Surgery, and undefined 2021.Computer vision in surgery. Elsevier, Accessed: Jun. 01, 2024. [Online]. Available: https://www.sciencedirect.com/science/article/pii/S0039606020307492.

Wen, X. et al. 2020. Microbial indicators and their use for monitoring drinking water quality—A review. mdpi.com, Accessed: Jun. 01, 2024. [Online]. Available: https://www.mdpi.com/2071-1050/12/6/2249.

Yang, Y. 2023. H. L.-E. Informatics, and undefined 2023. Spatiotemporal dynamic decoupling states of eco-environmental quality and land-use carbon emissions: A case study of Qingdao City, China. Elsevier, Accessed: Jun. 01, 2024. [Online]. Available: https://www.sciencedirect.com/science/article/pii/S1574954123000213?casa_token=SQDAHUIqwoAAAAAA:aF2bd3VapC8N4J_ULo8vUp4FyjLLUq2Ffl7fhgU4Rm_ky94di_QxJjUuqdNKdOyipUemWkghDW_Z.

Yuh, Y., Tracz, W. and Matthews, H. 2023. S. T.-E. informatics, and undefined 2023. Application of machine learning approaches for land cover monitoring in northern Cameroon. Elsevier, Accessed: Jun. 01, 2024. [Online]. Available: https://www.sciencedirect.com/science/article/pii/S1574954122004058.

Zhong, S. et al. 2021. Machine Learning: New ideas and tools in environmental science and engineering. *Environ. Sci. Technol.*, 55(19): 12741–12754, Oct. 2021, doi: 10.1021/ACS.EST.1C01339.

Zhong, S., Zhang, K., Bagheri, M., Burken, J.G., Gu, A., Li, B., Ma, X., Marrone, B.L., Ren, Z.J., Schrier, J. and Shi, W. 2021. Machine Learning: New ideas and tools in environmental science and engineering. ACS Publications, Accessed: Jun. 01, 2024. [Online]. Available: https://pubs.acs.org/doi/abs/10.1021/acs.est.1c01339.

Chapter 9

Tracing Plant Growth Patterns

Employing Artificial Intelligence and Computer Vision for Explicit Mapping

Monica Sankat*, *Ram Kumar** and *Kumari Shivangi

Introduction

Understanding and improving plant development patterns is an essential endeavor in agriculture and horticulture that has a substantial impact on sustainability, resource management, and production. Manual measurements and observations are frequently used in traditional techniques of tracking and analyzing plant growth, which can be labor-intensive, time-consuming, and prone to human error. A revolutionary method for precisely and effectively mapping plant growth patterns is presented by the combination of Artificial Intelligence (AI) and Computer Vision (CV), as technological breakthroughs continue to impact numerous sectors. This chapter explores the importance of applying Artificial Intelligence and Computer Vision technologies in agriculture, emphasizing how they could transform plant growth analysis and providing a thorough study of their uses, advantages, and potential future developments.

The complicated process of plant growth is controlled by several elements, such as nutrient levels, soil quality, water availability, genetics, and environmental circumstances. In the past, manual techniques have been used by scientists and farmers to track these variables and evaluate the health of their plants. Though somewhat successful, these approaches frequently have drawbacks like subjectivity, constrained scalability, and the incapacity to deliver real-time data. Furthermore, the growing world population and the ensuing need for increased agricultural productivity call for more accurate and effective methods of plant monitoring and growth optimization.

School of Computing Science and Engineering, VIT Bhopal University, Kothrikalan, Sehore, Madhya Pradesh, 466114.

* Corresponding author: drramkumar.research@gmail.com

Let us introduce computer vision and artificial intelligence. When paired with Computer Vision's ability to interpret visual input, Artificial Intelligence's ability to process massive volumes of data and spot patterns presents a potent solution to these problems. Large datasets can be analyzed by Artificial Intelligence algorithms, especially machine learning and deep learning models, to find nuances in patterns and relationships that human observers might miss. These technologies can be used to track several growth metrics in plants, identify early symptoms of illness or stress, and offer useful information to improve crop management techniques.

Phenotyping, or the measuring of an organism's physical and biochemical characteristics, is one of the main uses of Artificial Intelligence and Computer Vision in plant growth mapping. High-resolution photos of plants can be taken by automated phenotyping systems using Computer Vision technology. These images are subsequently subjected to Artificial Intelligence algorithms to extract specific information about plant morphology, including leaf size, shape, color, and texture. These systems can run constantly, giving researchers and farmers access to real-time data that supports their decision-making regarding plant health and growth circumstances. For example, autonomous vehicles such as drones that are fitted with cameras and sensors can drive across agricultural fields, take pictures and gather information about the health of the plants, the state of the soil, and the surrounding environment (Zhang et al., 2017). After that, this data is processed by Artificial Intelligence systems to provide detailed maps of plant growth trends. With the help of these maps, which may identify stress points for plants as well as regions where they are doing well, specific interventions like modifying irrigation schedules, adding fertilizer, or putting insect control measures in place can be made.

Artificial Intelligence and Computer Vision technology can also improve plant disease identification and diagnosis. To minimize productivity losses and prevent extensive crop damage, early disease diagnosis is essential. Visual inspection is a common component of traditional illness diagnosis techniques, however it can be arbitrary and unreliable.

History of Plant Growth Mapping

Earlier plant growth mapping was done by observing changes in plant sizes, shape and color to assess growth stages and health and the tools used during these processes included rulers and measuring sticks but these provided rudimentary data and lacked accuracy and scope. Moving on new approaches were adopted which included documenting growth stages, environmental conditions and plant resources. Scientists such as Carl Linnaeus and Charles Darwin proposed models and taxonomies providing a diverse understanding of plant development. Furthermore, more accurate plant growth quantification was made possible by the use of more exact measurement tools, such as micrometers, planimeters, and calipers (Islam et al., 2024). The systematic analysis of growth patterns and environmental impacts in agricultural research was made possible by the introduction of statistical tools. Massive amounts of surveillance of crop health and growth patterns were made possible by the development of remote sensing technologies including satellite images and aerial photography, which completely changed the field of plant growth mapping.

Moreover, plant growth mapping became much more accurate and straightforward in the late 20th century with the switch from analog to digital imagery. Plant development was precisely predicted and mapped under various situations using computer models that included physiological and environmental data, made possible by the use of digital cameras and scanners to provide comprehensive, high-resolution images. Machine learning and artificial intelligence (AI), which were first used in plant growth mapping in the early 20th century, are becoming more and more popular. Large datasets can be analyzed, growth trends can be found, and future growth can be accurately predicted thanks to this technology. With Artificial Intelligence combined with the Internet of Things, computer vision technology has become a crucial component of plant growth modeling, enabling the automated analysis of plant photos, growth stage tracking, and real-time health indicators.

Plant growth simulation developments of the present and future center on precision agriculture, where artificial intelligence (AI) and machine learning (ML) are key tools for improving plant growth through precise and consistent environmental condition monitoring. Plant growth simulation's long-term objectives include sustainability and the development of technologies that enhance resource efficiency, lessen environmental effects, and guarantee food safety (Islam et al., 2024). Further advancements in the utilization of sensing technologies, including thermal imaging, spectrum analysis, and drone-based photography, hold promise for improving the ability to image plant development. History illustrates how agricultural practices have developed from basic restrictions to more effective, technologically advanced techniques. With the further advancement of AI, computer vision, and other cutting-edge technology, plant growth modeling will become increasingly precise, effective, and vital to the practice of sustainable agriculture.

Fundamentals of Artificial Intelligence and Computer Vision in Agriculture

Artificial Intelligence (AI) in agriculture incorporates leveraging calculations and machine learning models to overhaul developing sharpens, boost alter yields, and move forward resource organization. Precision development may be an essential application where Artificial Intelligence analyzes data from sensors, climate gauges, and slavish imagery to optimize field-level organization. Alter checking through Artificial Intelligence enables agriculturists to assess trim prosperity, recognize diseases early, and take preventive measures, along these lines to secure crops. Besides, Artificial Intelligence models predict soil quality by analyzing data from soil sensors, making a distinction between agriculturists selecting and fitting fertilizers and medicines. Prescient analytics fueled by Artificial Intelligence can figure trim yields, bug pervasions, and climate designs, allowing agriculturists to create taught choices nearly planting, water framework, and gathering. Other than that, mechanized devices such as autonomous tractors and collectors utilize Artificial Intelligence to perform developing assignments with tall precision, diminishing the requirement for manual labor and extending efficiency. Computer Vision (CV), a subset of AI, enables computers to disentangle and plan visual data from the environment. In cultivating, Computer Vision is critical for distinctive applications, such as alter and weed recognizing confirmation (Hesami and Jones, 2020). Computer Vision systems can identify between crops and weeds, allowing for more effective herbicide application, reduced chemical use, and the advancement of feasible development strategies. Plant disease area is another pivotal application where Computer Vision systems analyze pictures of plants to recognize illnesses at an early organize, allowing for lucky mediations and treatment. Besides, Computer Vision advancement makes a difference in common item counting and yield estimation by precisely evaluating the number of characteristic items on trees, making a contrast farmers expect and manage their harvests. Furthermore, Computer Vision-powered meanders and robots can inspect wide areas, providing real-time data on trim conditions, improvement plans, and prospective difficulties, allowing agriculturists to make data-driven decisions and progress toward overall proficiency.

Technologies and Tools for Plant Growth Mapping

Monitoring, measuring, and analyzing plant growth and health is done through the use of several technologies in plant growth mapping. While multispectral, hyperspectral, and thermal imaging technologies are used to analyze plant health and stress levels, remote sensing—which makes use of drones and satellite imagery—provides large-scale data. Soil moisture, nutritional levels, and climatic conditions are tracked by sensors positioned on the ground. 3D modeling examines structural alterations, while computer vision scans photos to identify growth patterns and diseases. Anomaly detection, automated categorization, and predictive analytics are provided by Artificial Intelligence and ML. Web of Things (IoT) devices allow for continuous data integration and monitoring, while Geographic Information Systems (GIS) incorporate spatial data. Robotics enables precision agriculture and automated monitoring, while cloud computing manages big databases and real-time analytics. Improved plant growth control is made possible by these coupled technologies, which improve farming methods.

Moreover, measurements of gene expression, cell division, cell expansion, and their rates and geographic distributions are crucial for providing a thorough characterization of the dynamic behaviors of the cells responsible for organ growth and development. Cell division and lineage patterns in shoot and root meristems were first identified by image-based research. The creation of imaging technologies to see cellular dynamics in living things resulted from this. Researchers can examine data and generate testable predictions by combining computer modeling and experimental methodologies. Predictive models that take system behavior into account may be developed using biological studies, such as z-stacks of confocal microscope pictures. After that, these models can produce counterintuitive predictions for more experimental investigation (Yang et al., 2020).

Deep Learning Applications in Plant Growth Monitoring

Multi-layer artificial neural networks (ANNs) are used in deep learning (DL), a subset of machine learning (ML), to model data. Since these models are data-driven, the parameters and outcomes they produce are predicated on the real-world instances they were trained on. DL may acquire data representations in several ways, including:

- **Supervised Learning:** Refers to predetermined results.
- **Unsupervised Learning:** This method concentrates on finding patterns in the data.

- **Semi-Supervised Learning:** Blends supervised and unsupervised techniques, weighing the advantages and disadvantages of each.
- **Reinforcement Learning:** Based on incentives and punishments, this approach employs trial-and-error techniques.

Plant development is difficult to generalize for traditional machine learning (ML) models like decision trees, k-nearest neighbors (k-NNs), and support vector machines (SVMs) because biological factors like physiology and genetics are so complicated (Sharma et al., 2023).

Dataset Acquisition and Processing

In plant growth mapping, there are several procedures and difficulties involved in the dataset gathering and processing. Field research, greenhouse trials, open databases, and remote sensing are some of the sources from which data is gathered. This data consists of genotypic information, phenotypic measures, environmental data, and high-resolution photographs. To maintain dataset integrity throughout processing, data cleaning, normalization, and annotation are required. The diversity and resilience of data are increased by data augmentation techniques including flipping, rotating, and creating synthetic data. While segmentation approaches separate individual plant sections and analyzes growth trends over time, feature extraction, both automatic and human, find important features. Multimodal data fusion improves analysis, whereas data integration merges sources to provide an extensive dataset. After that, the dataset is divided into training, validation, and testing sets, and its model dependability is checked using cross-validation before being ready for analysis.

IoT in Plant Growth Mapping

Plant growth mapping is greatly improved by the Internet of Things (IoT), which allows for continuous monitoring and data integration of several environmental and plant-specific characteristics. Field-based Internet of Things (IoT) sensors monitor temperature, humidity, nutrient levels, and soil moisture in real-time, sending the collected data to a central system for analysis. This consolidated data gathering offers a thorough understanding of the dynamics of plant development.

IoT systems that use real-time monitoring can quickly identify problems such as nutrient deficits or drought stress and may send out automatic notifications that allow for early response. To forecast future plant development, yield, and possible problems, machine learning evaluates the gathered data. This helps with anomaly identification and informed decision-making (Dawn et al., 2023).

By providing targeted fertilization, irrigation, and pest management that are suited to individual plant needs, optimizing resource consumption, and boosting sustainability, IoT makes precision agriculture possible.

Using computers or cellphones, farmers may remotely oversee their crops, and the Internet of Things (IoT) can automate equipment like greenhouse climate management and irrigation pumps.

More effective choices for cultivation, harvesting, and general crop management are supported by IoT's complete insights on plant health and growth, which are obtained through the integration and analysis of data from several sources. Analyzing historical data enables one to see trends, enhance procedures, and adjust to shifting circumstances. Overall, by facilitating the monitoring, control, and optimization of plant health and development, IoT enhances agriculture's production, sustainability, and efficiency.

Robotics in Plant Growth Mapping

Plant growth mapping is revolutionized by robotics since it automates observations, data collecting, and interventions. While fixed robots continuously monitor particular regions, mobile robots with cameras and sensors scan fields or greenhouses to collect detailed data on plant health and environmental conditions. In precision agriculture, robots utilize real-time data to administer pesticides, fertilizers, and water with extreme accuracy. Additionally, robots with vision capabilities suppress weeds and other pests, minimizing the need for chemical applications. To evaluate plant health and identify illnesses, they take high-resolution pictures and employ multispectral and hyperspectral imaging. Robots detect developmental changes over time and assess growth indicators for plant phenotyping. This automation boosts productivity, decreases human work, and may be expanded to accommodate different types of agricultural activities. Robots with sensors to improve growth conditions improve environmental monitoring.

Practical Applications of Plant Growth Mapping

Artificial intelligence (AI) has advanced at a faster pace recently because of big data breakthroughs. These innovations combine clever algorithms with large datasets, allowing Artificial Intelligence software to automatically identify patterns or characteristics within large datasets. Machine learning, which is utilized by computer systems to do certain tasks, is a type of artificial intelligence. Typically, two categories exist: supervised learning and unsupervised learning strategies. A matrix containing independent x and dependent y variables is used to represent augmented learning approaches. Y is frequently a vector of continuous values when employing regression in classification issues, and a scalar to represent the category labels.

Plant Image Acquisition

During this stage, pertinent plant photos are taken and automatically analyzed for categorization. These photos are captured by high-resolution digital cameras and are essentially collections of binary data. Additionally, smartphones are important since they can take pictures in a variety of file types, including jpg, png, and tif. Following the capture of all required photos, they go to the image preprocessing step for any necessary corrections and improvements. Techniques for image enhancement are used if the pictures don't match the requirements for processing (Perez-Sanz et al., 2017).

Precise illness categorization depends on accurate picture collection. Since the captured pictures are used to train machine learning models, their quality has a significant impact on the overall efficiency of the framework. Numerous well-known picture datasets have been developed by agricultural research.

Image Preprocessing

Processing the acquired pictures is essential at the first stage of image collection to resolve several difficulties, including noise, blur, poor lighting, and undesired backdrops. This pretreatment stage makes sure that the raw data is transformed into a format that can be used and that any distortions are removed, preparing it for automated illness classification techniques. Following that, the photos go through crucial feature extraction and segmentation processes.

Preprocessing preserves consistency in image resolution while optimizing the use of computational resources. Standardization, color scale modification, distortion removal, regularization of picture size, and noise reduction are important preprocessing techniques. Images are frequently resized to suit a predetermined color space and scaled to certain proportions for the best analysis. Utilizing a white background can improve visual clarity, according to studies.

In agricultural research, the HSV (Hue, Saturation, and Value) approach is a common preprocessing technique that closely resembles human visual capabilities. Techniques for masking and removing backgrounds are widely utilized to increase processing accuracy and efficiency. Colored photos are frequently converted into the HSI (Hue, Saturation, and Intensity) color space, with the H component being especially helpful for additional analysis (Barbedo, 2013).

High-pass filters are used to emphasize regions with noticeable intensity gradients, whereas low-pass filters are used to minimize high-frequency noise. The Fast Fourier Transform technique converts pictures into the spatial frequency domain, while the Laplacian filter enhances the sharpness of image outlines. Histograms are utilized to alter the power distribution and enhance image quality, while Gaussian smoothing channels are utilized to lessen noise.

It's critical to segment the picture of the contaminated leaf.

Image Segmentation

A key tool in agricultural research is segmentation, which is breaking a picture up into its constituent parts so that each item can be examined more closely and valuable characteristics may be extracted. This procedure aids in identifying a plant's afflicted and unaffected areas. To extract potentially useful characteristics, it is essential to segment preprocessed pictures to categorize sick leaves.

Traditional techniques for segmentation include clustering, region-based, edge detection, and thresholding. Thresholding is often used in remote sensing, detection, and classification to segment pictures depending on pixel intensity levels. Global, variable, and adaptive thresholding are some techniques it uses. Using methods like the Sobel operator, Canny edge detector, and Laplacian of Gaussian (LoG) filter, edge detection divides a picture according to its edges. By applying techniques like Region Growing and Region Splitting, region-based segmentation separates the picture into areas according to pixel similarities in intensity, color, and form. Using common algorithms like K-means and fuzzy C-means, clustering groups' pixels based on texture, color, or other attributes (Yang et al., 2023).

However complex photos with precise features are typically a challenge for older approaches. Segmentation techniques based on deep learning have been proven to be more effective than conventional approaches. By categorizing every pixel in the image, semantic segmentation separates the image into useful areas. CNN and other deep learning methods improve the accuracy of segmentation. A sophisticated type of semantic segmentation, instance segmentation uses picture recognition to anticipate instances of object classes, enabling it to tackle complicated jobs. It is standard practice to employ methods like Tensor Mask, Deep Mask, RCNN, YOLO, and Instance Cut. Each approach has its own advantages and limitations, making the choice dependent on the specific application requirements.

Case Study on Plant Growth Mapping

Automated Tomato Disease Detection

Scientifically named Solanum Lycopersicon, tomatoes are an important agricultural commodity that is cultivated widely in Asia and around the world because of their high nutritional content, which includes high quantities of potassium, beta-carotene, and vitamins E and C. Tomatoes, although quite popular, are extremely susceptible to a wide range of diseases brought on by bacteria, microorganisms, and pests. Therefore, to guarantee crop health and output, automated disease detection has emerged as a crucial field of study. Various approaches have been investigated for this reason. With the help of the Plant Village dataset, Francis and Deisy (2019) created a CNN model with four convolutional layers and two dense layers. The model demonstrated an astounding 87% accuracy rate in differentiating between healthy and diseased tomato and apple leaves. Similarly to this, Basavaiah and Anthony (2020) used texture, color, and form variables from 200 photos to evaluate several machine learning techniques. The Random Forest model demonstrated a 94% accuracy rate. Probabilistic neural networks (PNN) and KNN were used by K and Rao (2019) to identify and classify illnesses from 600 photos, including Verticillium wilt, powdery mildew, and Septoria leaf spot. PNN outperformed KNN with an accuracy of 91.88%. Vadivel and Suguna (2022) employed the BPNN approach on 10,000 augmented photos and used the K-means method for feature extraction, resulting in an astounding 99.4% accuracy. By fine-tuning ResNet and Xception models on 4281 pictures, Chakravarthy and Raman (2020) achieved an amazing 99.95% accuracy in identifying early blight disease. In their analysis of many CNN architectures using 14,903 pictures of ten distinct tomato illnesses from the PlantVillage dataset, Kumar and Vani (2019) discovered that the VGG16 model outperformed other designs such as LeNet, Xception, and ResNet50, with the greatest accuracy of 99.25% (Jafar et al., 2024). These papers demonstrate the progress made in automated tomato disease diagnosis and the potential of different machine learning and deep learning models to improve crop health and agricultural productivity.

Automated Potato Disease Detection

A staple crop in every country, the potato suffers greatly from many illnesses, such as pink rot, common scab, black scurf, and black leg. Reversing these illnesses is essential to maintaining agricultural sustainability and food security. Numerous studies have explored the field of deep learning (DL) and machine learning (ML) techniques for automated disease detection in potato plants to address this. In a thorough analysis of machine learning techniques including ANN, SVM, and RF,

Patil et al. (2017) found that ANN was the most accurate, achieving an astounding 92% accuracy rate in the identification of sugarcane illnesses from leaf photos (Singh et al., 2024). With the help of textural features and the maximum-minimum color difference technique, Suttapakti and Bunpeng (2019) developed a novel method that produced an impressive 91.67% accuracy in disease classification. Convolutional neural networks (CNNs) were first used by Arshaghi et al. (2023) to categorize potato diseases. They achieved previously unheard-of accuracy rates of 100% and 99% across a range of disease groups, signifying a substantial advancement in automated disease identification. Arya and Rajeev (2019) also investigated the efficacy of CNN and AlexNet designs; AlexNet performed exceptionally well, achieving an astounding 98.33% accuracy rate in disease identification. Further highlighting the promise of DL in disease diagnosis, Mahum et al. (2023) developed an advanced DL technique including Efficient DenseNet and reweighted cross-entropy loss, attaining a testing set accuracy of 97.2%. Al-Amin et al. (2019) achieved accuracies of up to 98.33% in their development of a specialized deep learning architecture for detecting late and early blight in potato leaves, proving the effectiveness of DL models in disease detection. In addition, Sharma et al. (2021) achieved remarkable accuracy rates of 97.66% by using DL CNNs to predict potato illnesses, outperforming traditional ML classifiers. The conclusion of these investigations points to a positive direction for the incorporation of ML and DL methods into farming operations, providing long-term remedies for crop protection and disease control in potato production (Sinshaw et al., 2022). Table 1 summarizes various Artificial Intelligence techniques used in agriculture.

Table 1. Represents various techniques used in agriculture, their strengths and their limitations.

Techniques	**Strengths**	**Limitations**
CALEX	Can formulate scheduling guidelines for crop management activities.	Takes time.
PROLOG	Removes less used farm tools from the farm.	Location-specific.
ANN	Predicts crop yield.	Only captures weather as a factor for crop yield.
ROBOTICS - Demeter	Can harvest up to 40 hectares of crop.	Expensive: Uses a lot of fuel.
ROBOTICS	Has 80% success rate in harvesting crops.	Slow picking speed and accuracy.
ANN and Fuzzy Logic	Reduces insects that attack crops.	Shows inability to differentiate between crop and weed.
Computer Vision System (CVS)	Works at a high speed. Can multitask.	Dimension-based detection which may affect good species.
Expert system using rule-base in disease detection	Faster treatment as diseases are diagnosed faster. Cost effective based on its preventive approach.	Time consuming. Needs constant monitoring to check if pest has built immunity to the preventive measure.
Web-Based Expert System	High performance.	Internet and web based.
ANN, GIS	95% accuracy.	Internet-based. Some rural farmers will not have access.

Challenges and Limitations

The incorporation of Artificial Intelligence in plant growth analysis poses substantial challenges and limitations that must be overcome to ensure successful implementation. Data quality and quantity problems occur because there aren't enough big and good-quality datasets needed to train strong models, and the process of labeling data is both time-consuming and expensive. The diversity of agricultural data, encompassing different types and formats, poses challenges when attempting to integrate them into comprehensive models. Advanced Artificial Intelligence models, especially deep learning networks, are often intricate and demand significant computational power, which can result in overfitting and limited applicability. Furthermore, these models serve as "black boxes," making it challenging to understand their decision-making processes, which is essential for building trust among farmers and stakeholders (Kumar et al., 2024). The challenges of scalability and deployment arise from the need for substantial infrastructure, including high-performance computing and dependable internet connectivity, which are often scarce in rural regions (Pérez-López et al., 2007). The expense of implementation can be prohibitive, restricting accessibility for small-scale farmers. Broader systemic issues encompass ethical and privacy concerns related to data ownership, the ever-changing regulatory and policy landscape, and the environmental consequences of deploying Artificial Intelligence infrastructure. Overcoming these challenges necessitates a collaborative effort among Artificial Intelligence researchers, agricultural scientists, policymakers, and farmers to create sustainable, efficient, and trustworthy Artificial Intelligence solutions in the field of agriculture.

Future Scope of Plant Growth Mapping

The long-term scope of Artificial Intelligence and Computer Vision (CV) in agriculture is wide and transformative, adjusted to revolutionize how we approach developing, food era, and resource organization. As the world's populations rise and climate modification impacts agrarian effectiveness, Artificial Intelligence and Computer Vision propel will become essential gadgets for ensuring food security and prudent development sharpens (Chouhan et al., 2021). One major zone of future headway is free development, where Artificial Intelligence-driven robots and drifts perform errands such as planting, checking, and collecting with irrelevant human intervention. These free systems will utilize advanced Computer Vision methods to investigate ranges, recognize altered prosperity issues, and isolate crops and weeds, in this way extending capability and lessening labor costs (Noguchi et al., 1998). Additionally, Artificial Intelligence calculations will update the exactness of agribusiness by analyzing colossal datasets from IoT sensors, lackey imagery, and climate figures to form real-time choices on water framework, fertilization, and bother control. This data-driven approach will optimize resource utilization, driving higher yields and more temperate development sharpens.

Another critical future advancement is the integration of Artificial Intelligence and Computer Vision in bug and sickness organizations. Prescient analytics will engage early disclosure of bug perversions and plant diseases, allowing for luck and centered on trade that minimizes alter hurt and reduce the requirement for chemical pesticides (Chouhan et al., 2024). Moreover, AI-powered soil prosperity watching systems will allow farmers with nitty-dirty bits of information about soil composition and prosperity, coordinating them in actualizing practical soil organization strategies and advancing trim productivity. The long term will also see the rise of AI-driven grandstand deciding disobedience that predicts and promotes designs and asks, making a distinction between farmers making taught choices nearly altering choices and accumulating timing, hence maximizing their benefits (Efremova et al., 2019).

The application of Artificial Intelligence and Computer Vision in cultivating will increase to making advanced alter combinations that are more adaptable to climate change and ailment. By analyzing innate data and common conditions, Artificial Intelligence calculations will offer assistance in breeding crops that are not because they were high-yielding but moreover competent in withstanding unfavorable conditions (Chouhan et al., 2024). This will be especially pivotal in tending to the challenges posed by climate modification, ensuring food security for future periods.

In expansion, the long run will witness the extension of sharply developing organic frameworks, where interconnected Artificial Intelligence systems direct each viewpoint of the agrarian get-ready. These natural frameworks will utilize machine learning models to ceaselessly learn and alter to changing conditions, overhauling their decision-making capabilities over time (Salman et al., 2023). The allotment of blockchain development in conjunction with Artificial Intelligence will in addition ensure straightforwardness and traceability inside the food supply chain, from development to table, boosting client certainty in food security and quality.

Inside the space of approach and organization, Artificial Intelligence will play a significant portion in shaping attainable agrarian courses of action. By giving data-driven encounters and prescient models, Artificial Intelligence will offer assistance to policy makers in making controls that progress attainable, developing sharpens, resource preservation, and characteristic security (Khajuria et al., 2018). Long-standing time will also see extended collaboration between tech companies, and country businesses, and ask approximately teach, developing headway and enlivening the progression of Artificial Intelligence and Computer Vision advancements custom fitted for agribusiness.

Conclusion

In conclusion, the integration of Artificial Intelligence and computer vision into agriculture messengers is a cutting-edge time of advancement and efficiency, publicizing courses of action to many of the preeminent pressing challenges in cutting-edge development. As these advancements continue to advance, they will enable free developing systems, correct resource organization, and updated bug and disease control, driving higher yields and more doable sharpens. The advancement of flexible alter combinations and sharp developing natural frameworks will energize and fortify food security and normal conservation. By developing collaboration among development companies, and agrarian businesses, and asking almost any teacher, to be prepared to animate the determination of Artificial Intelligence and Computer Vision in cultivating, and driving basic changes in proficiency and productivity. Inevitably, the long run of farming lies in leveraging these cutting-edge developments to make a more flexible, beneficial, and conservative food-era system, able to meet the demands of a creating world people while securing our ordinary resources for future times. This transformative potential grows past proficiency and productivity picks up, touching on noteworthy points of common viability and money related common sense. The capacity of Artificial Intelligence and computer vision to analyze perpetual sums of data and provide vital bits of information will empower farmers to create better-informed choices, optimizing resource utilization and minimizing waste. Precision agribusiness, engaged by these progresses, will ensure that water, fertilizers, and pesticides are associated inside the foremost beneficial way, lessening characteristic influence and advancing maintainable sharpens.

References

Barbedo, J.G.A. 2013. Digital image processing techniques for detecting, quantifying and classifying plant diseases. https://springerplus.springeropen.com/articles/10.1186/2193-1801-2-660.

Chouhan, S.S., Singh, U.P., Sharma, U. and Jain, S. 2021. Leaf disease segmentation and classification of Jatropha Curcas L. and Pongamia Pinnata L. biofuel plants using computer vision based approaches. *Measurement*, 171: 108796, ISSN 0263-2241, https://doi.org/10.1016/j.measurement.2020.108796.

Chouhan, S.S., Singh, U.P. and Jain, S. 2024. Introduction to computer vision and drone technology. *In*: Chouhan, S.S., Singh, U.P. and Jain, S. (eds.). *Applications of Computer Vision and Drone Technology in Agriculture 4.0*. Springer, Singapore. https://doi.org/10.1007/978-981-99-8684-2_1.

Chouhan, S.S., Singh, U.P., Sharma, U. et al. 2024. Classification of different plant species using Deep Learning and Machine Learning algorithms. *Wireless Pers. Commun.*, 136: 2275–2298. https://doi.org/10.1007/s11277-024-11374-y.

Dawn, N., Ghosh, T., Ghosh, S., Saha, A., Mukherjee, P., Sarkar, S., Guha, S. and Sanyal, T. 2023. Implementation of Artificial Intelligence, Machine Learning, and Internet of Things (IoT) in revolutionizing Agriculture: A review on recent trends and challenges. https://www.researchgate.net/publication/370404876_Implementation_of_Artificial_Intelligence_Machine_Learning_and_Internet_of_Things_IoT_in_revolutionizing_Agriculture_A_review_on_recent_trends_and_challenges.

Efremova, N., West, D. and Zausaev, D. 2019. AI-based evaluation of the SDGs: The case of crop detection with earth observation data. https://arxiv.org/abs/1907.02813.

Hesami, M. and Jones, A.M.P. 2020. Application of artificial intelligence models and optimization algorithms in plant cell and tissue culture, September 2020. https://pubmed.ncbi.nlm.nih.gov/32984921/https://pubmed.ncbi.nlm.nih.gov/32984921/.

Islam, S., Nasim Reza, Md., Samsuzzaman, Ahmed, S., Cho, Y.J, Dong Hee Noh, D.H, Chung, S.-O. and Hong, S.J. 2024. Machine vision and artificial intelligence for plant growth stress detection and monitoring: A review. https://www.researchgate.net/profile/Sumaiya-Islam-9/publication/379810822_Machine_vision_and_artificial_intelligence_for_plant_growth_stress_detection_and_monitoring_A_review/links/661bc88266ba7e2359d6a779/Machine-vision-and-artificial-intelligence-for-plant-growth-stress-detection-and-monitoring-A-review.pdf.

Jafar, A., Bibi, N., Naqvi, R.A., Sadeghi-Niaraki, A. and Jeong, D. 2024. Revolutionizing agriculture with artificial intelligence: Plant disease detection methods, applications, and their limitations. https://www.frontiersin.org/journals/plant-science/articles/10.3389/fpls.2024.1356260/full.

Khajuria, R. Sunanda and Singh, S. 2018. A survey of existing leaf disease techniques using Artificial Neural Network. *International Journal of Computer Trends and Technology (IJCTT)*, 59(1): 52–62, ISSN:2231–2803, doi: 10.14445/22312803/IJCTT-V59P109, https://ijcttjournal.org/archives/ijctt-v59p109.

Kumar, R. et al. 2024. Hybrid approach of cotton disease detection for enhanced crop health and yield. In *IEEE Access*. https://ieeexplore.ieee.org/document/10589357.

Noguchi, N., Reid, J.F., Zhang, Q. and Tian, L.F. 1998. Vision intelligence for precision farming using fuzzy logic optimized genetic algorithm and Artificial Neural Network. https://www.

semanticscholar.org/paper/Vision-Intelligence-For-Precision-Farming-Using-And-Noguchi-Reid/25f44ac407418e8b053913454d673ecbaeddbc61.

Pérez-López, A.J., López-Nicolas, J.M., Núñez-Delicado, E., del Amorz, F.M. and Carbonell-Barrachina, A.A. 2007. Effects of agricultural practices on color, carotenoids composition, and minerals contents of sweet peppers, cv. Almuden. https://pubs.acs.org/doi/10.1021/jf071534n.

Perez-Sanz, F., Navarro, P.J. and Egea-Cortines, M. 2017. Plant phenomics: An overview of image acquisition technologies and image data analysis algorithms. https://academic.oup.com/gigascience/article/6/11/gix092/4316962?login=false.

Salman, Z., Muhammad, A., Piran, Md. J. and Han, D. 2023. Crop-saving with AI: Latest trends in deep learning techniques for plant pathology. https://www.frontiersin.org/journals/plant-science/articles/10.3389/fpls.2023.1224709/full.

Sharma, S., Verma, K. and Hardaha, P. 2023. Implementation of Artificial Intelligence in Agriculture. https://ojs.bonviewpress.com/index.php/JCCE/article/view/174.

Singh, S., Kumar, R. and Sharma, G. 2024. Deep learning for disease detection in sugarcane leaves. https://www.taylorfrancis.com/chapters/edit/10.1201/9781003430421-47/deep-learning-disease-detection-sugarcane-leaves-souravdeep-singh-ram-kumar-geeta-sharma?context=ubx&refId=22a6be5f-5ecf-4c1f-abf5-d3deec6cd5df.

Sinshaw, N.T., Assefa, B.G., Mohapatra, S.K. and Beyene, A.M. 2022. Applications of computer vision on automatic potato plant disease detection: A systematic literature review. https://pubmed.ncbi.nlm.nih.gov/36419507/.

Yang, T., Zhou, S., Xu, A., Ye, J. and Yin, J. 2023. An approach for plant leaf image segmentation based on YOLOV8 and the Improved DEEPLABV3+. https://www.mdpi.com/2223-7747/12/19/3438.

Yang, Y., Wu, T., Wang, S. and Li, H. 2020. Fractional evergreen forest cover mapping by MODIS time-series FEVC-CV methods at sub-pixel scales. https://www.sciencedirect.com/science/article/abs/pii/S0924271620300824.

Zhang, X., Huang, C., Wu, D., Qiao, F., Li, W., Duan, L., Wang, K., Xiao, Y., Chen, G., Liu, Q., Xiong, L. and Yang, W. 2017. High-throughput phenotyping and QTL mapping reveals the genetic architecture of maize plant growth. *Plant Physiology*, 173(3). https://academic.oup.com/plphys/article/173/3/1554/6115934.

Chapter 10

AI-Driven Circular Economy

Innovations in Agro and Food Waste Management

*Sachin Rathour** and *Evans Kemboi*

Introduction

The circular economy strives to reduce waste and resource use efficiently, transforming from the traditional linear model of production that follows a 'take-make-dispose' approach (Edirisinghe, 2024). The circular economy represents a shift towards a more sustainable approach. In contrast, the circular economy is characterized by its restorative and regenerative design, aiming to close the loops of product lifecycles through increased resource efficiency and sustainability (Kirchherr, 2017).

The existing linear economic model is unsustainable, promoting the unjust use of finite natural resources and generating excessive waste, which severely harms the environment. In contrast, a circular economy (CE) offers a restorative, sustainable, and regenerative alternative and is gaining global traction. Among all the existing digital technologies, artificial intelligence (AI) is a key promoter for CE, significantly facilitating its adoption and implementation in practical applications. This chapter explores the intersection of AI and CE and discusses policies for implementing CE principles using AI. Here Fig. 1(a) and 1(b) illustrate the elaborative comparison between the linear economy and circular economy:

Ph.D. Research Scholar, Department of Agricultural Economics, Institute of Agricultural Sciences, Banaras Hindu University, Varanasi-221005, Uttar Pradesh, India.

* Corresponding author: sachinrt638@bhu.ac.in

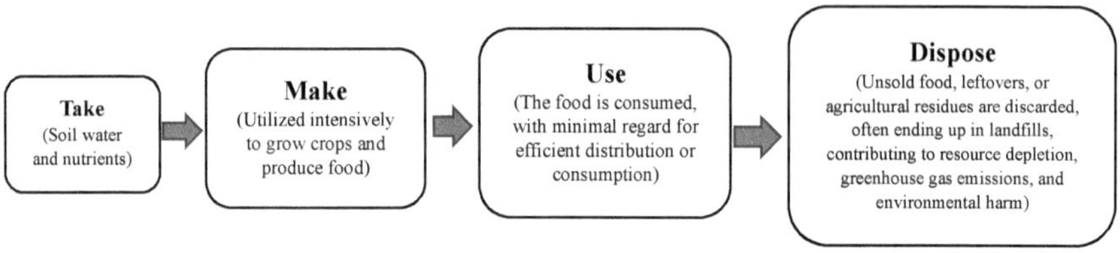

Fig. 1(a). Linear economy model depicting the "Take–Make–Use–Dispose".

Versus

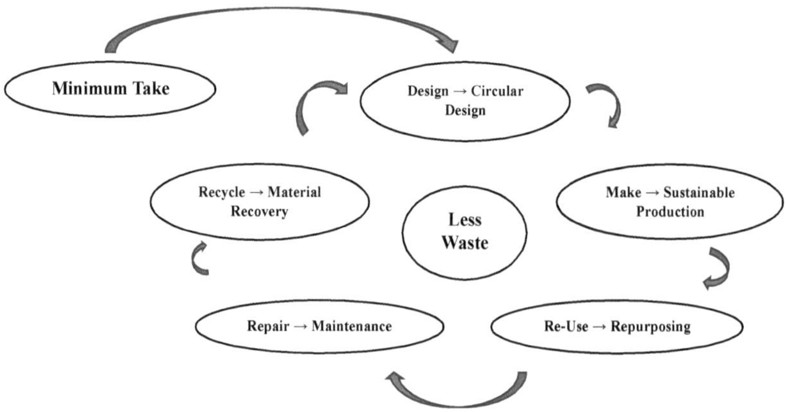

Fig. 1(b). Circular economy model depicting the "Design–Make–Reuse–Repair–Recycle".

Source: Fig. 1(a) and 1(b) illustrate a model designed by the authors.

Key Principles and Strategies: Reduce, Reuse, Recycle (3R)

The principles of reduce, reuse, and recycle (3R) are foundational to the circular economy, aiming to produce a sustainable system where resources are employed efficiently and waste is reduced significantly.

The principle of reducing involves minimizing the number of resources used and waste generated in the first place. For example; Companies reduce packaging materials to decrease waste; and use energy-efficient appliances and machinery. The principle of reusing involves using items more than once, either for the same purpose or for a different function, to extend their lifecycle. Let us say, reusable shopping bags, water bottles, and containers; converting and refurbishing and reselling used electronics and the principle of recycling involves the processing of used materials into new products to prevent the waste of potentially usable materials. For example, recycling paper products into new paper items; and converting used plastics into new plastic products or alternative materials.

Application of 3R (Reduce, Reuse, Recycle) with AI

Artificial intelligence has become a transformative force for enhancing the application of the 3R principles—Reduce, Reuse, and Recycle—by providing advanced tools and

```
A[Start] --> B[Reduce]
B --> C [Predictive Maintenance]
B --> D [Energy Management]
B --> E [Smart Agriculture]
A --> F [Reuse]
F --> G [Product Lifecycle Management]
F --> H [Shared Economy Platforms]
F --> I [Smart Sorting and Refurbishment]
A --> J [Recycle]
J --> K [Automated Sorting Systems]
J --> L [Material Recovery]
J --> M [Recycling Process Optimization]
```

Fig. 2. Flow chart of AI in the circular economy.

solutions for sustainable practices (Vasarhelyi, 2022). Here are modern applications of AI in each of these principles:

Table 1. AI applications in the circular economy.

AI Technique	Application Area	Specific Use Case	Benefits
Machine learning	Predictive maintenance	Predicting equipment failures in manufacturing	Reduces downtime, extends equipment life
Computer vision	Quality control	Automated inspection of products on the assembly line	Ensures product quality, reduces waste
Natural language processing	Customer engagement	Chatbots for customer queries on sustainable products	Enhances customer experience, promotes sustainable choices
IoT	Smart grids	Real-time energy management	Optimizes energy use, reduces waste
Robotics	Waste management	Autonomous robots for sorting recyclables	Improves sorting accuracy, increases recycling rates

Source: (Stahel, 2016).

```
+------------------+        +------------------+        +------------------
+        +------------------+
|   Resource       |        |   Design &       |        |   Production &
|        |  End-of-Life     |
|  Optimization    |------->|  Manufacturing   |------->|   Distribution
|------->|  Management      |
+------------------+        +------------------+        +------------------
+        +------------------+
          |                              |                              |
|
          v                              v                              v
v
+------------------+       +------------------+       +------------------+
+------------------+
| Machine Learning |    | Computer Vision   |    | IoT              |    |
Robotics         |
|  - Predictive    |    |  - Quality        |    |  - Asset Tracking|    | -
Automated       |
|    Maintenance   |    |    Control         |    |  - Smart Grids   |    |
Sorting         |
+------------------+       +------------------+       +------------------+
+------------------+
```

Fig. 3. Conceptual diagram showing how AI tools are applied within the circular economy.

To visualize the impact of AI on circular economy practices, let's consider a hypothetical graph showing resource efficiency improvement over time with and without AI implementation:

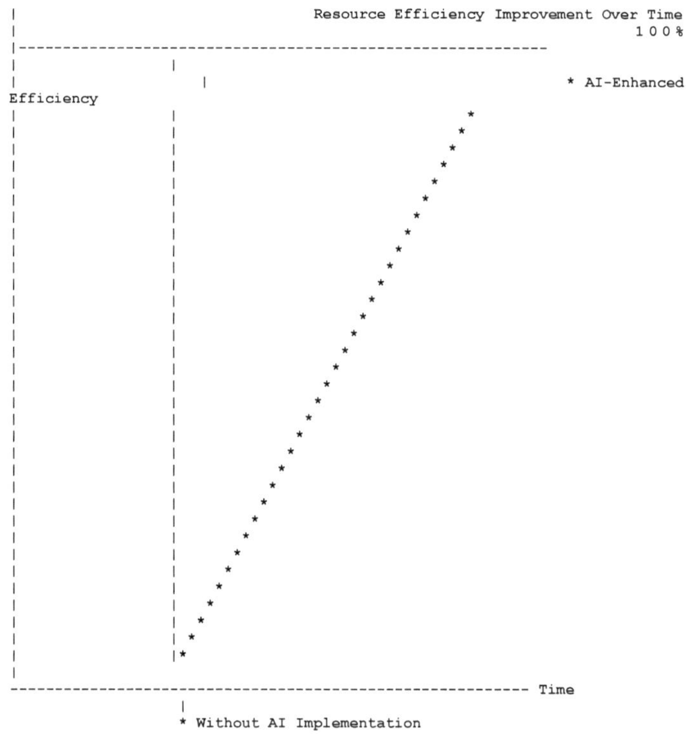

Fig. 4. A graphical representation of AI's impact on the circular economy.

Impact of AI on Waste-to-Value Chain Management

AI technology has transformative potential for waste-to-value chains, significantly enhancing the efficiency, effectiveness, and economic viability of converting waste materials into valuable resources (European Commission, 2020).

Value Creation, Delivery Mechanisms, and Capture Strategies

In the context of the circular economy (CE), value creation, delivery mechanisms, and capture strategies are crucial components that enable businesses to generate, deliver, and sustain value while maintaining resource efficiency and minimizing waste. Let's delve into each component with a focus on how AI enhances these aspects. Figure 5, is a flowchart showing how AI operated activities are employed in the creation of the waste to value chain.

```
A [Waste Collection] --> B [Smart Bins]
 A --> C [Optimized Logistics]
 B --> D [Automated Sorting]
 C --> E [Inventory Management]
 D --> F [Material Recovery]
 F --> G [E-Waste Recycling]
 F --> H [Plastic Recycling]
 E --> I [Waste-to-Energy]
 I --> J [Process Optimization]
 I --> K [Predictive Maintenance]
```

Fig. 5. Flowchart of AI in waste-to-value chain.

Table 2. How AI enhances value creation, delivery mechanisms, and capture strategies.

<table>
<tr><td colspan="2">A. Value creation: Value creation in a circular economy involves designing products, services, and processes that maximize resource efficiency and sustainability.</td></tr>
<tr><td colspan="2">B. Delivery mechanisms: Delivery mechanisms in a circular economy refer to how value is delivered to customers, focusing on sustainable and efficient processes. AI enhances delivery mechanisms by:</td></tr>
<tr><td>Efficient supply chain management</td><td>Supply chain optimization: AI optimizes logistics, reducing transportation emissions and costs (McKinsey, 2020).
Inventory management: Machine learning algorithms predict inventory needs, minimizing waste and ensuring timely delivery.</td></tr>
<tr><td>Product-as-a-Service (PaaS) Models</td><td>Usage monitoring: AI tracks product usage and condition, ensuring timely maintenance and upgrades.
Dynamic Pricing Models: AI adjusts pricing based on usage patterns and demand, maximizing resource utilization.</td></tr>
<tr><td>Enhanced recycling and reverse logistics</td><td>Automated sorting systems: AI-powered robots sort waste more efficiently, improving recycling rates.
Reverse logistics optimization: AI plans and manages the return of used products for recycling or refurbishing.</td></tr>
<tr><td colspan="2">C. Capture strategies: Capture strategies in a circular economy involve mechanisms to retain and reinvest value created, ensuring sustainability and profitability. AI enhances capture strategies by:</td></tr>
<tr><td>Data-driven insights</td><td>Consumer behavior analysis: AI analyzes customer data to identify trends and preferences, informing sustainable product development (Forbes, 2019; HBR, 2022).
Market demand forecasting: AI predicts market demand, allowing companies to align production with sustainability goals.</td></tr>
<tr><td>Revenue models</td><td>Subscription services: AI helps manage subscription models for products-as-a-service, providing steady revenue streams.
Dynamic pricing: AI regulates prices based on market conditions and customer behavior, optimizing revenue capture.</td></tr>
<tr><td>Efficiency gains</td><td>Operational efficiency: AI streamlines operations, reducing costs and waste, and thereby improving profitability.
Energy management: AI optimizes energy use in manufacturing and operations, reducing costs and environmental impact.</td></tr>
</table>

Here is a visual representation of how AI integrates with value creation, delivery mechanisms, and capture strategies in the circular economy:

```
+-------------------+         +-------------------+          +-------------------
+         +-------------------+
|    Value Creation |------->|    Delivery         |------->|    Value Capture
|------->|    Sustainability  |
|                    |          |    Mechanisms        |          |    Strategies
|         |    and Profitability|
+-------------------+         +-------------------+          +-------------------
+         +-------------------+
            |                                  |                                |
|
            v                                  v                                v
v
+-------------------+        +-------------------+        +-------------------+
+-------------------+
| Machine Learning |     | Supply Chain        |     | Data-Driven        |     |
Operational         |
|   - Resource        |     | Optimization        |     | Insights           |     |
Efficiency          |
|     Optimization  |     |   - AI Algorithms |     |   - Consumer Data |     |   -
AI Analytics   |
+-------------------+        +-------------------+        +-------------------+
+-------------------+
| Generative Design|     | PaaS Models          |     | Revenue Models    |     |
Energy Management|
|   - Durable         |     |   - Usage            |     |   - Subscription  |     |   -
Smart Grids    |
|     Products        |     |     Monitoring     |     |     Services        |     |   -
Energy           |
+-------------------+     +-------------------+     +-------------------+     |
Optimization   |
| Predictive          |     | Enhanced             |     | Efficiency Gains |
+-------------------+
|  Maintenance       |    |  Recycling          |    |  - Cost Reduction|
|  - AI Sensors      |    |  - AI Sorting       |    |  - Waste Minim.  |
+-------------------+    +-------------------+    +-------------------+
```

Fig. 6. AI-enhanced circular economy framework.

Designing Business Models for the Circular Economy—Toward a Conceptual Framework

To address how circular economy principles may be integrated into a business model and identify the universally applicable components of a circular business model, a narrative conceptual review was conducted. The process was divided into three steps (see Fig. 7):

i. Identifying the current state of knowledge on circular business models (CBMs).
ii. Categorizing the initial body of literature based on business model structure components.
iii. Synthesizing and developing a framework for a circular business.

Agro-based Waste Management

Agro-based waste management is a critical aspect of sustainable agriculture and environmental protection while maximizing resource utilization. Globally, the generation of agricultural waste from crop residue, agro-industries, animal production, and other sources presents substantial challenges. These wastes, if not adequately

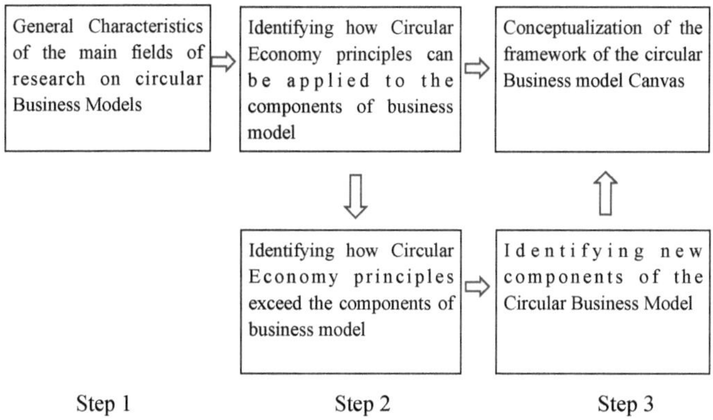

Fig. 7. Flow chart illustrates the concept of creating a business model framework for the circular economy (Lewandowski, 2016).

managed, can cause pollution, environmental damage, and health risks (Sivakumar, 2022). Implementing circular agricultural production models that prioritize waste reduction, reuse, and recycling is critical for maximizing agricultural waste's economic benefits while minimizing its negative environmental impact (Ogbu, 2023).

Types of Agro-based Waste

Agro-based wastes may be divided into three broad categories, namely crop residues, animal manures and agrochemical residues.

1) *Crop residues*: Crop residues include the stalks, leaves, husks, and roots left in the field after the harvest of crops like wheat, rice, maize, and sugarcane. According to Sharma, 2020a, farmers use 20% of rice straw and burn the other 50% for crop preparation. The availability of these residues can be substantial, for instance, rice straw constitutes about 50% of the total biomass of the rice plant (VanHung, 2020). Improper management of crop residues, such as open burning, leads to severe air pollution, contributing to greenhouse gas emissions and particulate matter. Conversely, leaving residues on fields can improve soil organic matter and moisture retention, reducing soil erosion and improving fertility. However, through the use of AI algorithms and real-time data, innovative approaches can be developed to effectively utilize agro-based waste and crop residues, contributing to sustainable agricultural practices and establishing food security in the face of evolving environmental conditions and growing food demand (Ziesche, 2023).

2) *Animal/livestock excreta*: Animal excreta includes the faeces and urine or other bodily waste products from livestock, such as cows, pigs, and chickens. It is rich in nutrients like nitrogen, phosphorus, and potassium, making it a valuable fertilizer. Bashir and Mohan (2023) demonstrated that sound manure management as a fertiliser can boost environmentally friendly farming and agricultural productivity. Moreover, it can be processed into biochar, a recognized method for sustainable land reclamation and enhancing plant growth on degraded soils. While manure is a potent fertilizer, improper management can lead to nutrient runoff, water pollution, and greenhouse gas emissions. The advancements in technology enable the conversion of agro-based

wastes into value-added products like biofertilizers, biofuels, and industrial enzymes, further promoting sustainability in agricultural processes and enhancing overall environmental health.

3) *Agro-chemical residues*: Agro-chemical residues include leftover pesticides, herbicides, and fertilizers that remain in the soil, water, and crops after application. These chemicals are essential for modern agriculture but can pose significant environmental and health risks. Excessive or incorrect use of agrochemicals can degrade soil, contaminate water, and harm non-target creatures such as beneficial insects and soil macrobiotics. The extensive use of agrochemicals has been reviewed for its wide-ranging impacts on human health, emphasizing the need for measures to minimize harm and a shift towards sustainable agriculture practices to mitigate these detrimental impacts (Carvalho, 2017; Ali, 2021).

Methods and Benefits of Agro-waste Management

Proper management of Agro-waste is crucial for sustainable agriculture, as it can lead to increased soil fertility, improved water retention, enhanced nutrient cycling, and overall soil health. Agro-waste management encompasses several methods, viz., composting, biochar production, anaerobic digestion, biogas production, incineration and pyrolysis. By implementing environmentally friendly practices like composting and recycling agro-waste, the agricultural sector may reduce its environmental footprint, enhance resource efficiency, and promote sustainable development for the future.

Composting

It is the biological decomposition of organic matter under controlled aerobic conditions, transforming Agro-waste into nutrient-rich compost. It involves the collection of agro-waste, followed by layering and regular turning to maintain aeration and moisture levels. Produces high-quality organic fertilizer, reduces landfill waste, and improves soil health by enhancing its nutrient content and structure.

Anaerobic Digestion

It is a series of biological processes of breaking down organic materials in the absence of oxygen, producing biogas (a mixture of methane and carbon dioxide) and digester (a nutrient-rich slurry). This method is extensively adopted for organic waste remediation that generates both organic fertilizer and bioenergy (Choi, 2021). Agro-waste is fed into a sealed digester where microbes decompose it anaerobically. It generates renewable energy in the form of biogas, reduces greenhouse gas emissions, and produces a nutrient-rich by-product usable as fertilizer.

Biochar Production

It is a stable configuration of carbon produced by pyrolyzing organic material in a low-oxygen environment. Agro-waste is processed by heating at high temperatures in the absence of oxygen, resulting in the formation of biochar. Sequesters carbon, improves soil fertility, enhances water retention, and reduces nutrient leaching.

Biogas Production

It involves the anaerobic digestion of organic waste to produce methane-rich biogas. Similar to anaerobic digestion, organic waste is placed in a digester where it undergoes microbial decomposition to release biogas, which provides a renewable energy source, decreases greenhouse gas emissions, and reduces reliance on fossil fuels.

Incineration and Pyrolysis

Incineration is the combustion of organic waste to produce energy, while pyrolysis involves thermal decomposition in the absence of oxygen. Agro-waste is burned at high temperatures in incineration or decomposed thermally in pyrolysis units. Reduces waste volume significantly, generates energy, and can produce by-products like ash or bio-oil.

In general, the benefits and advantages of Agro-waste processing and refining methods are believed to be cost-effective and eco-friendly, as well as it could contribute positively toward more sustainable waste management and a less-carbon-intensive energy utilization path (Hoang, 2023).

Biomass Valorization

Nzihou (2010), defined valorization as the conversion of waste and biomass to energy, fuels and other useful materials, with a particular focus on environmental indicators and sustainability goals, i.e., the process of converting Agro-waste into valuable products like biofuels, chemicals, and materials. Organic materials such as food waste, grass, and manure are being converted into VFAs, carboxylic acids, bioenergy, and bioplastics to help minimize waste and foster a circular economy (Onwucha, 2023).

AI-based Technologies for Agro-waste Conversion

Current developments in AI-based technologies for Agro-waste conversion have shown significant potential for sustainable agriculture by improving waste management and resource efficiency (Taneja, 2023). Machine Learning Algorithms are utilized in optimizing anaerobic digestion processes, predicting compost quality, and enhancing biochar production (Cruz, 2022). It improves efficiency, reduces operational costs, and increases yield and quality of end-products. Image Processing facilitates monitoring crop residue levels, and detecting contamination in waste streams thus, enabling precise and timely interventions, reducing waste and enhancing resource recovery. Predictive Analytics are applicable in forecasting waste generation patterns, and optimizing waste collection and processing schedules and therefore, enhance planning and resource allocation, and minimize waste management costs. AI plays an important role in addressing food waste issues by monitoring and optimizing food production and supply chains, redistributing excess food, and supporting circular economy initiatives. By leveraging AI in Agro-waste conversion, the agricultural sector can reduce waste, enhance sustainability, and contribute to a more efficient and equitable food system.

Utilization of IoT in Improving Waste Management

The utilization of IoT technology significantly enhances the efficiency and effectiveness of waste management practices in urban areas by providing real-time monitoring and data-driven insights. The Internet of Things (IoT) is being used to improve Agro-waste management (Koshariya, 2023). Sensors facilitate monitoring temperature, moisture, and pH levels in compost piles, detecting gas emissions from manure. Provides real-time data, enhances process control, and improves safety and compliance. Smart waste bins may be used to improve waste management by providing early automatic identification of waste at the bin level (Ali, 2020). Smart bins are furnished with sensors to detect fill levels, compact waste, and signal collection needs. Reduces collection frequency and costs, and improves efficiency in waste management. Moreover, remote monitoring through tracking the performance of waste treatment facilities, and monitoring environmental impact facilitates timely maintenance, ensures regulatory compliance, and enhances decision-making.

Comparative Economic Analysis Between Traditional and AI-based Tools

AI-based tools in Agro-waste management offer significant economic and operational advantages over traditional methods. While the initial investment is higher, the extended benefits for the long term such as reduced operational costs, increased efficiency, and lower environmental impact make AI-based tools a more sustainable and profitable choice. The ability to scale operations, adapt to market changes, and achieve higher quality outputs further strengthens the case for AI integration in Agro-waste management.

Table 3. Summary comparative of traditional and AI-based tools.

Particulars	Traditional Methods	AI-based Tools
Initial Investment	Generally lower upfront costs but may require frequent maintenance and manual labor.	Higher initial costs due to technology implementation but offer long-term savings.
Operational Costs	Higher due to manual labor and less efficient processes.	Lower due to automation, optimization, and predictive maintenance.
Waste Reduction	Moderate waste reduction based on manual sorting and processing.	Significant waste reduction through precise and optimized AI algorithms.
Labor Requirements	High labor requirement for manual operations.	Reduced labor requirements due to automation and machine learning.
Processing Speed	Slower due to manual and semi-automated processes.	Faster due to real-time data processing and AI-driven optimizations.
Scalability	Limited scalability without significant labor increases.	High scalability with minimal additional costs.
Maintenance Costs	Regular maintenance is required for manual and semi-automated tools.	Lower maintenance costs due to predictive maintenance capabilities.

Food Waste Reduction and Recovery

Food waste management is a critical component of sustainable development, addressing both environmental and economic challenges. Reduction and recovery of food waste not only minimize the burden on landfills but also reclaim valuable resources. Effective strategies include reducing food loss at the source, redistributing surplus food to those in need, and recycling waste into energy or compost. Leveraging innovative technologies and practices, such as artificial intelligence and circular economy principles, can significantly enhance the efficiency and efficacy of food waste management efforts. AI tools, such as predictive analytics, machine learning algorithms, and IoT-enabled monitoring systems, can optimize supply chains, forecast demand more accurately, and identify inefficiencies in real-time, leading to more effective reduction and recovery of food waste, and ultimately contributing to a more sustainable future (Jagtap, 2021; Garcia-Garcia, 2017).

Strategies for Minimizing Food Waste Using AI Tools and Techniques

Food waste is considered as waste which hasn't been utilized in any form and has been altered from the food supply chain. Food waste not only affects a country; it affects the planet on a large scale. Globally, approximately 13 percent of food produced is lost between harvest and retail, while an estimated 17 percent of total global food production is wasted in households, in food service and retail together (UN, 2020). As per UNEPs' (United Nations Environment Programme) food wastage index report, India produces 40 percent of the food wasted is parallel to nearly 92,000 crore/year. This is similar to nearly one percent of the GDP which is depleted in the form of food wastage in India. 68.7 million tonnes of food are wasted annually in Indian homes, each person in India wastes 55 kg of food per year (UNEP, 2021). The FSSAI (Food Safety and Standards Authority of India) reported that 1/3rd of all food produced in India gets wasted or spoilt before it is even consumed or utilized in any form. The reasons why food waste occurs and persists are seasonality, perishability and the consumer's changing demands (Sharma, 2020b).

The Scope of Global Food Waste

This scientific diagram provides a detailed visualization of the annual household food waste in various countries. It illustrates both the total food waste (in tonnes) and the estimated food waste per capita (in kilograms).

Key observations: India accounts for the second highest level of food waste after China (Onyeaka, 2023), this may be due to their large population. Australia has the highest food waste per capita, indicating significant individual food waste. Russia has the lowest per capita food waste, showing more efficient food (see Fig. 8).

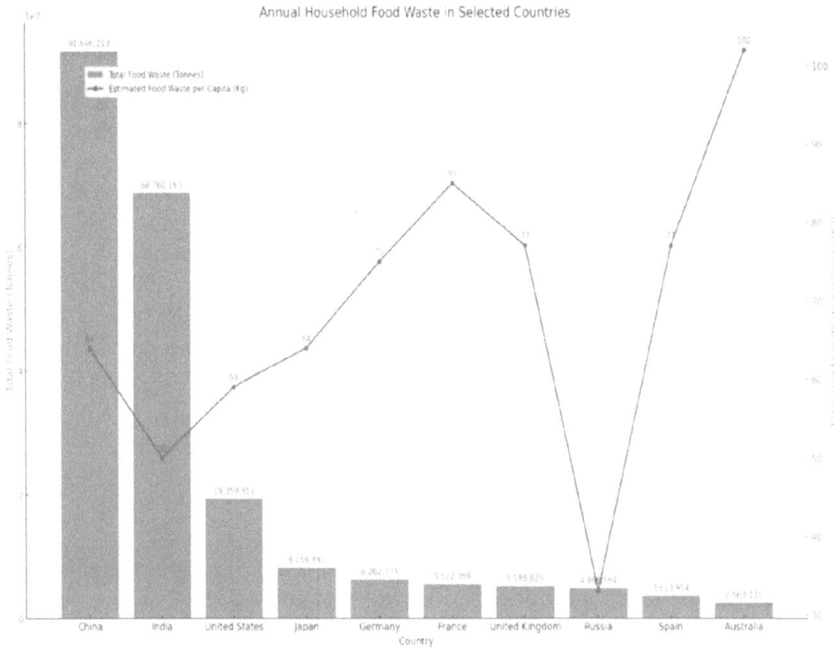

Fig. 8. The Extent of Global Food Waste: Annual Household Food Waste figures in Selected Countries (data source: Onyeaka, 2023).

Source: Visual representation (graph) created by the author using Python.

Importance of Minimizing Food Waste

Food is important to human survival; its waste hinders the sustainability of our food systems. Food waste not only negatively impacts food security and its availability, but also contributes to increasing the cost of food material. Production of food starts from the farm and moves through the system up until it reaches our plate. Before going to waste, the production of food utilizes resources such as water, land, energy, capital, and labor. Additionally, this food that goes to waste ends up in landfills which leads to greenhouse gas emissions, contributing to climate change.

Role of AI Tools and Techniques in Reducing Food Waste

In the recent past, a surge has been witnessed in the development of AI to assist humans in diverse domains (Furman, 2019). Artificial intelligence is known to mimic human intellect and intelligence. The inclusion and amalgamation of AI into the food industry hold huge potential for addressing food waste, supporting the circular economy, and enhancing sustainability in food production. There are various AI-based technologies for possible use to reduce food wastage and its valorization (Table 4).

Table 4. Role of AI-based technology in reducing food wastes.

S.no.	Technology	Applications	Benefits	References
1.	Integrated system Bayesian network and machine learning algorithms model	Determines household food wastage produced and discarded at the household level	Reduces food waste through consumer learning of food waste at household level helping in reducing food waste	Anggraeni, 2021
2.	Computer vision technology	Winnow Vision system uses a modern deep-learning-based image-classification algorithm to automate waste information gathering	Eliminates the need for manual identification of the items wasted Winnow Vision can automatically classify food items with an accuracy level surpassing humans	Nu, 2024. Winnow Solutions (UK)
3.	Predictive machine learning	Four AI-ML based techniques were tested to reduce food wastage in small grocery stores	Reduced food waste from 6,169.39 kg and 653.27 kg to 117.64 kg	Nascimento, 2022
4.	Machine with advanced sensors and machine learning	Identifies and segregates waste materials in the beverage industry	Reduces the human labor and provides precision	Hassan, 2023
5.	AI-driven waste-sorting robots	Segregate waste different types of waste into recyclable and non-recyclable	Reduces landfill waste, promotes circular economy	Ali and Shehata, 2021
6.	AI and blockchain	Enhances traceability in the food supply chain, reduces food fraud	Reducing food fraud which may lead to significant health complications	Dedeoglu, 2023
7.	IoT based valorisation system	Real time monitoring systems for measuring biogas, pH, humidity, temperature, etc.	Managing the food waste valorization process	Said, 2023
8.	Machine learning	Comparison between demand forecasts and real demand	To reduce food waste from 14 to 52% in catering services	Rodrigues, 2024

Artificial intelligence and optimization algorithms for smart warehouse management to reduce food waste. Predictive analysis helps in predicting the food demand and reducing waste in the supply chain. It also helps in understanding market preferences. Predictive models have been used to estimate the expected shelf-life of the food product (Ferreira, 2022). Brightloom is a company that uses AI and predictive analytics for consumer preferences by optimizing menu offerings and pricing of food retailers. It helps retailers in predicting exact demand reducing overproduction which reduces food waste, and also ensuring that the demands are positively aligned with the customer demand (Brightloom, 2023).

Conclusion

To summarize, AI is critical for promoting the circular economy since it optimizes resource management, improves waste sorting, and enables predictive maintenance. In agriculture, artificial intelligence improves supply chain efficiency, lowering food waste and boosting sustainability. Overall, artificial intelligence promotes sustainable development and environmental stewardship, making it a critical enabler of the circular economy.

Declaration

Ethical Responsibilities of Authors

"All authors have read, understood, and have complied as applicable with the statement on "Ethical responsibilities of Authors" as found in the Instructions for Authors".

Ethical Approval

Not applicable.

Use of AI Tool

The corresponding author declares that no AI tool has been used to prepare this manuscript.

Conflict of Interest

On behalf of all the authors the corresponding author states that there is no conflict of interest to disclose.

Competing Interests

Not applicable.

Author's Contributions

Author[1]: Conceptualization of chapter, Conducted the initial literature review, and supervision, Reviewed and edited the manuscript for intellectual content, and oversaw the entire project; **Author**[2]: Developed the theoretical framework, provided critical feedback, and revised the manuscript.

Funding

No funding has been received.

References

Ali, I. and Aboelmaged, M.G.S. 2022. Implementation of supply chain 4.0 in the food and beverage industry: Perceived drivers and barriers. *International Journal of Productivity and Performance Management*, 71: 1426–1443.

Ali, S., Ullah, M.I., Sajjad, A., Shakeel, Q. and Hussain, A. 2021. Environmental and health effects of pesticide residues. pp. 311–336. *In*: Inamuddin, Ahamed, M.I. and Lichtfouse, E. (eds.). *Sustainable Agriculture Reviews* 48. Springer, Cham.

Ali, T., Irfan, M., Alwadie, A.S. and Glowacz, A. 2020. IoT-based smart waste bin monitoring and municipal solid waste management system for smart cities. *Arabian Journal for Science and Engineering*, 45: 10185–10198.

Anggraeni, M.C., Silaban, C.A., Anggreainy, M.S. and Cahyadi, E. 2021. Role of Artificial Intelligence in the management of food waste. In *2021 2nd International Conference on Artificial Intelligence and Data Sciences (AiDAS)* (pp. 1–6). IEEE.

Bashir, S. and Mohan, A. 2023. An institutional approach to manure recycling. *In*: Jawaid, M. and Khan, A. (eds.). *Manure Technology and Sustainable Development*. Sustainable Materials and Technology. Springer, Singapore.

Carvalho, F.P. 2017. Pesticides, environment, and food safety. *Food and Energy Security*, 6: 48–60.

Choi, G., Kim, H. and Lee, C. 2021. Long-term monitoring of a thermal hydrolysis-anaerobic co-digestion plant treating high-strength organic wastes: Process performance and microbial community dynamics. *Bioresource Technology*, 319: 124138.

Cruz, I.A., Chuenchart, W., Long, F., Surendra, K.C., Andrade, L.R.S., Bilal, M., Liu, H., Figueiredo, R.T., Khanal, S.K. and Ferreira, L.F.R. 2022. Application of machine learning in anaerobic digestion: Perspectives and challenges. *Bioresource Technology*, 345: 126433.

Dedeoglu, V., Malik, S., Ramachandran, G., Pal, S. and Jurdak, R. 2023. Blockchain meets edge-AI for food supply chain traceability and provenance. *In Comprehensive Analytical Chemistry*, 101: 251–275.

Edirisinghe, L.G.L.M., de Alwis, A.A.P., Wijayasundara, M. and Hemali, N.A. 2024. Quantifying circularity factor of waste: Assessing the circular economy potential of industrial zones. *Cleaner Environmental Systems*, 12: 100160.

European Commission [EU], Circular Economy Action Plan 2020. https://environment.ec.europa.eu/strategy/circular-economy-action-plan_en.

Ferreira, C. and Gonçalves, G. 2022. Remaining useful life prediction and challenges: A literature review on the use of Machine Learning methods. *Journal of Manufacturing Systems*, 63: 550–562.

Forbes. 2019. AI's Role in Sustainable Consumerism. https://scholar.google.co.in/scholar?q=Forbes.+(2019).+AI%27s+Role+in+Sustainable+Consumerism.&hl=en&as_sdt=0&as_vis=1&oi=scholart.

Furman, J. and Seamans, R. 2019. AI and the economy. *Innovation Policy and the Economy*, 19: 161–191.

Garcia-Garcia, G., Woolley, E., Rahimifard, S., Colwill, J., White, R. and Needham, L.A. 2017. Methodology for sustainable management of food waste 2017. *Waste and Biomass Valorization*, 8: 2209–2227.

Harvard Business Review [HBR], How AI Can Foster Sustainable Consumer Behavior. 2020. https://scholar.google.co.in/scholar?q=Harvard+Business+Review+(HBR).+(2020).+%22How+AI+Can+Foster+Sustainable+Consumer+Behavior.&hl=en&as_sdt=0&as_vis=1&oi=scholar.

Hassan, M., Wahab, N.A.B.A. and Nor, R.B.M. 2023. The role of artificial intelligence in waste reduction in the beverage industry: A comprehensive strategy for enhanced sustainability and efficiency. *AI, IoT and the Fourth Industrial Revolution Review*, 13: 1–8.

Hoang, A.T., Pandey, A., Huang, Z., Nižetić, S., Le, A.T. and Nguyen, X.P. 2023. Biofuels an option for agro-waste management. In *Environmental Sustainability of Biofuels*, 27–47.

Jagtap, S., Trollman, H., Trollman, F., Garcia-Garcia, G., Parra-López, C., Duong, L. and Martindale, W. 2021. The role of artificial intelligence in circular economy: Automating the recovery of food waste. *Sustainable Production and Consumption*, 27: 101–111.

Kirchherr, J., Reike, D. and Hekkert, M. 2017. Conceptualizing the circular economy: An analysis of 114 definitions. *Resources, Conservation and Recycling*, 127: 221–232.

Koshariya, A.K., Khatoon, S., Marathe, A.M., Suba, G.M., Baral, D. and Boopathi, S. 2023. Agricultural waste management systems using Artificial Intelligence techniques. In *AI-Enabled Social Robotics in Human Care Services*, 236–258.

Lewandowski, M. 2016. Designing the business models for circular economy—Towards the conceptual framework. *Sustainability*, 8: 43.

McKinsey and Company, AI in Supply Chain Management. 2020. Access, https://scholar.google.co.in/scholar?q=McKinsey+%26+Company.+(2020).+%22AI+in+Supply+Chain+Management.%22&hl=en&as_sdt=0&as_vis=1&oi=scholart.

Nascimento, A.M., Queiroz, A., de Melo, V.V. and Meirelles, F.S. 2022. Applying Artificial Intelligence to Reduce Food Waste in Small Grocery Stores. https://aisel.aisnet.org/isla2022/6/.

Nu, Y., Belavina, E. and Girotra, K. 2024. Using Artificial Intelligence to reduce food waste. Available at SSRN 4826777.

Nzihou, A. 2010. Toward the valorization of waste and biomass. *Waste Biomass Valor.*, 1: 3–7.

Ogbu, C.C. and Okey, S.N. 2023. Agro-industrial waste management: The circular and bioeconomic perspective. *In*: Ahmad, F. and Muhammad Sultan, M. (eds.). *Agricultural Waste-New Insights.* IntechOpen. doi: 10.5772/intechopen.109181.

Onwucha, C.N., Talabi, J.O., Ajayi, S.O., Ehi-Eromosele, C.O. and Ajanaku, K.O. 2023. Valorization of biomass using deep eutectic solvent: A short review. In *IOP Conference Series: Earth and Environmental Science*, 1197: 012002.

Onyeaka, H., Tamasiga, P., Nwauzoma, U.M., Miri, T., Juliet, U.C., Nwaiwu, O. and Akinsemolu, A.A. 2023. Using Artificial Intelligence to tackle food waste and enhance the circular economy: Maximising resource efficiency and minimising environmental impact: A review. *Sustainability*, 15: 10482.

Rodrigues, M., Miguéis, V., Freitas, S. and Machado, T. 2024. Machine learning models for short-term demand forecasting in food catering services: A solution to reduce food waste. *Journal of Cleaner Production*, 435: 140265.

Said, Z., Sharma, P., Nhuong, Q.T.B., Bora, B.J., Lichtfouse, E., Khalid, H.M. and Hoang, A.T. 2023. Intelligent approaches for sustainable management and valorisation of food waste. *Bioresource Technology*, 128952.

Sharma, A., Georgi, M., Tregubenko, M., Tselykh, A. and Tselykh, A. 2022. Enabling smart agriculture by implementing artificial intelligence and embedded sensing. *Computers and Industrial Engineering*, 165: 107936.

Sharma, A., Singh, G. and Arya, S.K. 2020a. Biofuel from rice straw. *Journal of Cleaner Production*, 277: 124101.

Sharma, A., Singh, P.V. and Musunur, L.P. 2020b. Artificial Intelligence and Robotics for Reducing Waste in the Food Supply Chain: Systematic Literature Review, Theoretical Framework, and Research Agenda. *OSF Preprints*.

Sivakumar, D., Srikanth, P., Ramteke, P.W. and Nouri, J. 2022. Agricultural waste management generated by agro-based industries using biotechnology tools. *Global Journal of Environmental Science and Management*, 8: 281–296.

Stahel, W. 2016. The circular economy. *Nature*, 531: 435–438.

Taneja, A., Nair, G., Joshi, M., Sharma, S., Sharma, S., Jambrak, A.R. and Phimolsiripol, Y. 2023. Artificial intelligence: Implications for the agri-food sector. *Agronomy*, 13: 1397.

United Nations [UN], International Day of Awareness of Food Loss and Waste. 2020. https://www.un.org/en/observances/end-food-waste-day.

United Nations Environment Programme [UNEP], Food Waste Index Report. 2021. https://www.unep.org/resources/report/unep-food-waste-index-report-2021.

Van Hung, N., Maguyon-Detras, M.C., Migo, M.V., Quilloy, R., Balingbing, C., Chivenge, P. and Gummert, M. 2020. Rice straw overview: Availability, properties, and management practices. *Sustainable Rice Straw Management*, 1–13.

Vasarhelyi, K. 2022. AI and Robotics in Recycling. Recycling Today https://www.colorado.edu/ecenter/2022/04/06/ai-robotics-recycling.

Ziesche, S., Agarwal, S., Nagaraju, U., Prestes, E. and Singha, N. 2023. Role of Artificial Intelligence in advancing sustainable development goals in the agriculture Sector, 152. *In*: Mazzi, F. and Floridi, L. (eds.). *The Ethics of Artificial Intelligence for the Sustainable Development Goals*. Philosophical Studies Series.

Chapter 11

Advancements in Machine Learning for Water Quality Assessment

***Rashmi Yadav,*[1] *Ajay N Satpute,*[2] *Ajay Kushwah,*[3,*] *Jitendra Kumar,*[4] *Susama Sudhishri*[5] and *Bandi Nageswar*[6]**

Introduction

Life on earth cannot exist without water. It is a finite supply and an incredibly valuable natural asset. Although the earth's surface is predominantly composed of water, only a small portion of it is suitable for use making this resource highly scarce. Approximately 1.1 billion people across the globe do not have access to clean drinking water, and it is anticipated that nearly two-thirds of all nations will face water stress by 2025 (Ross Michael, 2016). The prudent utilization of the nation's water resources is necessary to fulfil the expanding demands of our emerging economy. Development, preservation, and management of water resources are therefore essential. The rapid

[1] Research Scholar, Water Technology Centre, ICAR-Indian Agricultural Research Institute, Pusa, New Delhi - 110-012, India.

[2] Scientist, ICAR-Indian Grassland and Fodder Research Institute, Jhansi, Uttar Pradesh, 284003, India.

[3] Research Scholar, Division of Agricultural Engineering, ICAR-Indian Agricultural Research Institute, Pusa, New Delhi - 110-012, India.

[4] Scientist, Division of Irrigation and Drainage Engineering ICAR-Central Soil Salinity Research Institute, Karnal - 132 001, Haryana, India.

[5] Professor and Principal Scientist, Water Technology Centre, ICAR-Indian Agricultural Research Institute, Pusa, New Delhi - 110-012, India.

[6] Teaching Associate, Regional Agriculture Research Station, Anakapalle, 531-001, Andhra Pradesh, India.

* Corresponding author: ajaykushwah98@gmail.com

growth of the economy has resulted in the generation of wastewater that contains one must understand various pollutants. This situation poses significant risks to the health of natural water environments (Johnson et al., 1997; Pratap et al., 2023). Consequently, multiple measures to control water pollution have been created and implemented. As mentioned previously, having a sufficient quantity of water with acceptable quality is crucial for survival. Consequently, it is crucial to exercise careful and wise management of this precious and limited resource. However, ensuring and maintaining acceptable water quality poses a challenge when it comes to managing water resources (Mukate et al., 2019). One must understand avoid and manage water contamination, it is imperative that one understands the problems and trends related to water pollution. Additionally, several illnesses may be transmitted by water, which emphasizes the significance of regularly checking the water quality. Water quality is usually assessed by collecting samples from various sites regularly and analyzing them in labs. To evaluate water quality in different water bodies or processes, however, the conventional approach of collecting samples manually and laboratory testing can be ineffective, expensive, and time-consuming (Kedia, 2015). Nowadays, computer vision and drone technology are utilized in agriculture (Chouhan et al., 2024a). As such, intelligent systems are increasingly being used for water quality assessment, particularly in scenarios requiring real-time data (Alshaltone et al., 2021). Deep learning (DL) and machine learning (ML) algorithms are used for a variety of tasks, including classifying different plant species (Chouhan et al., 2024b), cotton crops (Kumar et al., 2024), leaf disease segmentation (Chouhan et al., 2021), and leaf disease detection using artificial neural networks (Khajuria et al., 2018). One such technology is a machine learning which on the virtue of its remarkable accuracy, adaptable customization, and smooth scalability, has emerged as an advanced approach for processing and analysing data in many different fields (Chen and Liu, 2014). Deep data analysis is the foundation of machine learning approaches, which allow for the identification of patterns and subsequent adaptations (Bagheri et al., 2019). Across a wide range of research, machine learning has enormous potential for assessing, classifying, and predicting water quality indicator (Nasir et al., 2022). Advances in machine learning techniques over the past several years have encouraged more researchers to think that vast volume of data may be efficiently collected and processed to satisfy the complicated and wide-ranging requirements of water quality assessment (Zhu et al., 2022) (Sun and Scanlon, 2019), For example, ML models accurately simulate the flow of pollutants and hydrological processes when large enough datasets are available (Ehteram et al., 2019). The inclusion of sensors, such as photosensors that determine certain color wavelengths or capacitance fluctuations that identify dissolved water pollutants, is advantageous for water quality detection parameters (Nasir et al., 2019). When these outcomes are effectively handled by AI, a substantial volume of data is generated quickly, precisely, and consistently. Moreover, the capacity of machine learning to handle intricate nonlinear relational data makes it easier to comprehend underlying systems (Messaoud et al., 2020). As a result, machine learning is likely to produce more precise evaluation results in water quality analysis and assessment, despite its complexity (Wang et al., 2021).

Overview on Machine Learning

A subtype of artificial intelligence (AI) called machine learning enable systems or computers to acquire knowledge through the analysis of data or personal encounters. Instead of depending just on pre-programmed rules or equations (Rani et al., 2022). Processing data, choosing appropriate algorithms, training models, and testing them are all steps that must be taken before machine learning is applied in real-world situations (Zhu et al., 2022). Before employing any machine learning (ML) algorithms, users must assemble data and ensure its significance. Subsequently, they choose a suitable machine learning method that aligns with their intended goal. The chosen technique is then used on the training set to produce a model that accurately depicts the dataset. The correctness and effectiveness of the model are evaluated by utilizing testing data. If the model's quality is not satisfactory, users might experiment with alternative methodologies, expand the dataset, or enhance the data's quality. The model is deployed to accomplish its intended goal after it performs satisfactorily. Categorically, machine learning algorithms are grouped into three primary categories: supervised learning, unsupervised learning, and Reinforcement learning (Berry et al., 2019; Mosaffa et al., 2022). In supervised learning; models are trained on labeled datasets with predetermined output. By analyzing input data, the model learns to anticipate desired outputs and is improved until it reaches the required accuracy (Han et al., 2011). It uses techniques including support vector machine, linear regression, and neural networks, which are often employed, mainly for classification and regression applications (Sarker et al., 2020). Unsupervised learning focuses mostly on grouping and association tasks by using unlabeled data to find hidden patterns and group related data (Han et al., 2011). Algorithms like hierarchical clustering and k-means are commonly used. Reinforcement learning uses response in the structure of incentives or penalties to repeatedly make judgements that lead algorithms through unpredictable settings. Since data is generated by the algorithm's interaction with the environment, predefined data is not given (H. Jhaveri et al., 2021). Well-known algorithms are SARSA, Q-learning, and Deep Deterministic Policy Gradient (DDPG) Algorithm, and more (Rani et al., 2022). However, it is less commonly applied in the realm of water environments compared to the other two classes of machine learning.

Advanced Machine Learning Models for Evaluation of the Quality of the Water

Accurately predicting the Water Quality Index (WQI) has been contingent significantly on integrating artificial intelligence (AI) forecasting models, notably machine learning classifiers like Random Forest, SVM, XGBoost, KNN, and Decision Trees.

Random Forest

The array-based Random Forest (RF) method is a classification technique that consists of many decision trees and operates as a supervised machine learning methodology (Fig. 1). As the number of trees in the forest rises, RF, as an ensemble approach, limits and reduces generalization mistakes by using the best results from several decision

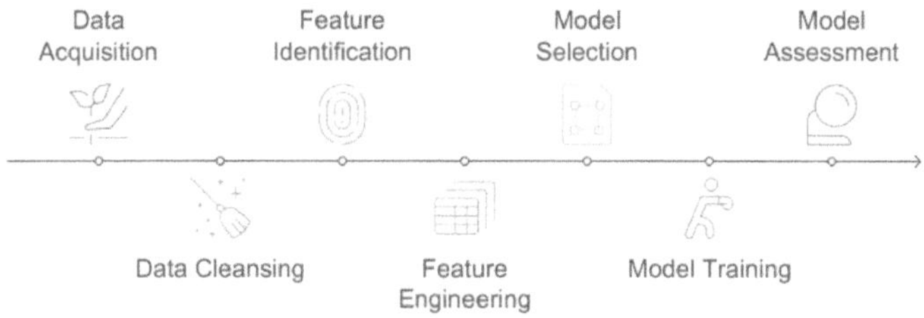

Fig. 1. Protocols for developing a machine learning model to forecast water quality factors.

trees (Forests and Breiman, 1999). The RF algorithm is better than the CART model because of its high accuracy, capacity to handle big datasets, lack of requirement for decision tree pruning, and ability to prevent over fitting. Furthermore, RF can evaluate the relative significance of every variable in the prediction process. It's crucial to remember that RF predictions in regression tasks are limited to the training data's range, which reduces the precision of the model in predicting extreme values.

Support Vector Machine (SVM)

SVM aims to find the optimal hyperplane that separates data into distinct classes. (Fan et al., 2018) Support Vector Machine (SVM), a supervised machine learning technique, is used for both regression and classification tasks, excelling particularly in regression. Identifying the ideal hyperplane in an N-dimensional space to divide data points into different classes inside the feature space is its main goal. The hyperplane's layout aims to optimize the separation between each class's closest points. The number of features dictates the hyperplane's dimensionality. SVM performs effectively in high-dimensional spaces. However, understanding and interpreting the final SVM model, variable weights, and individual impacts can be challenging due to the black-box nature of the technique.

Extreme Gradient Boosting (XGBoost)

The traditional gradient boosting decision tree methodology is not the same as XGBoost, an upgraded decision tree approach. XGBoost adds the second-order Taylor expansion to the loss function, using both first and second-order derivatives, whereas regular GBDT only uses first-order derivative information for the optimization issue (Chen and Guestrin, 2016). A regularisation term is also included in the loss function to manage the model's complexity and avoid over fitting. Through boosting, XGBoost enhances the performance of underachieving students. While XGBoost's iterative refining process and SVM's structural complexity increase their predictive potential, they also bring uncertainty, particularly when it comes to identifying complex patterns and non-linear correlations.

Artificial Neural Network (ANN)

An information processing system called an Artificial Neural Network (ANN) is made to resemble the connections and functionality of real neurons in the human brain. When tweaked (trained), ANNs may represent intricate, non-linear functions with a large number of parameters whose output nearly resembles the result obtained from a known dataset. For effective training, ANNs need a large quantity of historical data. Once sufficiently trained, however, they should be able to provide correct outputs for inputs that have never been seen before (Palani et al., 2008). ANN processes inputs by multiplying them by a sequence of weights, adding a constant value, and processing the resultant data through a transfer function, as seen in the figure. The input layer, middle layer, and output layer make up the three layers that make up a neural network. The input layer is in charge of supplying the network with inputs (Fig. 2). The majority of calculation is done by the intermediate layer, which may consist of one or more hidden layers. Within the centre layer, these concealed layers function as sublayers. The output layer compiles the outcomes of the middle layer's parallel calculations. Each neuron's output can be compared with observed data or utilized as an input for neurons in a layer below it. Through a training procedure, the network's weights and biases are modified (Haghiabi et al., 2016).

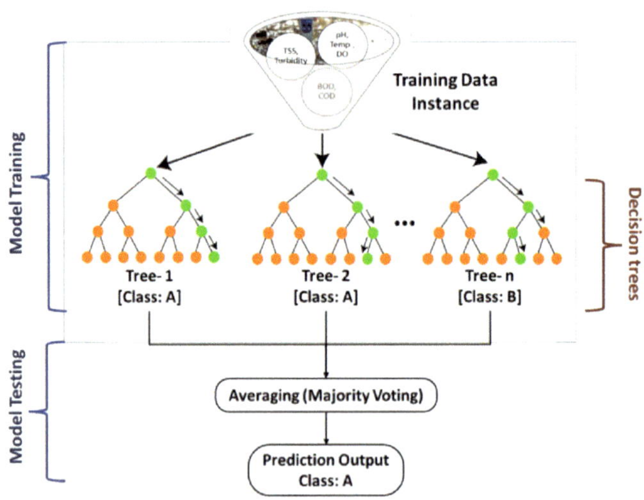

Fig. 2. A typical operation of Random forest.

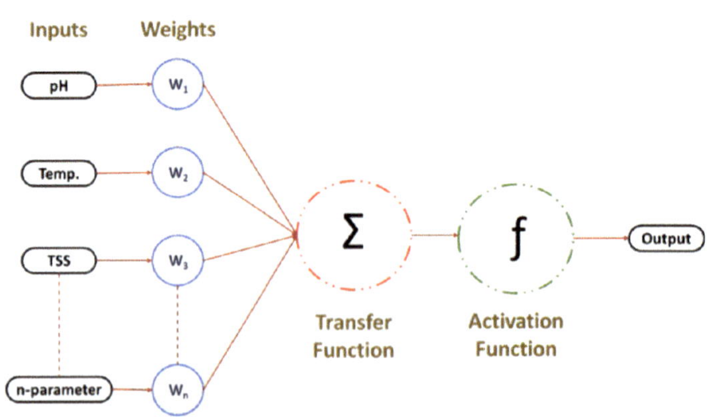

Fig. 3. A typical operation in neuron neural network.

K-Nearest Neighbors (KNN) Model

The principle of KNN uses the majority of neighbors to classify data by locating the closest neighboring points and classifying them accordingly. There are several ways to break a tie in this situation. Nevertheless, because KNN runs over all of the training data iteratively, determining the nearest neighbor each time, is not appropriate for huge datasets (Beyer et al., 1999). The Euclidean distance function (Di) is utilized in the following ways to get the feature vector's closest neighbor

$$D_i = \sqrt{(x_1 - x_2) + (y_1 - y_2)^2} \tag{1}$$

where, x_1, x_2, y_1, and y_2 are parameters for input data.

Decision Tree (DT)

The Decision Tree (DT) is an uncomplicated, fundamental method that bases judgments on pertinent input variable values. After utilizing entropy to determine the root parameter, the weights of the other factors are assessed. DT bases its judgments on a range of values derived from certain qualities, arranging all variable selections into an orderly framework. Research has demonstrated that decision tree models function effectively with imbalanced data. Nonetheless, group techniques for instance, Gradient Boosting (GB) and Random Forest (RF) consistently beat individual decision trees (Malek et al., 2022). Decision-tree-based models have the advantages of being very efficient, robust in lieu of missing values, and competent to handle both standard attributes along data. When it comes to short-term forecasting, decision-tree-based techniques are more appropriate than other machine learning algorithms and frequently provide faster computation speed (Lu and Ma, 2020).

Machine Learning for Different Water Quality Parameter Assessment

As machine learning provides insights into causative variables and complex variable interactions, it turns out to be a valuable tool for forecasting water quality (Najah Ahmed et al., 2019). In many different disciplines, a variety of machine learning models are widely used (Tikhamarine et al., 2020). For instance, to anticipate water quality, Liao and Sun (Liao and Sun, 2010) combined decision tree techniques with artificial neural networks (ANN). Four machine learning techniques were used in Shafi et al.'s classification of water quality: SVM, Neural Networks (NN), Deep Neural Networks (DNN), and KNN across various water bodies (Shafi et al., 2018). It was noted that DNN had the greatest accuracy of these methods, at 93%. The existence of molecular oxygen in the aquatic environment, or dissolved oxygen, is influenced by several elements including salinity, temperature, and air pressure.

Biochemical Oxygen Demand Prediction

Aquatic organic contamination can be detected using the commonly used biochemical oxygen demand (BOD) test. It calculates how much oxygen aerobic bacteria need in the direction of breaking down organic matter in a water sample over a given length of time (33). The United Kingdom's Royal Commission on Sewage Disposal first suggested the current technique for calculating biological oxygen demand, or BOD5, in 1912. (United States Geological Survey (USGS)) BOD5 calculates the amount of oxygen used per litre of water during five days of incubation at 20°C (Zaitsev et al., 2018). The most important factors influencing the prediction accuracy among the 17 water quality measures were determined to be temperature, pH, BOD, and phosphorus content (Siljic Tomic et al., 2018). By using information from past BOD5 readings, machine learning models can predict the BOD5 levels with greater accuracy (Ooi et al., 2022). The amount of biological oxygen demand (B.O.D.) in the Johor River in Malaysia was predicted using a support vector Regression and an Artificial Neural Network (ANN) to anticipate BOD values for traditional water reclamation plants (WRPs) (Zhu et al., 2018). Ma et al. (2020) utilized a combination of Deep Matrix Factorization and Deep Neural Networks to effectively predict BOD values. The Indian River Narmada's water quality was estimated using a decision tree model that predicted six water quality parameters: pH, temperature, ammonia-nitrogen (NH_3-N), nitrate-nitrogen (NO_3-N), BOD (biochemical oxygen demand), and COD (chemical oxygen demand) (Jaloree et al., 2014). The study by (Cipullo et al., 2019) used the random forest (RF) approach to estimate the bio available chemical concentration in soil sample tests with superior prediction accuracy. Numerous research has also presented other machine learning techniques, such as random forest regression (RFR), M5 trees, and boosted regression trees (BRT) (Yuchi et al., 2019). More sophisticated activation functions, which enable greater model convergence than the sigmoid function usually employed in conventional ANNs, are credited with this improvement, as well as the numerous layers amid the input and output in DNNs, which simplify the training process (Schmidhuber, 2015).

Dissolved Oxygen Prediction

An essential quality of surface water indicator is dissolved oxygen (DO), it symbolizes the overall condition of the aquatic environment and its capacity to support aquatic life. Heat, salinity, and air pressure are some of the variables that affect dissolved oxygen, which is a representation of oxygen molecules in the aqueous environment (Zaitsev et al., 2018). Fish and other aquatic life in water bodies depend on D.O. The wind and aerating action reflect the water's dissolved oxygen content. The stability of aquatic creatures and fish life depends on the D.O. level, which needs to be maintained within a certain range. The higher the D.O. level, the better the conditions for aquatic life and fish survival. D.O. is utilized as an indicator to show the condition of any aquatic system, and while it is in the water column, it is vital to microorganisms (Chen and Liu, 2014). A linear polynomial neural network (PNN) model was utilized to predict the concentrations of DO in the Danube River (Siljic Tomic et al., 2018). A different squad of researchers (Solanki et al., 2015) presented a deep learning network model that outperformed supervised learning-based techniques in terms of accuracy and performance when it came to predicting the pH and dissolved oxygen contents of water. The viability of employing machine learning to forecast dissolved oxygen levels has been shown by the successful implementation of earlier scenarios. ANNs (such as

BPNN, MLP, and FFNN) have been successfully used among other computational approaches for DO prediction, forecasting, modeling, and estimating (Wen et al., 2013). While it is possible to estimate dissolved oxygen using machine learning, gathering data on water quality in difficult-to-reach locations is still a challenge.

Water Salinity Prediction

Predicting climatic fluctuations and comprehending ocean dynamics depend on precise estimation of the ocean subsurface salinity (Qi et al., 2024). Salinity has a major effect on the physical, chemical, and biological activities that take place in the ocean. Estuaries are critical for material exchange between land and sea, with salinity indicating freshwater mixing. Influenced by river inputs, tides, and weather, salinity helps assess water quality and ecosystem health in estuaries (Guillou et al., 2023). The distribution of salinity is essential for the development and reproduction of marine life Seawater's density is strongly impacted by its salt concentration, which in turn affects circulation and stratification (Schmidt et al., 2018). An efficient machine learning and regression model was developed to estimate future salt levels based on past data and predict the salinity of water quality (Chawla et al., 2023). Support Vector Regression delivered the most accurate surface salinity estimates, particularly enhancing projections derived from complex numerical models in low-salinity periods. This demonstrates the potential benefits of applying algorithms for machine learning alongside traditional physical models (Guillou et al., 2023).

Water Turbidity Prediction

The quantity of light scattered by suspended particles in water, often referred to as suspensions, is measured as turbidity, which may be operationally defined as the percentage that can be eliminated by a 0.22 μm (or 0.45 μm) pore size filter (Kirk, 1985). Turbidity in water affects aquatic life, and measuring pollution levels requires a precise understanding of it. An algorithm based on neural networks is being used in the marine environment to track turbidity (Kumar et al., 2022). Wang et al., 2021 showed that using an artificial neural network with tide levels as inputs and NTU readings as outputs accurately predicts turbidity. Key data includes tide levels from the two cycles before the forecast period (Wang et al., 2021). High-turbidity zones were successfully predicted by a machine learning technique utilizing meteorological and satellite TSS data, with random forest regression producing the best results. The region of elevated turbidity from the day before, the wind, and the tide level were important predictions; rainfall had little bearing (Nagayama and Tonooka, 2023). However, forecasting turbidity is still a topic for future research since the causes of turbidity occurrence and dispersion might differ.

Algal Bloom Prediction

The growing nutrient release into aquatic environments has raised serious concerns about eutrophication and harmful algal blooms. The natural water balance has been upset by rapid industrialization, growing urbanization, overcrowding, and climate change. This has resulted in significant nutrient loads entering aquatic ecosystems and a decline in the quality and habitats of freshwater (Ly et al., 2019). Freshwater algal blooms are regarded as an undesirable ecological outcome of eutrophication also concluded that using The Han River's eutrophication and algal blooms have been successfully predicted using machine learning (ML) techniques, which have also shed light on the fundamental principles governing these processes (Ly et al., 2021). Three models of deep learning—multilayer perceptrons (MLP), RNN, and LSTM—were employed to forecast chlorophyll-a (Chl-a), an indicator of algal activity, and their performance was evaluated using RMSE and ordinary least squares (OLS) regression analysis. LSTM regression proved to be the most accurate for predicting harmful algal blooms compared to OLS regression (Lee and Lee, 2018). Similarly, (Kim et al., 2022) demonstrated four models of ML—k-Nearest Neighbour, Decision Tree, LR, and SVM—to forecast cyanobacterial bloom occurrences. By using logCyano as a feature, which is the logarithm of cyanobacterial cell density these models achieved in general fine precision. Growing machine learning models have shown important connections among microbiological populations, nutrients, and algal bloom dynamics, moreover, they have tremendous promise in fields where chemical, physical, and biological, data are easily accessible. These models might help determine the significance of groundwater contributions by identifying signals from groundwater in algal blooms. All things considered, machine learning techniques can improve our capacity to quickly and precisely determine the part that groundwater plays in algal blooms (Sheik et al., 2024).

Machine Learning Across Different Water Scenarios

A potent technique for tackling a variety of problems in water management systems and treatment is machine learning. Artificial intelligence approaches are transforming the way we manage water resources, from hydrologic modeling to real-time monitoring and predictive analysis. Recently, there has been a lot of interest in the use of machine learning in water resource management. The dire need for viable options is highlighted by the deterioration of water quality in metropolitan areas, which is mostly attributable to industrial and municipal effluent from human activities. The ML utilization in water quality research has gained prominence due to significant advancements in analysis and prediction techniques. With integrated and periodic water quality monitoring data serving as standards for water system management, data collecting is the key first step in the development of models for machine learning. Despite being widely used, standard environmental monitoring techniques have difficulties when it comes to *in situ* monitoring because of practical constraints (Li et al., 2020). Poor model performance can result from missing, inaccurate, or damaged data brought on by machinery malfunctions or human mistakes, underscoring the need for data cleaning in machine learning applications. There are several approaches to data cleaning, such as removing problematic data, imputation based on average or median, or using approaches for matrix completion and machine learning to improve raw data (Donders et al., 2006). Prediction accuracy in machine learning applications is influenced by selecting the kind of model and the caliber of the training set (Zhu et al., 2022). Artificial Neural Networks (ANN) and Support Vector Machines

(SVM) excel in predicting water quality components, with SVM often demonstrating superior accuracy and generalization compared to ANN (Wang et al., 2013). SVM's effectiveness is attributed to its stability in optimizing model parameters and setting an upper bound on generalization error, minimizing errors more effectively than ANN (Kim, 2003). Real-time monitoring or predictive modeling is crucial for managing river systems effectively, with Bootstrapped Wavelet Neural Networks (BWNN) and Long Short-Term Memory (LSTM) networks demonstrating proficiency in handling variable and non-seasonal time-series data on water quality (Zhi et al., 2021). Machine learning techniques including Support Vector Machines (SVM), Decision Trees (DT), Random Forests (RF), and Artificial Neural Networks (ANN) have been widely used in groundwater quality assessments in recent years. A comparative study of five algorithms for data mining—the typical DT, RF, chi-square automatic interaction detector, and iterative dichotomizer 3—was carried out to determine the critical factors affecting the quality of groundwater in semiarid regions and how they affect high-quality groundwater areas in Tabriz City, Iran (Jeihouni et al., 2020). Moreover, the groundwater quality maps that are produced via the application of geographic information system techniques are a typical way to identify groundwater contamination with greater accuracy (Lee et al., 2019). Machine learning is essential to wastewater treatment because it helps with the prediction, operation, and administration of wastewater treatment facilities, technological optimization, and monitoring of water quality. Before treatment, a water quality evaluation is required for both industrial and household wastewater due to the variety of contaminants present (Nishat et al., 2023). Soft sensors built on black box models provide real-time online monitoring, which improves the monitoring of indicators such as *E. coli* (Foschi et al., 2021). Artificial Neural Networks (ANN) and soft sensors work together to overcome maintenance and operating issues in wastewater treatment facilities, allowing for real-time ammonia and chlorine monitoring. Numerous factors affect the effluent quality of wastewater treatment plants, and machine learning helps with maintenance and operation management. While deep learning improves measurement accuracy in flow-measurement sensors, XGBoost and RF detect contaminants and their source spots in wastewater networks (Buras and Solano Donado, 2021). Machine learning applications in wastewater treatment facilities have chances to save expenses and enhance overall performance (Wang et al., 2021).

Challenges in Predicting Water Quality Using ML

Machine learning (ML) represents a formidable computational paradigm for addressing multifaceted issues within the domain of water resources, offering applications ranging from the optimization of water resource allocation to the prediction of water quality and the management of recurrent water scarcity. Notwithstanding its potential, the deployment of ML algorithms for water quality assessment in this sector is impeded by a plethora of challenges. Those challenges are described below. The utilization of machine learning (ML) algorithms for any designated application necessitates the availability of a substantial corpus of high-fidelity data. ML models are calibrated based on the acquired datasets, enabling them to discern patterns or trends within the data. Such identified patterns are instrumental in the prediction of variables. After model calibration, validation processes are undertaken, and only upon achieving a predefined threshold of accuracy are these models deemed suitable for deployment in real-world field applications. However, the financial exigencies associated with data procurement serve as a deterrent to the widespread adoption of ML algorithms within the domain of water resources. This challenge is further exacerbated by technological constraints. Moreover, even in instances where data is accessible, it may not invariably conform to the requisite resolution (Zhu et al., 2022).

Machine learning-derived models are predominantly empirical and are tailored to the specific conditions for which they were formulated. The intrinsic complexity and heterogeneity of real-world aquatic systems present significant challenges to the generalizability of water quality predictions. Moreover, the environmental dynamics are distinct across different water bodies, thereby limiting the applicability of a model to its system of origin. This limitation poses a substantial impediment to the broad-scale deployment of ML methodologies. To mitigate this issue, models necessitate recalibration when applied to disparate systems from those they were originally designed for, incurring additional resource expenditures. Consequently, reliance on 'black-box' solutions should be eschewed in favor of approaches that yield process-based scientific insights (Schafer et al., 2022). The initial stage in the development of any machine learning (ML) model necessitates the judicious selection of pertinent input parameters. It is crucial to select a subset of parameters affecting the water quality that encapsulate sufficient predictive information regarding water quality. This selection is paramount, as an excess of input variables may result in redundancy, whereas an insufficient number could lead to inaccuracies in the model. Optimal variable selection can enhance model accuracy, as inappropriate choices may adversely affect model performance. Furthermore, each ML model exhibits unique responses to different sets of input variables and data patterns. Therefore, it is imperative to identify variables that augment the efficacy of a specific ML model. The methodologies such as the partial autocorrelation function (PACF), autocorrelation function (ACF), correlation coefficient, and cross-correlation function, are instrumental in discerning the salient predictors (Asadollah et al., 2021). Better data acquisition through sophisticated sensors, enhancing the applicability of the model to a universal scale and an interdisciplinary approach are to be considered to address the above-mentioned challenges.

Conclusion

The evaluation of water quality, which has historically relied on labor-intensive approaches, gains a great deal from machine learning (ML) techniques since they improve accuracy, efficiency, and affordability. Applications of machine learning include controlling shortages and allocating water resources optimally. Notwithstanding, several obstacles persist, such as the requirement for extensive and superior quality data sets, limitations in terms of finances and technology, and the intricacy of actual aquatic systems. Resource requirements rise when machine learning models require to be recalibrated for diverse contexts. Preventing errors and duplication, and choosing the appropriate input parameters are essential for an effective machine learning implementation. Key predictors can be found using methods such as the correlation coefficient, autocorrelation function, and partial autocorrelation function (PACF, ACF). Interdisciplinary techniques and sophisticated sensor data gathering are essential to addressing these issues. In conclusion, while machine learning (ML) has the potential to manage water quality, it is critical to overcome implementation, generalization, and data issues. Improved parameter selection and data collection will increase the applicability and dependability of ML models in a variety of aquatic settings.

Declaration

Ethical Responsibilities of Authors

"All authors have read understood, and have complied as applicable with the statement on "Ethical responsibilities of Authors" as found in the Instructions for Authors".

Ethical Approval

Not applicable.

Use of AI Tool

The corresponding author declares that no AI tool has been used to prepare this manuscript.

Conflict of Interest

On behalf of all the authors the corresponding author states that there is no conflict of interest to disclose.

Competing Interests

Not applicable.

Authors' Contributions

RY: Conceptualization, Methodology, Writing First Draft, Writing - Review & Editing, **AS**: Investigation, Formal analysis, Visualization, Software, Data Curation, **AK:** Writing - Original Draft, Formal analysis, Methodology, **JK**: Formal analysis, Visualization, Review & Editing, Supervision, **SS**: Review & Editing, Supervision, **BN**: Review & Editing, Formal analysis.

Funding

No funding has been received.

References

Alshaltone, O., Nasir, N., Barneih, F., Majali, E.A. and Al-Shammaa, A. 2021. Multi Sensing Platform. *14th International Conference on Developments in e-Systems Engineering (DeSE)*, 174–179. https://doi.org/10.1109/DeSE54285.2021.9719474.

Asadollah, S.B.H.S., Sharafati, A., Motta, D. and Yaseen, Z.M. 2021. River water quality index prediction and uncertainty analysis: A comparative study of machine learning models. *Journal of Environmental Chemical Engineering*, 9(1): 104599. https://doi.org/10.1016/j.jece.2020.104599.

Bagheri, M., Akbari, A. and Mirbagheri, S.A. 2019. Advanced control of membrane fouling in filtration systems using artificial intelligence and machine learning techniques: A critical review. *Process Safety and Environmental Protection*, 123: 229–252. https://doi.org/10.1016/j.psep.2019.01.013.

Berry, M.W., Mohammed, A.H. and Yap, B.W. 2019. *Supervised and Unsupervised Learning for Data Science*. Springer.

Beyer, K., Goldstein, J., Ramakri, R. and Shaft, U. 1999. When is "nearest neighbor" meaningful? In: *International Conference on Database Theory*. Springer, 217–235.

Buras, M.P. and Solano Donado, F. 2021. Identifying and estimating the location of sources of industrial pollution in the sewage network. *Sensors*, 21(10): 3426. https://doi.org/10.3390/s21103426.

Chawla, P., Cao, X., Fu, Y., Hu, C., Wang, M., Wang, S. and Gao, J.Z. 2023. Water quality prediction of salton sea using machine learning and big data techniques. *International Journal of Environmental Analytical Chemistry*, 103(18): 6835–6858. https://doi.org/10.1080/03067319.2021.1963713.

Chen, T. and Guestrin, C. 2016. XGBoost: A Scalable Tree Boosting System. *Proceedings of the 22nd ACM SIGKDD International Conference on Knowledge Discovery and Data Mining*, 785–794. https://doi.org/10.1145/2939672.2939785.

Chen, W.-B. and Liu, W.-C. 2014. Artificial neural network modeling of dissolved oxygen in reservoir. *Environmental Monitoring and Assessment*, 186(2): 1203–1217. https://doi.org/10.1007/s10661-013-3450-6.

Chouhan, S.S., Singh, U.P., Sharma, U. and Jain, S. 2021. Leaf disease segmentation and classification of Jatropha Curcas L. and Pongamia Pinnata L. biofuel plants usingcomputer vision based approaches. *Measurement*, 171: 108796, ISSN 0263-2241. https://doi.org/10.1016/j.measurement.2020.108796.

Chouhan, S.S., Singh, U.P. and Jain, S. 2024a. Introduction to computer vision and drone technology. *In*: Chouhan, S.S., Singh, U.P. and Jain, S. (eds.). *Applications of Computer Vision and Drone Technology in Agriculture 4.0*. Springer, Singapore. https://doi.org/10.1007/978-981-99-8684-2_1.

Chouhan, S.S., Singh, U.P., Sharma, U. et al. 2024b. Classification of different plant species using Deep Learning and Machine Learning algorithms. *Wireless Pers. Commun.*, 136: 2275–2298. https://doi.org/10.1007/s11277-024-11374-y.

Cipullo, S., Snapir, B., Prpich, G., Campo, P. and Coulon, F. 2019. Prediction of bioavailability and toxicity of complex chemical mixtures through machine learning models. *Chemosphere*, 215: 388–395.

Donders, A.R.T., Van Der Heijden, G.J.M.G., Stijnen, T. and Moons, K.G.M. 2006. Review: A gentle introduction to imputation of missing values. *Journal of Clinical Epidemiology*, 59(10): 1087–1091. https://doi.org/10.1016/j.jclinepi.2006.01.014.

Ehteram Ghotbi, Kisi Najah Ahmed, Hayder Ming Fai, Krishnan Abdulmohsin Afan and EL-Shafie. 2019. Investigation on the potential to integrate different artificial intelligence models with metaheuristic algorithms for improving river suspended sediment predictions. *Applied Sciences*, 9(19): 4149. https://doi.org/10.3390/app9194149.

Fan, J., Wang, X., Wu, L., Zhou, H., Zhang, F., Yu, X., Lu, X. and Xiang, Y. 2018. Comparison of support vector machine and extreme gradient boosting for predicting daily global solar radiation using temperature and precipitation in humid subtropical climates: A case study in China. *Energy Conversion and Management*, 164: 102–111. https://doi.org/10.1016/j.enconman.2018.02.087.

Forests, R. and Breiman, L. 1999. In Statistics Department University of California Berkeley, 1–29.

Foschi, J., Turolla, A. and Antonelli, M. 2021. Soft sensor predictor of E. coli concentration based on conventional monitoring parameters for wastewater disinfection control. *Water Research*, 191: 116806. https://doi.org/10.1016/j.watres.2021.116806.

Guillou, N., Chapalain, G. and Petton, S. 2023. Predicting sea surface salinity in a tidal estuary with machine learning. *Oceanologia*, 65(2): 318–332. https://doi.org/10.1016/j.oceano.2022.07.007.

Haghiabi, A.H., Azamathulla, H.M. and Parsaie, A. (n.d.). Prediction of head loss on cascade weir using ANN and SVM. *SH Journal of Hydraulic Engineering*, 23: 102–110.

Han, J., Kamber, M. and Pei, J. 2011. Data Mining: Concepts and Techniques (3rd ed.).

Jaloree, S., Rajput, A. and Gour, S. 2014. Decision tree approach to build a model for water quality. *Binary Journal of Data Mining & Networking*, 4: 25–28.

Jeihouni, M., Toomanian, A. and Mansourian, A. 2020. Decision tree-based data mining and rule induction for identifying high quality groundwater zones to water supply management: A novel hybrid use of data mining and GIS. *Water Resources Management*, 34(1): 139–154. https://doi.org/10.1007/s11269-019-02447-w.

Jhaveri, R.H., Ramani, S.V., Srivastava, G., Gadekallu, T.R. and Aggarwal, V. 2021. Fault-resilience for bandwidth management in industrial software-defined networks. *IEEE Transactions on Network Science and Engineering*, 8(4): 3129–3139. https://doi.org/10.1109/TNSE.2021.3104499.

Johnson, D.C., Enriquez, C.E., Pepper, I.L., Davis, T.L., Gerba, C.P. and Rose, J.B. 1997. Survival of giardia, cryptosporidium, poliovirus and salmonella in marine waters. *Water Science and Technology*, 35(11-12): 261–268. https://doi.org/10.2166/wst.1997.0744.

Kedia, N. 2015. Water quality monitoring for rural areas—A sensor cloud based economical project. *2015 1st International Conference on Next Generation Computing Technologies (NGCT)*, 50–54. https://doi.org/10.1109/NGCT.2015.7375081.

Khajuria, R, Sunanda and Singh, S. 2018. A survey of existing leaf disease techniques using artificial neural network. *International Journal of Computer Trends and Technology (IJCTT)*, 59(1): 52–62. ISSN:2231-2803, doi:10.14445/22312803/IJCTT-V59P109.

Kim, J., Jonoski, A. and Solomatine, D.P. 2022. A classification-based machine learning approach to the prediction of cyanobacterial blooms in ChilgokWeir, South Korea. *Water*, 14(4): 542. https://doi.org/10.3390/w14040542.

Kim, K. 2003. Financial time series forecasting using support vector machines. *Neurocomputing*, 55(1-2): 307–319. https://doi.org/10.1016/S0925-2312(03)00372-2.

Kirk, J.T.O. 1985. Effects of suspensoids (turbidity) on penetration of solar radiation in aquatic ecosystems. *Hydrobiologia*, 125(1): 195–208. https://doi.org/10.1007/BF00045935.

Kumar, R. et al. 2024. Hybrid Approach of Cotton Disease Detection for Enhanced Crop Health and Yield in IEEE Access, doi: 10.1109/ACCESS.2024.3419906.

Kumar, L., Afzal, M.S. and Ahmad, A. 2022. Prediction of water turbidity in a marine environment using machine learning: A case study of Hong Kong. *Regional Studies in Marine Science*, 52: 102260. https://doi.org/10.1016/j.rsma.2022.102260.

Lee, K.-J., Yun, S.-T., Yu, S., Kim, K.-H., Lee, J.-H. and Lee, S.-H. 2019. The combined use of self-organizing map technique and fuzzy c-means clustering to evaluate urban groundwater quality in Seoul metropolitan city, South Korea. *Journal of Hydrology*, 569: 685–697. https://doi.org/10.1016/j.jhydrol.2018.12.031.

Lee, S. and Lee, D. 2018. Improved prediction of harmful algal blooms in four major South Korea's rivers using deep learning models. *International Journal of Environmental Research and Public Health*, 15(7): 1322. https://doi.org/10.3390/ijerph15071322.

Li, W., Fang, H., Qin, G., Tan, X., Huang, Z., Zeng, F., Du, H. and Li, S. 2020. Concentration estimation of dissolved oxygen in Pearl River Basin using input variable selection and machine learning techniques. *Science of the Total Environment*, 731: 139099. https://doi.org/10.1016/j.scitotenv.2020.139099.

Liao, H. and Sun, W. 2010. Forecasting and evaluating water quality of Chao Lake based on an improved decision tree method. *Procedia Environmental Sciences*, 2: 970–979. https://doi.org/10.1016/j.proenv.2010.10.109.

Lu, H. and Ma, X. 2020. Hybrid decision tree-based machine learning models for short-term water quality prediction. *Chemosphere*, 249: 126169. https://doi.org/10.1016/j.chemosphere.2020.126169.

Ly, Q.V., Lee, M.-H. and Hur, J. 2019. Using fluorescence surrogates to track algogenic dissolved organic matter (AOM) during growth and coagulation/flocculation processes of green algae. *Journal of Environmental Sciences*, 79: 311–320. https://doi.org/10.1016/j.jes.2018.12.006.

Ly, Q.V., Nguyen, X.C., Lê, N.C., Truong, T.-D., Hoang, T.-H.T., Park, T.J., Maqbool, T., Pyo, J., Cho, K.H., Lee, K.-S. and Hur, J. 2021. Application of Machine Learning for eutrophication analysis and algal bloom prediction in an urban river: A 10-year study of the Han River, South Korea. *Science of the Total Environment*, 797: 149040. https://doi.org/10.1016/j.scitotenv.2021.149040.

Ma, J., Ding, Y., Cheng, J.C.P., Jiang, F. and Xu, Z. 2020. Soft detection of 5-day BOD with sparse matrix in city harbor water using deep learning techniques. *Water Research*, 170: 115350. https://doi.org/10.1016/j.watres.2019.115350.

Malek, N.H.A., Wan Yaacob, W.F., Md Nasir, S.A. and Shaadan, N. 2022. Prediction of water quality classification of the Kelantan River Basin, Malaysia, using Machine Learning techniques. *Water*, 14(7): 1067. https://doi.org/10.3390/w14071067.

Messaoud, S., Bradai, A., Bukhari, S.H.R., Quang, P.T.A., Ahmed, O.B. and Atri, M. 2020. A survey on machine learning in Internet of Things: Algorithms, strategies, and applications. *Internet of Things*, 12: 100314. https://doi.org/10.1016/j.iot.2020.100314.

Mosaffa, H., Sadeghi, M., Mallakpour, I., Naghdyzadegan Jahromi, M. and Pourghasemi, H.R. 2022. Application of machine learning algorithms in hydrology. In *Computers in Earth and Environmental Sciences*, 585–591. Elsevier. https://doi.org/10.1016/B978-0-323-89861-4.00027-0.

Mukate, S., Wagh, V., Panaskar, D., Jacobs, J.A. and Sawant, A. 2019. Development of new integrated water quality index (IWQI) model to evaluate the drinking suitability of water. *Ecological Indicators*, 101: 348–354. https://doi.org/10.1016/j.ecolind.2019.01.034.

Nagayama, K. and Tonooka, H. 2023. Prediction of the area of high-turbidity water in the Yatsushiro Sea, Japan, using Machine Learning with satellite, meteorological, and oceanographic data. *Remote Sensing*, 15(6): 1652. https://doi.org/10.3390/rs15061652.

Najah Ahmed, A., Binti Othman, F., Abdulmohsin Afan, H., Khaleel Ibrahim, R., Ming Fai, C., Shabbir Hossain, M., Ehteram, M. and Elshafie, A. 2019. Machine learning methods for better water quality prediction. *Journal of Hydrology*, 578: 124084. https://doi.org/10.1016/j.jhydrol.2019.124084.

Nasir, N., Al Ahmad, M. and Murad, A.A. 2019. Capacitive detection and quantification of water suspended solids. *2019 International Conference on Electrical and Computing Technologies and Applications (ICECTA)*, 1–5. https://doi.org/10.1109/ICECTA48151.2019.8959628.

Nasir, N., Kansal, A., Alshaltone, O., Barneih, F., Sameer, M., Shanableh, A. and Al-Shamma'a, A. 2022. Water quality classification using machine learning algorithms. *Journal of Water Process Engineering*, 48: 102920. https://doi.org/10.1016/j.jwpe.2022.102920.

Nishat, A., Yusuf, M., Qadir, A., Ezaier, Y., Vambol, V., Ijaz Khan, M., Ben Moussa, S., Kamyab, H., Sehgal, S.S., Prakash, C., Yang, H.-H., Ibrahim, H. and Eldin, S.M. 2023. Wastewater treatment: A short assessment on available techniques. *Alexandria Engineering Journal*, 76: 505–516. https://doi.org/10.1016/j.aej.2023.06.054.

Ooi, K.S., Chen, Z., Poh, P.E. and Cui, J. 2022. BOD5 prediction using machine learning methods. *Water Supply*, 22(1): 1168–1183. https://doi.org/10.2166/ws.2021.202.

Palani, S., Liong, S.-Y. and Tkalich, P. 2008. An ANN application for water quality forecasting. *Marine Pollution Bulletin*, 56(9): 1586–1597. https://doi.org/10.1016/j.marpolbul.2008.05.021.

Pratap, B., Kumar, S., Nand, S., Azad, I., Bharagava, R.N., Romanholo Ferreira, L.F. and Dutta, V. 2023. Wastewater generation and treatment by various eco-friendly technologies: Possible health hazards and further reuse for environmental safety. *Chemosphere*, 313: 137547. https://doi.org/10.1016/j.chemosphere.2022.137547.

Qi, J., Sun, G., Xie, B., Li, D. and Yin, B. 2024. Deep learning to estimate ocean subsurface salinity structure in the Indian Ocean using satellite observations. *Journal of Oceanology and Limnology*, 42(2): 377–389. https://doi.org/10.1007/s00343-023-3063-z.

Rani, A., Kumar, N., Kumar, J., Kumar, J. and Sinha, N.K. 2022. Machine learning for soil moisture assessment. *In Deep Learning for Sustainable Agriculture*. Elsevier, 143–168. https://doi.org/10.1016/B978-0-323-85214-2.00001-X.

Ross Michael, P. 2016. Water Rights in Southeast Asia and India.

Sarker, I.H., Kayes, A.S.M., Badsha, S., Alqahtani, H., Watters, P. and Ng, A. 2020. Cybersecurity data science: An overview from machine learning perspective. *Journal of Big Data*, 7(1): 41. https://doi.org/10.1186/s40537-020-00318-5.

Schäfer, B., Beck, C., Rhys, H., Soteriou, H., Jennings, P., Beechey, A. and Heppell, C.M. 2022. Machine learning approach towards explaining water quality dynamics in an urbanised river. *Scientific Reports*, 12(1): 12346. https://doi.org/10.1038/s41598-022-16342-9.

Schmidhuber, J. 2015. Deep learning in neural networks: An overview. *Neural Networks*, 61: 85–117. https://doi.org/10.1016/j.neunet.2014.09.003.

Schmidt, H., Seitz, S., Hassel, E. and Wolf, H. 2018. The density–salinity relation of standard seawater. *Ocean Science*, 14(1): 15–40. https://doi.org/10.5194/os-14-15-2018.

Shafi, U., Mumtaz, R., Anwar, H., Qamar, A.M. and Khurshid, H. 2018. Surface water pollution detection using Internet of Things. *2018 15th International Conference on Smart Cities: Improving Quality of Life Using ICT & IoT (HONET-ICT)*, 92–96. https://doi.org/10.1109/HONET.2018.8551341.

Sheik, A.G., Kumar, A., Patnaik, R., Kumari, S. and Bux, F. 2024. Machine learning-based design and monitoring of algae blooms: Recent trends and future perspectives—A short review. *Critical Reviews in Environmental Science and Technology*, 54(7): 509–532. https://doi.org/10.1080/10643389.2023.2252313.

Siljic Tomic, A., Antanasijević, D., Ristić, M., Perić-Grujić, A. and Pocajt, V. 2018. A linear and non-linear polynomial neural network modeling of dissolved oxygen content in surface water: Inter- and extrapolation performance with inputs' significance analysis. *Science of The Total Environment*, 610-611: 1038–1046. https://doi.org/10.1016/j.scitotenv.2017.08.192.

Solanki, A., Agrawal, H. and Khare, K. 2015. Predictive analysis of water quality parameters using deep learning. *International Journal of Computer Applications*, 125(9): 29–34.

Sun, A.Y. and Scanlon, B.R. 2019. How can Big Data and machine learning benefit environment and water management: A survey of methods, applications, and future directions. *Environmental Research Letters*, 14(7): 073001. https://doi.org/10.1088/1748-9326/ab1b7d.

Tikhamarine, Y., Souag-Gamane, D., Ahmed, A.N., Sammen, S.Sh., Kisi, O., Huang, Y.F. and El-Shafie, A. 2020. Rainfall-runoff modelling using improved machine learning methods: Harris hawks optimizer vs. particle swarm optimization. *Journal of Hydrology*, 589: 125133. https://doi.org/10.1016/j.jhydrol.2020.125133.

United States Geological Survey (USGS) n.d. Biological Oxygen Demand (BOD) and Water. USGS, US Dept of the Interior, Washington DC. (n.d.). See: https://www.usgs.gov/special-topic/water-scienceschool/science/biological-oxygen-demand-bod-and-water.

Wang, D., Thunell, S., Lindberg, U., Jiang, L., Trygg, J., Tysklind, M. and Souihi, N. 2021. A machine learning framework to improve effluent quality control in wastewater treatment plants. *Science of the Total Environment*, 784: 147138. https://doi.org/10.1016/j.scitotenv.2021.147138.

Wang, R., Kim, J.-H. and Li, M.-H. 2021. Predicting stream water quality under different urban development pattern scenarios with an interpretable machine learning approach. *Science of the Total Environment*, 761: 144057. https://doi.org/10.1016/j.scitotenv.2020.144057.

Wang, Y., Chen, J., Cai, H., Yu, Q. and Zhou, Z. 2021. Predicting water turbidity in a macro-tidal coastal bay using machine learning approaches. *Estuarine, Coastal and Shelf Science*, 252: 107276. https://doi.org/10.1016/j.ecss.2021.107276.

Wang, Y., Zheng, T., Zhao, Y., Jiang, J., Wang, Y., Guo, L. and Wang, P. 2013. Monthly water quality forecasting and uncertainty assessment via bootstrapped wavelet neural networks under missing data for Harbin, China. *Environmental Science and Pollution Research*, 20(12): 8909–8923. https://doi.org/10.1007/s11356-013-1874-8.

Wen, X., Fang, J., Diao, M. and Zhang, C. 2013. Artificial neural network modeling of dissolved oxygen in the Heihe River, Northwestern China. *Environmental Monitoring and Assessment*, 185(5): 4361–4371. https://doi.org/10.1007/s10661-012-2874-8.

Yuchi, W., Gombojav, E., Boldbaatar, B., Galsuren, J., Enkhmaa, S., Beejin, B., Naidan, G., Ochir, C., Legtseg, B., Byambaa, T., Barn, P., Henderson, S.B., Janes, C.R., Lanphear, B.P., McCandless, L.C., Takaro, T.K., Venners, S.A., Webster, G.M. and Allen, R.W. 2019. Evaluation of random forest regression and multiple linear regression for predicting indoor fine particulate matter concentrations in a highly polluted city. *Environmental Pollution*, 245: 746–753. https://doi.org/10.1016/j.envpol.2018.11.034.

Zaitsev, N.K., Dvorkin, V.I., Melnikov, P.V. and Kozhukhova, A.E. 2018. A dissolved oxygen analyzer with an optical sensor. *Journal of Analytical Chemistry*, 73(1):102–108. https://doi.org/10.1134/S1061934818010136.

Zhi, W., Feng, D., Tsai, W.-P., Sterle, G., Harpold, A., Shen, C. and Li, L. 2021. From hydrometeorology to river water quality: Can a deep learning model predict dissolved oxygen at the continental scale? *Environmental Science & Technology*, 55(4): 2357–2368. https://doi.org/10.1021/acs.est.0c06783.

Zhu, J.-J., Kang, L. and Anderson, P.R. 2018. Predicting influent biochemical oxygen demand: Balancing energy demand and risk management. *Water Research*, 128: 304–313. https://doi.org/10.1016/j.watres.2017.10.053.

Zhu, M., Wang, J., Yang, X., Zhang, Y., Zhang, L., Ren, H., Wu, B. and Ye, L. 2022. A review of the application of machine learning in water quality evaluation. *Eco-Environment & Health*, 1(2): 107–116. https://doi.org/10.1016/j.eehl.2022.06.001.

Chapter 12

Enhancing Agricultural Support with AI in the Farmer ChatBot Framework

Jayanthi J,[1] *Arun Kumar K*[2,*] and *Vimal Kumar Semwal*[3]

Introduction

In today's rapidly evolving technological landscape, chatbots have emerged as versatile tools capable of understanding human language and executing predefined tasks. Stemming from Joseph Weizenbaum's pioneering creation of Eliza in 1966, these digital entities have undergone significant advancements since Alan Turing posed the fundamental question of machine intelligence. Today, they permeate various industries, serving as invaluable customer interaction and support resources. However, despite their widespread adoption, the effectiveness of chatbots varies considerably. The challenge lies not merely in their creation but in crafting chatbots that excel in functionality and user experience.

India's agricultural sector contributes significantly to the country's economic growth, accounting for approximately 16% of the GDP and employing around 52% of the population. Despite its importance, many farmers lack access to authentic and up-to-date farming information, hindering their decision-making processes. Traditional methods of disseminating information, such as field visits, are often hindered by

[1] School of Computer Science and Engineering, VIT Bhopal University, Kothrikalan, Sehore Madhya Pradesh – 466114. https://orcid.org/0000-0003-3276-2432.

[2] School of Bioscience, Engineering and Technology, VIT Bhopal University, Kothrikalan, Sehore Madhya Pradesh – 4661142. https://orcid.org/0000-0002-5552-5459.

[3] Africa Rice Center (AfricaRice), Nigeria Station, C/O IITA PMB-5320, Ibadan, Nigeria. Email: V.Semwal@cgiar.org

* Corresponding author: arunkumar.k@vitbhopal.ac.in

accessibility issues, resulting in inefficiencies and decreased crop yields (Singh et al., 2019). Agriculture, as the art and science of soil cultivation, stands as a cornerstone of national development. Despite the modernization of plant breeding techniques and the advent of agrochemicals such as pesticides and fertilizers, farmers often lag in leveraging the latest technological advancements. Therefore, there's a critical need to equip them with up-to-date information and techniques to ensure optimal agricultural productivity.

The proliferation of Information Technology (IT) in recent years, particularly in rural areas, presents an opportunity to address these challenges. With the exponential growth of mobile connectivity, especially in rural regions, mobile devices have become powerful tools for disseminating agricultural information. However, challenges such as language diversity and rural populations' hesitancy towards modern technologies persist, complicating the dissemination of accurate information (Jain, 2020). Chat-bot systems, serving as question-answering systems, bridge the gap between humans and machines, enabling users to query and receive accurate responses. In domains such as medicine and travel, significant research has been conducted on question-answering systems. However, in agriculture, retrieving precise answers from web data poses unique challenges, as websites often present irrelevant information alongside relevant responses. To address this, deep learning techniques, particularly Recurrent Neural Networks (RNNs), are increasingly employed (Singh et al., 2019). In this context, the utilization of chatbot systems holds promise as a solution to disseminate accurate agricultural information efficiently. By leveraging the ubiquity of mobile devices and advancements in natural language processing, chatbots can empower farmers with timely and relevant agricultural knowledge, thereby enhancing productivity and supporting rural livelihoods.

Natural Language Processing (NLP)

Natural Language Processing (NLP) is a branch of computer science that falls under the broader field of Artificial Intelligence (AI), with a primary focus on enabling computers to comprehend and interpret human language, both spoken and written, in a manner akin to humans (Jurafsky and Martin, 2019). This interdisciplinary field combines computational linguistics, which involves rule-based mode ling of human language, with statistical, machine learning, and deep learning models. NLP technologies encompass a range of tasks aimed at providing computers with the ability to understand human language in various forms, such as text or speech data, and to grasp its full meaning, including the speaker's or writer's intention and sentiment. These tasks include speech recognition, where software deciphers spoken commands or translates text between languages, and part-of-speech (POS) tagging, which identifies the grammatical role of each word in a sentence.

However, creating software that accurately discerns the intended meaning of text or voice data is highly challenging due to the inherent ambiguity of human language. From homonyms and homophones to idioms and metaphors, human language is replete with complexities that take years for humans to master. Programmers must meticulously teach NLP-driven applications to recognize and understand these nuances from the outset to ensure their usefulness (Jurafsky and Martin, 2019).

Word sense disambiguation is another critical NLP task, involving the selection of a word's meaning from its potential meanings based on semantic analysis and context. Named entity recognition (NER) identifies words or phrases as specific entities, such as locations or individuals, while sentiment analysis delves into intangible elements within the text, including attitudes and emotions.

These NLP tasks underpin various applications that range from voice-activated GPS devices and digital assistants to customer service chatbots and speech-to-text dictation programs (Jurafsky and Martin, 2019). Furthermore, NLP's utility extends beyond consumer conveniences to corporate solutions, where it enhances business operations, boosts worker productivity, and streamlines critical procedures. Thus, NLP continues to play a pivotal role in advancing human-computer interaction and driving technological innovation across diverse domains. Chatbots, powered by AI, NLP, and ML, come in two main types: task-oriented and data-driven/predictive. Task-oriented chatbots focus on specific tasks, offering structured, conversational responses using rules and minimal ML. On the other hand, data-driven/predictive chatbots, like virtual assistants, use NLP, NLU, and ML to provide personalized, interactive experiences, learning from user interactions to anticipate needs and make suggestions (Jurafsky and Martin, 2019). Examples include Apple's Siri and Amazon's Alexa.

NLTK, the Natural Language Toolkit, is a powerful platform for Python programming with human language data. It offers a wide range of text processing tools and access to various language resources like WordNet. NLTK is suitable for linguists, engineers, students, and industry users alike, providing comprehensive guides and documentation. It's free, open-source, and compatible with Windows, Mac OS X, and Linux (Bird et al., 2009).

Research Methodology

In this chapter, a Systematic Literature Review method is employed to investigate the applications of Chatbot for Agriculture purposes. The aim is to gather, analyze, and draw conclusions from published research papers relevant to the topic. The research question guides the inquiry, focusing the search to find relevant data and insights from existing literature. Table 1 research questions for systematic literature review of the agriculture sector.

The research chapters were identified through searches conducted in database journals using specific keyword strings related to the research questions outlined in Table 1. These searches employed combinations of keywords such as "Chatbot" AND "Agriculture" and "India" AND "Call center". The findings from these searches are presented in Table 2.

Exploring the Applications and Functionalities of Chatbots in the Farmer ChatBot (FCB) Framework

Chatbots in the Farmer ChatBot (FCB) (Fig. 1) framework are designed to offer agricultural support and advisory services tailored to the needs of farmers. These intelligent systems act as virtual assistants, capable of answering a wide range of queries related to farming, such as crop management, pest control, weather forecasts,

Table 1. Research questions.

ID	Research Questions	Purpose
RQ1	How are chatbots utilized in the Farmer ChatBot (FCB) in India?	To explore the specific applications and functionalities of chatbots within the FCB framework in India.
RQ2	What are the key features and capabilities of chatbots deployed in FCB for agricultural support?	To identify the essential functionalities and characteristics of chatbots tailored for providing agricultural assistance through FCB.
RQ3	What are the challenges faced in implementing chatbots within the FCB ecosystem for agricultural advisory services?	To understand the obstacles and difficulties encountered during the integration and operation of chatbots in FCB for agricultural advisory purposes.
RQ4	How effective are chatbots in enhancing the accessibility and efficiency of agricultural information delivery through FCB?	To assess the impact and effectiveness of chatbots in improving the accessibility, timeliness, and quality of agricultural information provided by FCB to farmers.
RQ5	What are the user perceptions and experiences regarding the use of chatbots in FCB for agricultural inquiries and support?	To investigate the attitudes, opinions, and satisfaction levels of farmers and stakeholders towards the use of chatbots as part of the FCB platform for agricultural assistance.

Table 2. Search result and number of articles.

S. no.	Journal - Databse	No. of Article
1.	IEEE Xplore	**3**
2.	ScienceDirect	**3**
3.	ACM Digital Library	**1**
4.	Springer	**6**
	Total	**13**

market prices, government schemes, and best agricultural practices. The chatbots are equipped with natural language processing (NLP) capabilities and machine learning algorithms, enabling them to understand and process agricultural terminology effectively. These chatbots are programmed to understand agricultural terminology and provide accurate information tailored to farmers' needs (Singh et al., 2019; Jain, 2020). By leveraging these technologies, the chatbots provide precise, context-aware responses and personalized recommendations. Furthermore, the chatbots are integrated with extensive agricultural knowledge bases and databases, ensuring that the information provided is accurate, reliable, and up-to-date. The FCB model strives to empower farmers by making agricultural knowledge easily accessible and improving decision-making in farming practices.

By utilizing chatbots, the Farmer ChatBot (FCB) framework aims to enhance accessibility to agricultural information and advisory services, especially in rural areas where access to traditional support systems may be limited. Chatbots offer farmers a convenient and efficient means of obtaining timely assistance and guidance, thereby empowering them to make informed decisions and improve agricultural practices.

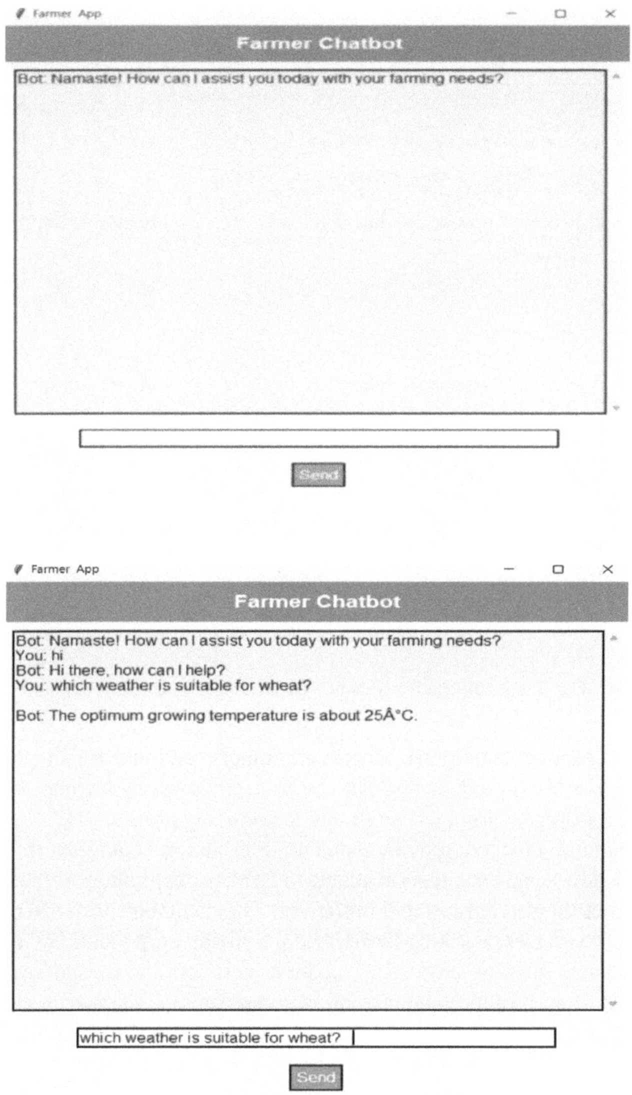

Fig. 1. Farmer ChatBot system – Forntend.

The applications and functionalities of chatbots within the Farmer ChatBot (FCB) framework encompass a diverse range of agricultural support services tailored to meet the specific needs of farmers. Firstly, chatbots play a crucial role in query resolution, providing immediate assistance on a wide array of agricultural topics such as crop cultivation techniques, pest management strategies, fertilizer usage, and soil health improvement measures. Additionally, they offer real-time weather updates, enabling farmers to effectively plan their agricultural activities and mitigate risks associated

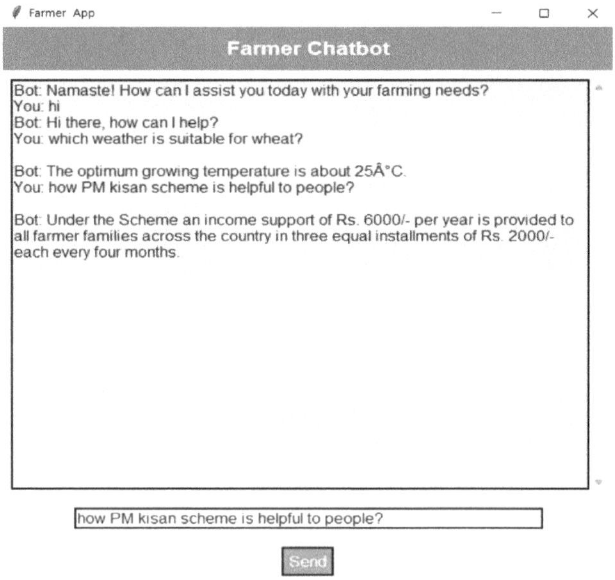

Fig. 2. Farmer ChatBot system – Government Scheme Information.

with adverse weather conditions. Moreover, farmers can inquire about current market prices for agricultural produce through chatbots, empowering them to make informed decisions regarding selling their crops and maximizing profits.

Furthermore, chatbots provide valuable information about government schemes (Fig. 2), subsidies, and incentives available to farmers, facilitating their participation in relevant agricultural programs and initiatives. These chatbots also offer personalized crop advisory services, assisting farmers in optimizing crop yields and reducing risks based on factors such as crop type, location, soil conditions, and seasonal trends. Additionally, they provide guidance on pest and disease management, recommend agricultural practices to improve productivity and sustainability and offer educational resources and training materials on modern farming practices, thereby empowering farmers with knowledge and skills to enhance their agricultural productivity and livelihoods.

Overall, the applications and functionalities of chatbots within the Farmer ChatBot (FCB) framework are instrumental in streamlining communication, delivering timely information, and supporting farmers in addressing various challenges and opportunities in agriculture.

Identifying Key Features of Chatbots for Agricultural Support in FCB

The chatbot deployed in the Farmer ChatBot (FCB) for agricultural support is designed with several key features to enhance user experience and accessibility.

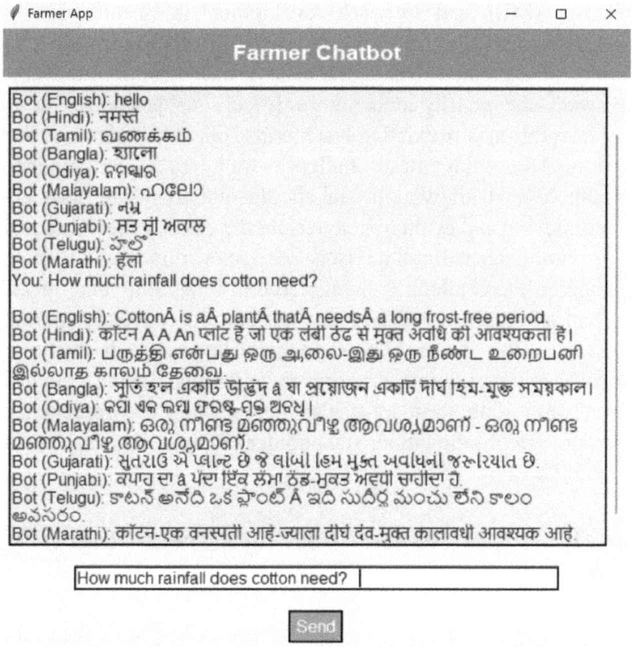

Fig. 3. Farmer ChatBot system – Multi Language Support.

Firstly, it is seamlessly connected with Google Translate for multilingual support, enabling users to interact in their preferred language. Additionally, the chatbot accepts text queries from users, ensuring enhanced accessibility for all users. Furthermore, it incorporates language translation capabilities, allowing it to translate user queries into the local language and provide responses in text formats. Initially launched in five languages during Phase I (English, Hindi, Tamil, Bangla, and Odiya), the chatbot's expansion plan includes the addition of five more languages (Malayalam, Gujarati, Punjabi, Telugu, and Marathi) (Fig. 3) with further accessibility planned for all 22 regional languages in subsequent stages. Through these features and expansion plans, the chatbot aims to cater to the diverse linguistic needs of farmers and stakeholders accessing agricultural support through the FCB platform.

Challenges in Implementing Chatbots for Agricultural Advisory Services within the FCB Ecosystem

Implementing chatbots within the Farmer ChatBot (FCB) ecosystem for agricultural advisory services presents several challenges that need to be addressed for effective deployment. One significant challenge is the linguistic diversity across India, requiring chatbots to understand and respond to queries in various languages and dialects. Additionally, ensuring accessibility and functionality in areas with limited internet connectivity is crucial. Technical expertise in natural language processing (NLP), machine learning (ML), and software development is essential for deploying and maintaining chatbots within the FCB framework. Furthermore, ensuring the quality and availability of agricultural data, addressing user adoption and acceptance, and addressing privacy and security concerns are critical considerations. Cultural sensitivity and seamless integration with existing systems also play vital roles in successful chatbot implementation. Addressing these challenges requires a comprehensive approach involving technological innovation, stakeholder collaboration, user education, and continuous monitoring and evaluation to ensure the effectiveness and sustainability of chatbots in providing agricultural advisory services within the FCB ecosystem.

Understanding the challenges encountered in integrating and operating chatbots within the Farmer ChatBot (FCB) for agricultural advisory purposes requires addressing several key areas. These include linguistic diversity, infrastructure integration, data security, internet connectivity reliability, user acceptance, and cultural sensitivities. Overcoming these obstacles necessitates a comprehensive approach involving technological innovation, stakeholder engagement, capacity building, and ongoing evaluation.

Assessing the Impact of Chatbots on Agricultural Information Delivery through FCB

The research chapter aims to examine how chatbots have influenced the delivery of agricultural information within the Farmer ChatBot (FCB) ecosystem and to what extent they have enhanced farmers' access to crucial information. The assessment may involve analyzing various metrics such as user engagement, response time, information accuracy, and user satisfaction to gauge the impact of chatbots on the efficiency and effectiveness of agricultural information delivery. Through this investigation, researchers can gain insights into the role of chatbots in transforming agricultural advisory services and informing decision-making processes among farmers.

Recent studies highlight the potential of deep learning and machine learning algorithms in classifying different plant species, showcasing advancements that could further optimize chatbot capabilities (Chouhan et al., 2024a). Moreover, computer vision and drone technology innovations offer promising applications in enhancing agricultural practices, underscoring the evolving landscape of AI in agriculture (Chouhan et al., 2024b; Kumar et al., 2023).

Exploring User Perceptions and Experiences of Chatbots in FCB for Agricultural Support

This point outlines the research focus of examining user perceptions and experiences regarding the utilization of chatbots within the Farmer ChatBot (FCB) platform for agricultural inquiries and support. It highlights the aim of delving into farmers' and stakeholders' attitudes, opinions, and satisfaction levels with chatbots, aiming to understand their overall experiences. Through this inquiry, researchers aim to gain insights into how users perceive chatbots within the Farmer ChatBot (FCB) framework and assess their satisfaction levels with the use of chatbots for agricultural assistance. Recent research on leaf disease segmentation and classification using computer vision provides further context on technological advancements influencing user interactions with chatbots (Chouhan et al., 2021; Khajuria et al., 2018).

Research Results

The development and implementation of the FarmerBot marks a significant leap in the integration of Artificial Intelligence (AI) and Machine Learning (ML) into agricultural practices. By utilizing advanced machine learning techniques, particularly TensorFlow, the chatbot is designed to provide real-time assistance to farmers by analyzing vast agricultural datasets. The system processes various queries related to farming, such as crop management, pest control, weather forecasting, and soil health, ensuring that farmers receive timely and relevant information to make informed decisions. A key feature of FarmerBot is its multilingual capability, leveraging the Google Translate API to make the chatbot accessible to farmers across diverse linguistic backgrounds, thereby bridging communication gaps and ensuring inclusivity. Furthermore, FarmerBot integrates Entity Recognition to accurately extract relevant details like crop types or pest issues from user queries, which helps in delivering tailored responses. The bot's performance is regularly evaluated using precision, recall, and accuracy metrics, ensuring high-quality outputs and minimizing errors. This continuous improvement process, which includes retraining the models with updated agricultural data, is vital for adapting to evolving farming trends and challenges. With the potential to incorporate additional features such as voice recognition, FarmerBot is poised to enhance user engagement and offer personalized, actionable insights to improve crop productivity, thereby empowering farmers to optimize their yields and improve their livelihoods. This innovation exemplifies how AI-driven solutions can revolutionize agricultural practices, making them more efficient, accessible, and sustainable.

The programming techniques employed in the development of FarmerBot combine a variety of state-of-the-art methodologies from natural language processing (NLP) and machine learning. At the core of the system is a deep learning model built with Keras and TensorFlow, enabling the chatbot to understand and classify user intents accurately. For query processing, CountVectorizer is used to convert text data into numerical representations, allowing the machine learning model to interpret and classify the input effectively. Additionally, Porter Stemmer from the NLTK library is utilized to preprocess user inputs by stemming words, thus standardizing variations of words to their root form for more efficient processing. The chatbot also leverages Entity Recognition to extract specific agricultural entities, such as crop types or pest names, which helps in personalizing responses. The integration of Google Translate API ensures seamless multilingual support, expanding accessibility for farmers in various regions. The overall system employs modular code architecture, where each function and process is encapsulated to ensure scalability and maintainability. The combination of these techniques allows FarmerBot to operate efficiently, provide accurate responses, and adapt to different languages and farming contexts, showcasing the power of modern AI and programming techniques in solving real-world problems.

A farmer query-related word cloud (Fig. 4) provides a visual representation of the most common and frequently asked terms by farmers when seeking assistance or

Fig. 4. Farmer ChatBot system – Word Cloud.

information related to agriculture. By analyzing farmer queries, such as those collected from sources like the Farmer ChatBot, the word cloud highlights critical topics like crop management, soil health, weather patterns, pest control, fertilizer usage, market prices, and government schemes. The size of each word in the cloud correlates with its frequency of occurrence, offering insights into the key challenges and concerns that farmers face. This word cloud serves as an effective tool to understand the most pressing agricultural issues, allowing developers and agricultural experts to focus on providing targeted solutions and responses that address the specific needs of farmers, thereby promoting more efficient and sustainable farming practices.

Conclusion

The integration of chatbots within the FCB framework presents significant opportunities to enhance agricultural support services in India. Key features such as multilingual support, audio/text queries, and language translation improve accessibility and user experience. However, challenges such as linguistic diversity, technical expertise, and user acceptance must be addressed for effective deployment. Despite these challenges, chatbots show promise in transforming agricultural advisory services, and empowering farmers with timely and relevant information. Future research should focus on addressing implementation challenges and evaluating the impact of chatbots on agricultural information delivery in FCB. Through continued innovation and stakeholder collaboration, chatbots can play a pivotal role in advancing agricultural productivity and supporting rural livelihoods in India.

Declaration

Ethical Responsibilities of Authors

"All authors have read, understood, and have complied as applicable with the statement on "Ethical responsibilities of Authors" as found in the Instructions for Authors".

Ethical Approval

Not applicable.

Use of AI Tool

Example 1: The corresponding author declares that no AI tool has been used to prepare this manuscript.

Conflict of Interest

On behalf of all the authors the corresponding author states that there is no conflict of interest to disclose.

Competing Interests

Not applicable.

Authors' contributions

Author[1]: Conceptualization, Methodology, Writing - Original Draft, Writing - Review & Editing, **Author[2]**: Investigation, Formal analysis, Visualization, Software, Supervision.

Funding

No funding has been received.

References

Chandolikar, N., Dale, C., Koli, T., Singh, M. and Narkhede, T. 2022, March. Agriculture assistant Chatbot using Artificial Neural Network. *In 2022 International Conference on Advanced Computing Technologies and Applications (ICACTA)* (pp. 1–5). IEEE.

Chouhan, S.S., Singh, U.P. and Jain, S. 2024. Introduction to computer vision and drone technology. *In*: Chouhan, S.S., Singh, U.P. and Jain, S. (eds.). *Applications of Computer Vision and Drone Technology in Agriculture 4.0.* Springer, Singapore. https://doi.org/10.1007/978-981-99-8684-2_1.

Chouhan, S.S., Singh, U.P., Sharma, U. and Jain, S. 2021. Leaf disease segmentation and classification of Jatropha Curcas L. and Pongamia Pinnata L. biofuel plants using computer vision based approaches. *Measurement*, 171: 108796, ISSN 0263-2241. https://doi.org/10.1016/j.measurement.2020.108796.

Chouhan, S.S., Singh, U.P., Sharma, U. et al. 2024. Classification of different plant species using Deep Learning and Machine Learning algorithms. *Wireless Pers. Commun.*, 136: 2275–2298. https://doi.org/10.1007/s11277-024-11374-y.

Jain, N., Jain, P., Kayal, P., Sahit, J., Pachpande, S. and Choudhari, J. 2019. AgriBot: Agriculture-specific question answer system.

Khajuria, R., Sunanda and Singh, S. 2018. A survey of existing leaf disease techniques using Artificial Neural Network. *International Journal of Computer Trends and Technology (IJCTT)*, 59(1): 52–62, May 2018. ISSN:2231-2803, doi:10.14445/22312803/IJCTT-V59P109, (https://ijcttjournal.org/archives/ijctt-v59p109).

Kiruthika, U., Raja, S.K.S., Balaji, V. and Raman, C.J. 2020. E-agriculture for direct marketing of food crops using Chatbots. *2020 International Conference on Power, Energy, Control and Transmission Systems (ICPECTS), Chennai, India*, pp. 1–4, doi: 10.1109/ICPECTS49113.2020.9337024.

Kumar, R. et al. 2024. Hybrid Approach of Cotton Disease Detection for Enhanced Crop Health and Yield. In *IEEE Access*, https://ieeexplore.ieee.org/document/10589357.

Lamba, T. and Keshri, R. 2022, August. Chatbot farming system. In *Proceedings of the Third International Conference on Information Management and Machine Intelligence: ICIMMI 2021* (pp. 311–318). Singapore: Springer Nature Singapore.

Maduri, P.K., Dhiman, P., Shukla, M.R., Anand, S. and Singh, S.P. 2021, December. Farmers agriculture assistance Chatbot. In *2021 3rd International Conference on Advances in Computing, Communication Control and Networking (ICAC3N)* (pp. 1884–1889). IEEE.

Maduri, P.K., Dhiman, P., Shukla, M.R., Anand, S. and Singh, S.P. 2021, December. Farmers agriculture assistance Chatbot. In *2021 3rd International Conference on Advances in Computing, Communication Control and Networking (ICAC3N)* (pp. 1884–1889). IEEE.

Momaya, M., Khanna, A., Sadavarte, J. and Sankhe, M. 2021, June. Krushi–the farmer chatbot. In *2021 International Conference on Communication information and Computing Technology (ICCICT)* (pp. 1–6). IEEE.

Niranjan, P.Y., Rajpurohit, V.S. and Malgi, R. 2019, July. A survey on chat-bot system for agriculture domain. In *2019 1st International Conference on Advances in Information Technology (ICAIT)* (pp. 99–103). IEEE.

Sawant, D., Jaiswal, A., Singh, J. and Shah, P. 2019, July. AgriBot-An intelligent interactive interface to assist farmers in agricultural activities. In *2019 IEEE Bombay Section Signature Conference (IBSSC)* (pp. 1–6). IEEE.

Sivakumar, S.A., Shankar, B.M., Anuradha, B., Karan, K.A., Karthik, A., Karthik, R. and Kumar, J.R.R. 2023. Artificial Intelligence based agricultural Chatbot and virtual assistant for delivery of harvested crops. *International Journal of Intelligent Systems and Applications in Engineering*, 11(8s): 576–583.

Chapter 13

Federated Learning

A Game-Changer in Agricultural Decision-Making and Precision Farming

Sisir Kumar Rajbongshi,[1,*] *Rajdeep Ghosh,*[1] *Kshirod Sarmah*[1] and *Satyajit Sarmah*[2]

Introduction

Agriculture, the backbone of civilization, has undergone remarkable transformations over the centuries, driven by technological advancements aimed at enhancing productivity, sustainability, and resilience. In the contemporary landscape, the convergence of data science and agriculture has sparked a new era of innovation, promising unprecedented insights into crop management, resource allocation, and decision-making processes. At the forefront of this revolution lies Federated Learning (FL), a decentralized machine learning paradigm poised to revolutionize agricultural practices and usher in a new era of precision farming.

The traditional approach to agricultural data analysis often entails centralizing vast datasets from multiple sources for analysis, presenting significant challenges related to data privacy, security, and scalability. However, in an era characterized by heightened concerns surrounding data protection and privacy, centralization poses

[1] Department of Computer Science, Pandit Deendayal Upadhyaya Adarsha Mahavidyalaya, Amjonga, Goalpara, Assam, India, Pin: 783124. https://orcid.org/0000-0003-1874-6882, https://orcid.org/0000-0002-8045-0137, https://orcid.org/0009-0006-8780-5689.

[2] Department of Information Technology, Gauhati University, Guwahati-781014. https://orcid.org/0000-0002-4032-5278.

* Corresponding author: sisirrajbongshi@gmail.com

formidable barriers to collaboration and knowledge sharing among stakeholders. Federated Learning offers a compelling solution to this conundrum by enabling collaborative model training across distributed data sources while preserving data privacy and security.

In essence, Federated Learning operates on the principle of decentralized model training, where individual data sources, such as farms, agricultural research institutions, and IoT sensors, retain control over their respective data. Instead of aggregating data in a central repository, FL leverages local computation to train machine learning models collaboratively. This distributed approach not only addresses concerns regarding data privacy and security but also minimizes communication overhead and reduces the reliance on centralized infrastructure.

The application of Federated Learning in agriculture holds immense promise across various domains, ranging from crop yield prediction and disease detection to resource optimization and environmental monitoring. By leveraging the collective intelligence embedded within distributed datasets, FL empowers agricultural stakeholders to make informed decisions, optimize resource allocation, and enhance productivity while minimizing environmental impact.

One of the key areas where Federated Learning demonstrates its transformative potential is in precision farming, a data-driven approach to agricultural management that aims to optimize inputs, maximize yields, and minimize waste. Traditionally, precision farming relies on the integration of data from various sources, including satellite imagery, weather forecasts, soil sensors, and crop yield monitors, to generate actionable insights for farmers. However, the centralized nature of data analysis in conventional precision farming models often poses challenges in terms of data access, privacy, and scalability.

Federated Learning offers a compelling alternative to traditional precision farming methods by enabling collaborative model training across distributed data sources. By leveraging FL, farmers can harness the collective intelligence embedded within local datasets to improve the accuracy and robustness of predictive models while preserving data privacy and security. This decentralized approach not only enhances the scalability and efficiency of precision farming systems but also fosters greater collaboration and knowledge sharing among agricultural stakeholders.

In this chapter, we explore the transformative potential of Federated Learning as a game-changer in agricultural decision-making and precision farming. We provide an overview of the fundamental principles of FL and its application in agricultural contexts, highlighting its capacity to optimize crop management practices, improve yield forecasts, and mitigate environmental impact. To further investigate the potential and application of Federated Learning in agriculture, we propose the following research questions:

i. **Data Privacy and Security:**
    - How can Federated Learning mitigate data privacy concerns in agricultural data sharing?
    - What are the key security challenges associated with implementing FL in agricultural ecosystems?

ii. **Implementation and Scalability:**
   - What are the practical challenges of deploying Federated Learning across diverse agricultural settings?
   - How can the scalability of Federated Learning systems be enhanced to support large-scale agricultural operations?

iii. **Performance and Accuracy:**
   - How does the performance of Federated Learning models compare to traditional centralized models in predicting crop yields and detecting diseases?
   - What techniques can be employed to improve the accuracy and robustness of FL models in the context of agriculture?

iv. **Collaboration and Knowledge Sharing:**
   - How can Federated Learning facilitate greater collaboration and knowledge sharing among farmers, researchers, and agricultural stakeholders?
   - What are the barriers to adopting FL in collaborative agricultural research and how can they be overcome?

v. **Environmental and Economic Impact:**
   - How can Federated Learning contribute to sustainable farming practices and reduce the environmental impact of agriculture?
   - What economic benefits can be realized through the adoption of Federated Learning in precision farming?

By addressing these questions, this research aims to provide a comprehensive understanding of the potential, challenges, and benefits of Federated Learning in transforming agricultural practices and promoting sustainable farming.

Literature Review

Machine learning (ML) uses data and algorithms to learn, and gradually improve the performance of a system as suggested by Samuel (1959) and Alpaydin (2020). Benos (2021) suggested that machine learning methods can be applied in agriculture allowing for more efficient and precise farming since machine learning can help establish knowledge-based farming systems. ML algorithms can analyze large datasets of agricultural data, such as crop yields, soil, and weather data, to identify patterns and gain knowledge of agricultural data. Liakos (2018) analyzed the use of ML for different agricultural areas: water management, soil management, livestock production and management, and crop management. Deep learning (DL) is a subset of machine learning (ML). LeCun (2015) and Schmidhuber (2022) recommended that DL extends classical ML by incorporating depth and complexity into models, which enables learning through multiple levels of abstraction. DL consists of multiple processing layers that enable hierarchical learning as suggested by Schmidhuber (2015). Jain (1999) in turn proposed that recurrent neural networks (RNNs) model temporal dynamic behavior. They contain loops, and connections between nodes create cycles, allowing the outputs from some nodes to affect subsequent inputs to the same nodes. A comparison of the classification and regression performance of deep learning in agriculture with other existing popular techniques is presented by

Kamilaris (2018). Their findings prove that deep learning significantly improves the performance of classification and prediction problems, providing high accuracy.

Wójtowicz (2016) stated that agriculture generates significant real-time data through low-cost and low-energy-consuming sensors and devices. Ground surveys in agriculture are expensive and typically cover only small areas or a selection of sample farms. In contrast, remote sensing offers cost-effective monitoring in agriculture. These devices generate real-time data and rely on open internet communication. This also leads to minimizing the risks associated with security and data privacy violations. Kaur (2022) stated that data privacy is also crucial for farmers, not only for large organizations.

One solution to the data security problem involves cooperation among multiple data owners to train and use shared machine learning models while preserving all training data locally. This is achievable with an emerging machine learning technology called federated learning (FL) as stated by Konečný (2016). FL was introduced by McMahan (2017), and the first aggregation algorithm, FedAvg, was introduced by Kairouz (2021). Mothukuri 2021 suggested that FL is a type of machine learning in which different clients or users collaborate to train a model while maintaining decentralized control of their data. It also reduces the risk of privacy violations as stated by Zhang (2016). FL trains machine learning models across multiple decentralized clients using only local data samples, avoiding the need to exchange data with other clients or a central server.

Chouhan (2024a) explores computer vision and drone technology in agriculture, covering concepts, theories, applications, and future advancements. It aims to provide readers with fundamental insights and practical adaptability of these technologies, starting with computer vision methodologies and then introducing drone technology. In a significant advancement for agricultural technology, Chouhan (2024b) has developed a deep learning model aimed at the real-time classification of different plant species. Leveraging the Internet of Things (IoT) practices, the model efficiently identifies and classifies plant species from images captured in real-time. Kumar (2024) presents a machine learning-based system for early disease detection in cotton crops through leaf image analysis. By evaluating various models, including Random Forest, SVM, and an Ensemble model, the research aims to enhance crop yield and quality by accurately classifying cotton leaves as "Healthy" or "Diseased". Efforts to produce bio-fuel from plants like Jatropha Curcas L. and Pongamia Pinnata L. face challenges due to biotic factors affecting plant growth. To address this, a Computer Vision methodology using a Hybrid Neural Network with Superpixel clustering for disease diagnosis is proposed by Chouhan (2021). The system achieved high accuracy in disease classification, with segmentation results showing average specificity of 0.9534 and 0.9795, sensitivity of 0.9637 and 0.9805, and classification accuracy of 0.9857, 0.9095, and 0.9607 for both plants.

Agricultural productivity is crucial for the Indian economy, but leaf diseases significantly impact farmers. Traditional disease detection methods are inefficient at larger scales, making automated techniques essential for efficiency and accuracy. Khajuria (2018) highlights the advancements in artificial neural networks (ANNs) and machine learning for leaf disease detection, segmentation, and identification, emphasizing the improvements brought by modern sensors and imaging techniques in enhancing ANN models.

Application of FL in Agriculture

Analyzing the application of Federated Learning (FL) in agricultural contexts involves examining its potential, challenges, opportunities, and implications for various aspects of farming practices. In this analysis, we delve into FL's role in enhancing crop management, improving yield predictions, optimizing resource allocation, mitigating environmental impact, and addressing privacy concerns in agricultural data analysis.

- **Enhancing Crop Management:**

Federated Learning offers a decentralized approach to crop management by leveraging insights from distributed datasets without compromising data privacy. Farmers can collaborate with agricultural researchers, extension services, and other stakeholders to collectively analyze crop performance, identify trends, and make informed decisions regarding planting strategies, pest control measures, and irrigation management. FL facilitates the integration of data from diverse sources, including weather forecasts, soil sensors, satellite imagery, and historical crop yield data, enabling farmers to optimize inputs, maximize yields, and minimize risks associated with environmental variability and climate change.

- **Improving Yield Predictions:**

The predictive capabilities of Federated Learning can revolutionize yield forecasting in agriculture. By aggregating data from multiple farms, regions, and agro-climatic zones, FL models can capture the complex interactions between environmental factors, crop genetics, and management practices, enabling more accurate and reliable yield predictions. Farmers can use these predictions to anticipate market demands, plan harvest schedules, and optimize post-harvest logistics, thereby improving profitability and reducing food waste. Moreover, FL facilitates continuous learning and adaptation, allowing predictive models to evolve as new data becomes available, leading to more robust and reliable yield forecasts.

- **Optimizing Resource Allocation:**

Federated Learning enables farmers to optimize resource allocation by providing insights into the spatial and temporal variability of soil fertility, moisture levels, and crop health across their fields. By integrating data from soil sensors, drones, and IoT devices, FL models can generate spatially explicit maps of soil properties, nutrient availability, and pest infestations, enabling precision application of fertilizers, pesticides, and water resources. This targeted approach to resource management not only improves crop yields and quality but also minimizes input costs, reduces environmental impact, and enhances sustainability in agriculture.

- **Mitigating Environmental Impact:**

Federated Learning has the potential to mitigate environmental impact by promoting sustainable farming practices and reducing resource wastage. By optimizing irrigation scheduling, minimizing chemical inputs, and adopting precision agriculture techniques, farmers can conserve water, reduce soil erosion, and minimize pollution runoff, thereby preserving ecosystem health and biodiversity. FL models can also support conservation efforts by identifying areas of high ecological value, prioritizing habitat restoration projects, and monitoring the effectiveness of land management practices in mitigating climate change impacts.

- **Addressing Privacy Concerns:**

One of the key advantages of Federated Learning is its ability to address privacy concerns associated with agricultural data analysis. By keeping data localized and performing model training on-device or at the edge, FL ensures that sensitive information remains secure and confidential. Farmers retain full control over their data, deciding which insights to share with collaborators while preserving the privacy of proprietary information. This decentralized approach to data analysis not only enhances trust and transparency among stakeholders but also fosters greater participation and collaboration in agricultural research and innovation.

In conclusion, Federated Learning has the potential to revolutionize agricultural decision-making and precision farming by leveraging distributed data sources, preserving privacy, and promoting collaboration among stakeholders. By enhancing crop management, improving yield predictions, optimizing resource allocation, mitigating environmental impact, and addressing privacy concerns, FL can contribute to sustainable agriculture and food security in a rapidly changing world. However, realizing the full potential of FL in agriculture requires addressing technical challenges, fostering interdisciplinary collaboration, and promoting the adoption of decentralized data analysis approaches across the agricultural value chain.

Challenges

The profound impact of Federated Learning (FL) as a catalyst for agricultural innovation stems from its ability to address critical challenges facing the agricultural sector while unlocking new opportunities for efficiency, sustainability, and resilience. FL revolutionizes traditional approaches to data analysis by enabling collaborative model training across distributed datasets while preserving data privacy and security. This decentralized paradigm fosters greater collaboration and knowledge sharing among agricultural stakeholders, leading to transformative advancements in various aspects of farming practices:

- **Data Privacy and Security:**

FL addresses concerns surrounding data privacy and security by keeping sensitive information localized and performing model training on-device or at the edge. This ensures that proprietary data remains secure and confidential, enhancing trust and transparency among stakeholders. By preserving data privacy, FL facilitates greater participation and collaboration in agricultural research and innovation, leading to the development of more robust and reliable predictive models for crop management, yield forecasting, and resource allocation.

- **Collaborative Decision-Making:**

FL enables collaborative decision-making among farmers, researchers, extension services, and other stakeholders by facilitating the aggregation of insights from diverse datasets without the need for centralizing sensitive information. This collaborative approach fosters greater synergy and coordination in addressing complex agricultural challenges, such as pest outbreaks, soil degradation, and water scarcity. By leveraging collective intelligence, FL empowers agricultural stakeholders to make informed decisions, optimize resource allocation, and enhance productivity while minimizing environmental impact.

- **Precision Agriculture:**

FL holds immense promise for precision agriculture, a data-driven approach to farming that aims to optimize inputs, maximize yields, and minimize environmental impact. By integrating data from various sources, including weather forecasts, soil sensors, satellite imagery, and historical crop yield data, FL models can generate spatially explicit maps for precision application of fertilizers, pesticides, and water resources. This targeted approach to resource management improves crop yields and quality while minimizing input costs and environmental risks, leading to more sustainable and resilient farming practices.

- **Real-Time Data Analysis:**

FL facilitates real-time data analysis and adaptive decision support systems in agriculture by enabling model training on distributed datasets in near real-time. This enables farmers to respond rapidly to changing environmental conditions, market demands, and pest outbreaks, enhancing resilience and agility in agricultural production systems. By harnessing the power of FL, farmers can make timely and informed decisions regarding planting strategies, pest control measures, and irrigation management, leading to improved productivity and profitability.

- **Innovation and Research:**

FL drives innovation and research in agriculture by providing a platform for collaborative experimentation, data analysis, and model development. Agricultural researchers can leverage FL to access diverse datasets from multiple sources, enabling them to conduct large-scale studies, develop predictive models, and identify novel insights into crop genetics, environmental interactions, and management practices. This collaborative approach accelerates the pace of innovation in agriculture, leading to the development of new technologies, practices, and solutions for addressing emerging challenges and opportunities in farming.

- **Scalability and Adaptability:**

FL offers scalability and adaptability in agricultural data analysis by accommodating the diverse and dynamic nature of farming systems. The decentralized nature of FL allows for the integration of data from a wide range of sources, including smallholder farms, large-scale operations, and research institutions, enabling comprehensive analyzes of agricultural landscapes at various scales. This scalability facilitates the development of tailored solutions and recommendations for farmers based on their specific needs, resources, and environmental conditions. Additionally, FL's adaptability enables models to evolve as new data becomes available, ensuring that predictive insights remain relevant and reliable in the face of changing environmental conditions, market dynamics, and technological advancements.

- **Resilience to Data Silos and Connectivity Issues:**

FL addresses challenges associated with data silos and connectivity issues in agricultural data analysis by enabling model training on distributed datasets without the need for centralized infrastructure or continuous internet connectivity. This resilience allows farmers in remote or resource-constrained areas to participate in collaborative data analysis efforts, contributing valuable insights to collective decision-making processes. Additionally, FL's ability to operate in low-bandwidth or intermittent connectivity environments ensures continuity and reliability in data analysis, even in areas with limited access to high-speed internet or advanced computing resources.

- **Empowerment of Small-holder Farmers:**

FL empowers small-holder farmers by providing access to cutting-edge technologies and data-driven insights previously only available to larger agricultural enterprises. By participating in collaborative data analysis efforts facilitated by FL, small-holder farmers can gain access to valuable information, resources, and market opportunities that enhance their productivity, profitability, and resilience. Additionally, FL enables small-holder farmers to contribute their localized knowledge and expertise to collective decision-making processes, fostering greater inclusiveness and equity in agricultural innovation efforts.

- **Adoption of Sustainable Farming Practices:**

FL promotes the adoption of sustainable farming practices by providing farmers with actionable insights and recommendations for optimizing resource use, minimizing environmental impact, and enhancing ecosystem resilience. By integrating data from environmental sensors, satellite imagery, and historical records, FL models can identify opportunities for improving soil health, conserving water, reducing greenhouse gas emissions, and preserving biodiversity. This data-driven approach to sustainability enables farmers to make informed decisions that balance economic viability with environmental stewardship, ensuring the long-term viability of agricultural production systems.

- **Global Collaboration and Knowledge Sharing:**

FL fosters global collaboration and knowledge sharing in agriculture by facilitating the exchange of data, expertise, and best practices among farmers, researchers, policymakers, and agricultural organizations worldwide. Through collaborative data analysis efforts facilitated by FL, stakeholders can leverage the collective wisdom of diverse perspectives and experiences to address shared challenges, such as food security, climate change adaptation, and rural development. This global collaboration accelerates the pace of agricultural innovation, enabling the rapid dissemination and adoption of transformative technologies and practices that benefit farmers and communities around the world.

In conclusion, Federated Learning serves as a catalyst for agricultural innovation by addressing critical challenges, fostering collaboration, enabling precision agriculture, facilitating real-time data analysis, and driving innovation and research.

By leveraging the power of FL, agricultural stakeholders can unlock new opportunities for efficiency, sustainability, and resilience in farming practices, leading to a more productive, profitable, and sustainable agricultural sector.

Conceptual Model: FL as a Catalyst for Agricultural Innovation

Let's illustrate a conceptual model that depicts how Federated Learning (FL) catalyzes agricultural innovation:

- **Input Data:**

Diverse datasets: These may include data collected from farms, such as crop yield records, soil nutrient levels, and irrigation schedules. Additionally, data from external sources like weather forecasts, satellite imagery, and market prices can provide valuable insights.

Examples: Soil sensors installed in fields collect real-time data on soil moisture and nutrient levels. Weather stations provide meteorological data such as temperature, humidity, and rainfall. Satellite imagery offers high-resolution images for monitoring crop health and detecting changes in land use over time.

- **Federated Learning Framework:**

Decentralized model training: Federated Learning allows machine learning models to be trained collaboratively on local devices without centralizing sensitive data. Each device contributes its insights to the model while keeping data secure and private.

Examples: Farmers' smartphones or IoT devices run machine learning algorithms to analyze local data, such as crop images or sensor readings, and send model updates to a central server. The server aggregates these updates to improve the overall model without accessing individual farmers' data.

- **Model Training and Optimization:**

Collaborative analysis: Federated Learning enables models to extract insights from distributed datasets, capturing patterns and correlations while respecting data privacy constraints.

Examples: Machine learning models trained through Federated Learning can predict crop yields based on historical data from multiple farms, identify pest infestations by analyzing images of affected crops, or optimize irrigation schedules using soil moisture sensor readings.

- **Predictive Analytics and Decision Support:**

Forecasting and recommendations: Predictive models generated through Federated Learning provide farmers with actionable insights and recommendations for optimizing farming practices.

Examples: A predictive model trained on data from multiple farms can forecast crop yields for the upcoming season, allowing farmers to adjust planting strategies or allocate resources accordingly. Decision support systems powered by Federated Learning can recommend specific pest control measures based on localized data, reducing the need for indiscriminate pesticide use.

- **Feedback Loop and Continuous Learning:**

Adaptation to changing conditions: Federated Learning models evolve based on new data and insights, ensuring their relevance and reliability in dynamic agricultural environments.

Examples: As new data becomes available, such as updated weather forecasts or soil sensor readings, Federated Learning models retrain themselves to incorporate these insights. For example, a yield prediction model may adjust its forecasts based on real-time weather data to account for unexpected weather events like droughts or storms.

- **Impact on Agricultural Innovation:**

Addressing challenges: Federated Learning drives innovation in agriculture by addressing critical challenges such as data privacy, resource optimization, and sustainability.

Examples: Precision agriculture techniques enabled by Federated Learning, such as variable rate irrigation or targeted pest management, optimize resource use and reduce environmental impact. Sustainable farming practices, such as no-till farming or cover cropping, are promoted through data-driven insights generated by Federated Learning models.

- **Empowerment of Stakeholders:**

Access to technology and insights: Federated Learning empowers farmers, researchers, policymakers, and agricultural organizations by providing access to cutting-edge technologies and data-driven insights.

Examples: Smallholder farmers in remote areas can benefit from Federated Learning models that provide localized recommendations for improving crop yields or managing pest outbreaks. Researchers can collaborate on data analysis efforts facilitated by Federated Learning to develop new solutions for addressing global agricultural challenges.

- **Global Collaboration and Knowledge Sharing:**

Exchange of data and expertise: Federated Learning fosters global collaboration and knowledge sharing in agriculture, enabling stakeholders to share data, expertise, and best practices.

Examples: Agricultural organizations collaborate on international research projects facilitated by Federated Learning to develop region-specific solutions for improving crop resilience or mitigating climate change impacts. Farmers participate in global knowledge-sharing networks that leverage Federated Learning to exchange insights and experiences on sustainable farming practices.

These examples illustrate how Federated Learning serves as a catalyst for agricultural innovation by enabling collaborative data analysis, fostering global collaboration, and empowering stakeholders to make informed decisions that enhance productivity, sustainability, and resilience in farming practices.

To minimize environmental footprints through dynamic data analysis and adaptive decision-making frameworks, we need to integrate real-time data from various sources, analyze it dynamically, and adapt our decisions accordingly to optimize resource use and reduce negative environmental impacts. This process involves formulating mathematical statements that model environmental factors, defining objectives for minimizing environmental footprints and developing algorithms that dynamically adjust decisions based on incoming data.

Let's break down the key components and provide mathematical statements for each:

- **Modelling Environmental Factors:**

Mathematical statements: Environmental factors such as water usage, energy consumption, greenhouse gas emissions, and soil erosion can be modeled using mathematical equations. For example, the impact of irrigation on water usage can be represented by a function that relates irrigation levels to crop water requirements and soil moisture content.

Objective: The objective is to minimize the environmental impact of agricultural activities, such as reducing water usage, energy consumption, and emissions while maximizing resource efficiency and sustainability.

- **Defining Objectives:**

Mathematical statements: Objectives can be formulated as optimization problems that minimize environmental footprints subject to constraints on resource availability, regulatory requirements, and economic feasibility. For example, the objective function may minimize the total water usage subject to constraints on crop yield, soil moisture levels, and irrigation capacity.

Objective function: The objective function quantifies the environmental impact of agricultural activities, such as water usage, energy consumption, and emissions, using mathematical expressions that reflect their contributions to environmental footprints.

- **Developing Adaptive Decision-Making Algorithms:**

Mathematical statements: Adaptive decision-making algorithms dynamically adjust agricultural practices based on real-time data and environmental conditions. These algorithms use mathematical models to predict future outcomes, evaluate the environmental impact of different actions, and optimize decisions to minimize environmental footprints.

- **Integration of Real-Time Data:**

Mathematical statements: Real-time data from sensors, satellites, weather stations, and other sources are integrated into mathematical models to update environmental parameters, predict future states, and optimize decisions. These models use mathematical techniques such as data assimilation, state estimation, and parameter identification to incorporate new data into existing models.

Hurdles and Prospects

Implementing Federated Learning (FL) in agricultural landscapes presents both hurdles and prospects. These challenges revolve around data diversity, privacy dilemmas, connectivity issues, and the need for infrastructure development. However, the potential benefits include improved data analysis, enhanced privacy protection, and increased collaboration among stakeholders. Let's explore these aspects in more detail:

- **Data Diversity:**

Hurdles: Agricultural data is often diverse, coming from various sources such as farms, research institutions, weather stations, and satellite imagery. Integrating these heterogeneous datasets poses challenges in terms of data standardization, interoperability, and quality assurance. Additionally, data may be sparse or incomplete, particularly in remote or resource-constrained areas, making it difficult to train robust machine learning models.

Prospects: FL enables collaborative model training across distributed datasets, allowing for the aggregation of insights from diverse sources while preserving data privacy. By leveraging FL, stakeholders can harness the collective intelligence embedded within localized datasets to improve data analysis and decision-making in agriculture.

- **Privacy Dilemmas:**

Hurdles: Agricultural data is often sensitive and proprietary, containing information about crop yields, soil properties, farm management practices, and market strategies. Sharing this data for collaborative analysis raises concerns about data privacy, security, and confidentiality. Farmers may be reluctant to participate in FL initiatives if they perceive a risk of data misuse, unauthorized access, or loss of competitive advantage.

Prospects: FL addresses privacy dilemmas by keeping data localized and performing model training on-device or at the edge. This ensures that sensitive information remains secure and confidential, enhancing trust and transparency among stakeholders. FL also allows farmers to retain control over their data, deciding which insights to share with collaborators while preserving the privacy of proprietary information.

- **Connectivity Issues:**

Hurdles: Rural areas often lack robust internet connectivity, hindering the adoption of FL and other data-driven technologies in agriculture. Limited access to high-speed internet, mobile networks, and computing infrastructure prevents farmers from participating in collaborative data analysis efforts and accessing real-time insights.

Prospects: FL operates in low-bandwidth or intermittent connectivity environments, making it suitable for agricultural landscapes with limited internet access. By performing model training on-device or at the edge, FL reduces reliance on centralized infrastructure and enables farmers to analyze data locally, even in remote areas with poor connectivity.

- **Infrastructure Development:**

Hurdles: Implementing FL in agricultural landscapes requires investment in infrastructure, including hardware, software, and data management systems. Farmers may lack the technical expertise and financial resources to deploy FL algorithms, acquire sensor technologies, and maintain computing infrastructure.

Prospects: Governments, research institutions, and agricultural organizations can support infrastructure development initiatives through funding, capacity-building programs, and technology transfer initiatives. By investing in FL infrastructure, stakeholders can democratize access to data-driven technologies, empower farmers, and promote sustainable agricultural development.

Conclusions

In conclusion, integrating FL in agricultural settings brings both challenges and opportunities. While problems such as data diversity, privacy concerns, connectivity constraints, and infrastructure development must be addressed, FL provides an opportunity for better data analysis, more privacy protection, and expanded stakeholder participation. By overcoming these barriers and capitalizing on FL's promise, stakeholders can harness the power of data-driven technology to address urgent issues and open up new prospects for agricultural innovation.

Declaration

Ethical Responsibilities of Authors

"All authors have read, understood, and have complied as applicable with the statement on "Ethical responsibilities of Authors" as found in the Instructions for Authors".

Ethical Approval

Not applicable.

Use of AI Tool

The corresponding author declares that no AI tool has been used to prepare this manuscript.

Conflict of Interest

On behalf of all the authors the corresponding author states that there is no conflict of interest to disclose.

Competing Interests

Not applicable.

Authors' Contributions

Author[1]: Conceptualization, Methodology, Writing - Original Draft, **Author[2]**: Writing - Review & Editing, Investigation, Formal analysis, Visualization **Author[3]**: Writing - Original Draft, Formal analysis, Methodology, **Author[4]:** Formal analysis, Supervision.

Funding

No funding has been received.

References

Alpaydin, E. 2020. *Introduction to Machine Learning*. MIT Press, Massachusetts, USA.

Benos, L., Tagarakis, A.C., Dolias, G., Berruto, R., Kateris, D. and Bochtis, D. 2021. Machine Learning in agriculture: A comprehensive updated review. *Sensors*, 21: 1–55.

Chouhan, S.S., Singh, U.P., Sharma, U. and Jain, S. 2021. Leaf disease segmentation and classification of Jatropha Curcas L. and Pongamia Pinnata L. biofuel plants using computer vision based approaches. *Measurement*, 171: 1–10.

Chouhan, S.S., Singh, U.P. and Jain, S. 2024a. Introduction to computer vision and drone technology. pp. 1–5. *In*: Chouhan, S.S., Singh, U.P. and Jain, S. (eds.). *Applications of Computer Vision and Drone Technology in Agriculture 4.0*. Springer, Singapore.

Chouhan, S.S., Singh, U.P., Sharma, U. and Jain, S. 2024b. Classification of different plant species using deep learning and machine learning algorithms. *Wireless Personal Communications*, 136: 2275–2298.

Jain, L.C. and Medsker. L.R. 1999. *Recurrent Neural Networks: Design and Applications*. CRC Press, Florida, USA.

Kairouz, P., McMahan, H.B. Avent, B., Bellet, A., Bennis, M., Bhagoji, A.N., Bonawit, K., Charles, Z., Cormode, G., Cummings, R. et al. 2021. Advances and open problems in federated learning. *Foundations and Trends in Machine Learning*, 14: 1–210.

Kamilaris, A. and Prenafeta-Boldú. F.X. 2018. Deep learning in agriculture: A survey. *Computers and Electronics in Agriculture*, 147: 70–90.

Kaur, J., Fard, S.M.H., Amiri-Zarandi, M. and Dara, R. 2022. Protecting farmers' data privacy and confidentiality: Recommendations and considerations. *Frontiers in Sustainable Food Systems*, 6: 1–9.

Khajuria, R., Sunanda and Singh, S. 2018. A survey of existing leaf disease techniques using Artificial Neural Network. *International Journal of Computer Trends and Technology*, 59: 52–62.

Konečný, J., McMahan, H.B., Yu, F.X., Suresh, A.T. and Bacon, D. 2016. Federated learning: Strategies for improving communication efficiency. *arXiv 1610.05492v2*: 1–10.

Kumar, R., Kumar, A., Bhatia, K., Nisar, K.S., Chouhan, S.S., Maratha, P. and Tiwari, A.K. 2024. Hybrid approach of cotton disease detection for enhanced crop health and yield. *IEEE Access*, 12: 132495–132507.

LeCun, Y., Bengio, Y. and Hinton, G. 2015. Deep Learning. *Nature*, 521: 436–444.

Liakos, K.G., Busato, P., Moshou, D., Pearson, S. and Bochtis, D. 2018. Machine Learning in agriculture: A review. *Sensors*, 18: 1–29.

McMahan, H.B., Moore, E., Ramage, D., Hampson, S. and Arcas, B.A. 2017. Communication efficient learning of deep networks from decentralized. pp. 1273–1282. In *Proceedings of the 20th International Conference on Artificial Intelligence and Statistics (AISTATS) 2017*, Fort Lauderdale, FL, USA, PMLR.

Mothukuri, V., Parizi, R.M., Pouriyeh, S., Huang, Y., Dehghantanha, A. and Srivastava, G. 2021. A survey on security and privacy of federated learning. *Future Generation Computer Systems*, 115: 619–640.

Samuel, A. 1959. Some studies in Machine Learning using the game of checkers. *IBM Journal of Research and Development*, 3: 210–229.

Schmidhuber, J. 2015. Deep learning in neural networks: An overview. *Neural Networks*, 61: 85–117.

Schmidhuber, J. 2022. Annotated history of modern AI and Deep Learning. *arXiv 2212.11279*: 1–75.

Wójtowicz, M., Wójtowicz, A. and Piekarczyk, J. 2016. Application of remote sensing methods in agriculture. *Communications in Biometry and Crop Science*, 11: 31–50.

Zhang, C., Xie, Y., Bai, H., Yu, B., Li, W. and Gao, Y. 2021. A survey on federated learning. *Knowledge-Based Systems*, 216: 1–11.

Chapter 14

Clean Streams, Clear Futures

AI Innovations in Water Quality Monitoring

Anuradha Devi

Introduction to Water Quality Monitoring

The process ensures that water resources are safe, available and sustainable through quality monitoring. This chapter gives a complete insight into the significance of water quality, the parameters to be measured for understanding it and conventional methods used in monitoring its causes, and all shortcomings. The concepts are well explained along with references from reliable sources, tables and detailed analysis of each aspect.

Importance of Water Quality

Human health, ecosystem sustainability and economic productivity all rely on water quality. Table 1 represents important uses of water including, drinking, irrigation for agriculture industrial processes and recreation. Water contaminated with these nutrients can cause major health concerns, such as waterborne infections, as well as harm aquatic life and ecosystems.

Guest faculty, Department of Chemical Engineering, GPCR, Rajgarh (MP).
Email: er.annu08@gmail.com, https://orcid.org/0000-0002-4407-0778

Table 1. Key uses of High-Quality Water (WHO, 2020).

Use	Description
Drinking Water	Safe for human consumption, free from harmful contaminants
Agriculture	Irrigation water that promotes healthy crop growth
Industry	Water used in manufacturing processes, cooling, and cleaning
Recreation	Water bodies used for swimming, boating, and fishing

Water Quality Parameters

Water quality is determined by measuring various physical, chemical, and biological parameters. Each parameter provides insight into the condition and usability of the water.

Physical Parameters:

- **Temperature**: Affects the biological and chemical processes in water.
- **Turbidity**: Indicates the clarity of water; high turbidity can be harmful to aquatic life.
- **Color**: Can indicate the presence of organic matter or pollutants.

Chemical Parameters:

- **pH**: Measures the acidity or alkalinity of water; important for aquatic life.
- **Dissolved Oxygen (DO)**: Essential for the survival of fish and other aquatic organisms.
- **Nutrients (Nitrate, Phosphate)**: High levels can lead to eutrophication, causing algal blooms.

Biological Parameters:

- **Coliform Bacteria**: Indicators of microbial contamination and potential presence of pathogens.
- **Biochemical Oxygen Demand (BOD)**: Indicates the amount of organic matter in water; high BOD can deplete oxygen levels.

The summary of common water quality parameters and their significance are given in Table 2.

Table 2. Common Water Quality Parameters and their significance (EPA, 2021).

Parameter	Type	Significance
Temperature	Physical	Affects biological and chemical processes
Turbidity	Physical	Indicates water clarity; impacts aquatic life
pH	Chemical	Indicates acidity or alkalinity; important for aquatic organisms
Dissolved Oxygen (DO)	Chemical	Essential for aquatic life
Nitrate	Chemical	High levels can cause eutrophication
Coliform Bacteria	Biological	Indicates microbial contamination

Traditional Water Quality Monitoring Methods

Traditional water quality monitoring involves periodic sampling and laboratory analysis (Table 3). These methods include:

- **Grab Sampling**: Collecting water samples at specific locations and times for laboratory analysis.
- **Composite Sampling**: Combining multiple samples taken at different times or locations to get an average measurement.
- **On-site Testing**: Using portable kits and meters to measure parameters like pH, temperature, and turbidity directly at the sampling site.

While these approaches provide accurate and complete information, they have limitations such as time and labor requirements, as well as the inability to deliver real-time data.

Table 3. Comparison of Traditional Water Quality Monitoring methods (AWWA, 2020).

Method	**Description**	**Advantages**	**Limitations**
Grab Sampling	Collecting samples at specific points	Accurate and detailed analysis	Time consuming, not real-time
Composite Sampling	Combining samples from different times/locations	Average measurement	Labor intensive, may miss temporal variations
On-site Testing	Using portable kits for immediate measurements	Quick results, portable	Limited to basic parameters, less accurate than lab analysis

Limitations of Traditional Methods

Despite their widespread use, traditional water quality monitoring methods have several limitations:

- **Temporal Limitations**: Periodic sampling may miss short-term pollution events or fluctuations in water quality.
- **Spatial Limitations**: Sampling at fixed locations may not capture the full variability of water quality in a water body.
- **Resource Intensive**: Requires significant human and financial resources for sampling, transportation, and laboratory analysis.
- **Delay in Results**: Laboratory analysis can take days to weeks, delaying the response to pollution events.

These limitations (Table 4) highlight the need for more efficient and real-time monitoring approaches, which is where AI-based systems can offer considerable benefits.

The State of Water Quality in India

Given India's vast population and diverse ecosystems, guaranteeing water quality is a unique challenge. In this segment, we elucidate the existing water quality in India,

Table 4. Limitations of Traditional Water Quality Monitoring methods (National Academy of Sciences, 2017).

Limitation	Description
Temporal Limitations	May miss short-term pollution events
Spatial Limitations	Fixed locations may not capture full variability
Resource Intensive	Requires significant human and financial resources
Delay in Results	Laboratory analysis can delay the response to pollution

including the status of its water resources, its challenges, and the implications for public health and the environment.

Overview of India's Water Resources

India finds itself in an enviable position of having such gigantic water belts like rivers, lakes and groundwater aquifers. Urbanization, industrialization and agricultural expansion are putting unsustainable pressure on the world's scarce water resources resulting in pollution and depletion. As per a report published by CPCB, 70% of India's water sources are threatened including the major rivers like Ganges and Yamuna (CPCB, 2020).

Current Challenges in Maintaining Water Quality

Despite efforts to improve water quality, India continues to grapple with various challenges, including industrial pollution, agricultural runoff, untreated sewage, and inadequate wastewater treatment infrastructure. These challenges are further compounded by ineffective regulatory enforcement, lack of public awareness, and limited resources for monitoring and management.

Impact of Poor Water Quality on Public Health and Environment

Poor water quality has far-reaching effects on human health, ecology, and socioeconomic development. Waterborne diseases including cholera, typhoid fever, and diarrhea are common in many parts of India, especially among underprivileged people with limited access to clean water and sanitation facilities. Furthermore, water pollution harms aquatic biodiversity, disrupts ecosystems, and jeopardizes livelihoods based on fishing and agriculture.

These water quality issues have far-reaching effects on public health and the ecosystem, as indicated in Table 5. High amounts of contaminants in water bodies can contaminate drinking water supplies, increasing the risk of waterborne diseases such diarrhea, typhoid, and hepatitis. Moreover, aquatic ecosystems suffer from the adverse effects of pollution, including habitat degradation, loss of biodiversity, and disruptions to ecosystem services such as water purification and nutrient cycling.

Table 5. Major pollutants in Indian water bodies (MeitY, 2021).

Pollutant	Sources	Effects
Industrial Waste	Factories, manufacturing units	Heavy metals, toxins
Agricultural Runoff	Pesticides, fertilizers	Nutrient pollution, algal blooms
Untreated Sewage	Urban areas, rural settlements	Pathogens, organic matter
Heavy Metals	Mining, industrial activities	Bioaccumulation, toxicity

Introduction to Artificial Intelligence

Artificial intelligence (AI) is transforming numerous industries, including water quality monitoring. With an emphasis on its potential in environmental monitoring and management of water quality, this section presents the fundamental concepts of artificial intelligence (AI), along with its types, applications, and evolution. Gaining a grasp of these foundational concepts is essential to understanding how artificial intelligence (AI) may solve problems with conventional water quality monitoring techniques.

What is Artificial Intelligence?

Artificial intelligence is the simulation of human intelligence in machines with decision-making, learning, and thinking capabilities. Artificial intelligence (AI) is an umbrella term for a variety of technologies that allow robots to carry out activities like speech recognition, visual perception, language translation, and decision-making that normally need human intelligence.

Key Concepts in AI:

The key concepts in artificial intelligence are summarized in Table 6.

- **Machine Learning (ML)**: A subset of AI that involves training algorithms to learn from and make predictions based on data.
- **Deep Learning (DL)**: A specialized form of ML that uses neural networks with many layers to analyze complex patterns in large datasets.
- **Natural Language Processing (NLP)**: Enables machines to understand and interpret human language.
- **Computer Vision**: The capability of machines to interpret and make decisions based on visual inputs.

Table 6. Key concepts in Artificial Intelligence (Russell and Norvig, 2020).

Concept	Description
Machine Learning (ML)	Algorithms that learn from data and make predictions
Deep Learning (DL)	Advanced ML using neural networks with multiple layers
Natural Language Processing (NLP)	Understanding and interpreting human language
Computer Vision	Analyzing and making decisions based on visual data

Evolution of AI

Since the mid-1900s, when artificial intelligence first emerged, a lot has changed. The initial motivation behind AI development was the development of algorithms designed to address certain challenges. Large datasets are now readily available, processing power has increased, and algorithm design innovations have made artificial intelligence (AI) a game-changing technology.

Key Milestones in AI Evolution:

- **1950s**: Alan Turing proposes the concept of a machine that can simulate any human intelligence.
- **1960s–1970s**: Development of early AI programs like ELIZA and Shakey the Robot.
- **1980s**: Introduction of neural networks and expert systems.
- **2000s**: Rise of big data and significant improvements in ML and DL algorithms.
- **2010s**: AI applications become mainstream with advancements in image and speech recognition (Russell and Norvig, 2020).

Types of AI

AI is broadly classified into three types based on its capabilities as shown in Table 7:

- **Narrow AI (Weak AI):** Designed to perform a narrow task, such as facial recognition or internet searches. It operates under a limited set of constraints and is currently the most common form of AI.
- **General AI (Strong AI):** Possesses the ability to understand, learn, and apply intelligence across a wide range of tasks at a level comparable to human beings. General AI is still theoretical and has not been realized.
- **Super-intelligent AI:** An AI that surpasses human intelligence across all fields. This type of AI is also theoretical and involves significant ethical and existential considerations.

Table 7. Types of Artificial Intelligence (Goodfellow et al., 2016).

Type	Description	Example Applications
Narrow AI (Weak AI)	Performs specific tasks within a limited scope	Facial recognition, spam filtering
General AI (Strong AI)	Exhibits human-like intelligence across diverse tasks	Theoretical, not yet realized
Super-intelligent AI	Surpasses human intelligence in all aspects	Theoretical, involves ethical considerations

AI in Environmental Monitoring

AI has the potential to transform environmental monitoring by providing real-time data analysis, predictive insights, and automated decision-making. AI systems can process vast amounts of environmental data from various sources, identify patterns, and predict future trends with high accuracy (IWA, 2020; EPA, 2021). Key applications of AI in environmental monitoring include:

- **Air Quality Monitoring**: AI algorithms analyze data from sensors and satellite imagery to track air pollution levels and predict air quality indices.
- **Water Quality Monitoring**: AI systems process data from water sensors to detect contaminants, predict pollution events, and recommend mitigation measures.
- **Climate Change Modeling**: AI helps in modeling and predicting climate patterns by analyzing historical weather data and simulating future scenarios.

AI in Water Quality Monitoring

AI technologies, particularly machine learning and deep learning, are being increasingly applied to water quality monitoring. These technologies enable the analysis of complex datasets from multiple sources, providing real-time insights and predictive analytics that enhance the efficiency and accuracy of water quality assessments.

Key Benefits of AI in Water Quality Monitoring (IIT, 2021; Grand View Research Inc., 2021):

- **Real-Time Monitoring**: AI systems can continuously analyze data from water quality sensors, providing instant alerts for any anomalies.
- **Predictive Analytics**: AI can forecast future water quality trends based on historical data, enabling proactive management.
- **Automated Decision-Making**: AI algorithms can recommend corrective actions in response to detected issues, reducing the need for human intervention.

AI in Water Quality Monitoring: A Paradigm Shift

Artificial Intelligence (AI) provides a variety of methods that can be used in water quality monitoring to improve the capacity to identify, anticipate, and address problems with water quality. This portion discusses particular AI methods and their applications in water quality monitoring systems, such as data analytics, machine learning, and deep learning.

Machine Learning Techniques

Machine Learning (ML) is a subset of AI that focuses on building systems that can learn from and make decisions based on data. In water quality monitoring, ML techniques analyze large datasets, identify patterns, and predict future water quality trends (Bishop, 2016; Murphy, 2012).

- **Supervised Learning**: This technique involves training an algorithm on a labeled dataset, where the input-output pairs are known. It is useful for predicting specific outcomes, such as contaminant levels in water.
  - **Example**: Predicting the concentration of nitrates in a river based on historical data of agricultural runoff and weather patterns.

- **Unsupervised Learning**: This technique deals with datasets where the output is not labelled. It is used for discovering hidden patterns or grouping data into clusters.
    - **Example**: Clustering different water samples based on their chemical properties to identify sources of pollution.
- **Reinforcement Learning**: This technique involves training algorithms to make a sequence of decisions by rewarding desired behaviors and penalizing undesired ones.
    - **Example**: Optimizing the operation of water treatment plants by continuously adjusting parameters to achieve the best water quality.

Deep Learning Techniques

Deep Learning (DL) is an advanced form of machine learning that uses neural networks with many layers to process complex data (Goodfellow et al., 2016; LeCun et al., 2015). DL is particularly effective in analyzing large volumes of water quality data and identifying intricate patterns that simpler algorithms might miss.

Data Analytics in Water Quality Monitoring

Data analytics involves processing and analyzing data to extract meaningful insights. In the context of water quality monitoring, data analytics helps in understanding trends, detecting anomalies, and making data-driven decisions (Davenport and Harris, 2007; Provost and Fawcett, 2013).

- **Descriptive Analytics**: Summarizes historical data to understand what has happened over a given period. This is often the first step in data analysis.
    - **Example**: Summarizing annual water quality data to identify periods of high pollution.
- **Predictive Analytics**: Uses historical data to predict future outcomes. It is critical for proactive water quality management.
    - **Example**: Predicting future contamination events based on past pollution patterns and environmental factors.
- **Prescriptive Analytics**: Suggests actions based on predictive analytics to achieve desired outcomes. It helps in decision-making processes.
    - **Example**: Recommending specific interventions to prevent predicted contamination events.

Case Studies: AI in Water Quality Monitoring

To illustrate the practical application of AI techniques in water quality monitoring, several case studies are discussed in this chapter. These examples highlight how AI has been successfully implemented to improve water quality management (IIT, 2021; Ministry of Urban Development, 2021).

- **Case Study 1: AI-Powered Monitoring of the Ganges River**
  - **Objective**: To monitor and improve water quality in the Ganges River using AI technologies.
  - **Methods**: Deployment of AI-enabled sensors along the river to collect real-time data on pollutants, followed by machine learning analysis to predict pollution trends and identify sources.
  - **Results**: Significant improvements in water quality, with timely identification and mitigation of pollution sources.
- **Case Study 2: Predictive Analytics in Groundwater Quality Management**
  - **Objective**: To predict and manage groundwater quality in agricultural regions.
  - **Methods**: Using historical data on groundwater quality and agricultural practices, machine learning models were developed to forecast future contamination risks.
  - **Results**: Enhanced ability to manage and mitigate groundwater contamination, ensuring safer drinking water for local communities.
- **Case Study 3: AI-Based Early Warning System for Urban Water Supplies**
  - **Objective**: To implement an early warning system for water quality in urban water supply networks.
  - **Methods**: Integration of AI algorithms with water quality sensors to provide real-time monitoring and early warnings of contamination.
  - **Results**: Improved response times to contamination events, reducing the risk of waterborne diseases in urban populations.

Key Components of AI-Based Water Quality Monitoring Systems

In order to create and implement AI-based water quality monitoring systems, several components, such as sensors, data gathering tools, and analytical algorithms, must be carefully considered. This section delves into the essential elements of artificial intelligence (AI) systems and their functions in augmenting the precision, dependability, and effectiveness of water quality surveillance.

Sensors and Data Collection Devices

Sensors play a crucial role in capturing data on water quality parameters such as temperature, pH, dissolved oxygen, and nutrient levels. Advances in sensor technology have led to the development of compact, affordable, and IoT-enabled sensors that can be deployed in diverse environmental settings, from rivers and lakes to wastewater treatment plants and distribution networks. Some examples of advanced sensors for water quality are given in Table 8.

Table 8. Examples of advanced Water Quality Sensors (CPCB, 2021; Wurtsbaugh et al., 2019).

Sensor Type	Measurement Parameters	Applications
Optical Sensors	Turbidity, Chlorophyll	Surface Water Monitoring
Fluorescence Sensors	Dissolved Organic Matter, Algal Biomass	Aquatic Ecosystem Monitoring
Spectrophotometers	Nutrients (Nitrate, Phosphate)	Agricultural Runoff Monitoring
Electrochemical Sensors	pH, Conductivity	Drinking Water Quality Monitoring

Machine Learning Algorithms for Data Analysis

Machine learning algorithms play a vital role in analyzing the vast amounts of data collected by sensors and identifying patterns, trends, and anomalies indicative of water quality changes. Supervised learning algorithms, such as decision trees and support vector machines, can be trained on historical data to predict water quality parameters based on environmental variables.

Unsupervised learning algorithms, such as clustering and anomaly detection, can identify patterns and anomalies in data, providing valuable insights into water quality dynamics. Commonly used machine learning algorithms for water quality analysis are shown in Table 9.

Table 9. Commonly used Machine Learning algorithms for Water Quality Analysis (Twardowski and Boss, 2011).

Algorithm	Description	Applications
Decision Trees	Tree-like models for classification and regression	Predicting water quality parameters
Support Vector Machines	Linear and non-linear classification models	Anomaly detection, pattern recognition
Random Forests	Ensemble learning method based on decision trees	Predictive modeling, feature selection
Neural Networks	Deep learning models inspired by the human brain	Image analysis, time series prediction

Implementation Challenges and Solutions

While the integration of AI into water quality monitoring offers significant benefits, there are several challenges to overcome. In this section, we explore these challenges and propose solutions to ensure the successful implementation of AI-based systems.

Addressing Technical Limitations

The technical difficulty involved in establishing AI-based water quality monitoring systems is one of the main obstacles. Specialised knowledge and resources are needed for the design and deployment of sensors, the creation of machine learning algorithms, and the integration of data from many sources. Furthermore, there are extra difficulties in guaranteeing the accuracy, scalability, and dependability of AI systems in real-world environmental scenarios.

Collaboration amongst multidisciplinary teams made up of engineers, data scientists, environmental scientists, and domain experts is crucial to addressing these technical restrictions. Teams can create strong AI-based systems that satisfy the unique requirements of water quality monitoring applications by utilising a variety of skill sets and knowledge areas. The use of AI technologies in the water sector can also be facilitated by funding R&D, technology infrastructure, and capacity-building initiatives.

Overcoming Regulatory and Policy Hurdles

Managing the legal and policy frameworks controlling environmental monitoring, data security, and privacy presents another difficulty in implementing AI-based water quality monitoring systems. Concerns about privacy about the gathering, storing, and distribution of private environmental data may impede cooperative research projects and data-sharing programmes. To guarantee the ethical and lawful application of AI technology in water quality monitoring, it is also necessary to address regulatory compliance requirements, including data quality standards, data ownership rights, and liability concerns.

Stakeholders must collaborate with legislators, regulators, and legal experts to create precise norms and criteria for the ethical application of AI in environmental monitoring to get over these legislative and regulatory roadblocks. The establishment of legislative frameworks and guidelines for AI-based water quality monitoring should be based on the fundamental concepts of transparency, accountability, and stakeholder involvement. Programmes for training and capacity building can also provide stakeholders with the tools they need to successfully manage legal requirements and compliance duties.

Ensuring Accessibility and Affordability

Under resource-constrained conditions, access to AI-based water quality monitoring systems may be restricted, especially in developing nations and rural areas. Adoption of AI solutions by local governments, community organizations, and small-scale water utilities may be hindered by the high upfront costs involved with technology procurement, installation, and maintenance. Widespread adoption may also be hampered by worries about the sustainability and affordability of AI-based systems, particularly continuing operational and technical support costs.

Innovative finance methods, such as grant funding, technology transfer agreements, and public-private partnerships, might be investigated to guarantee the affordability and accessibility of AI-based water quality monitoring technologies. Technology transfer, capacity building, and information exchange can be facilitated by cooperative activities involving governments, international organizations, academia, and the commercial sector, especially in underserved regions. The scalability and sustainability of AI-based water quality monitoring systems can also be improved by funding research and development projects that aim to lower the cost and complexity of AI technologies. Examples of these projects include the creation of inexpensive sensors, open-source software platforms, and cloud-based analytics services.

Conclusion

An extensive overview of the use of artificial intelligence (AI) in water quality monitoring can be found in "Clean Streams, Clear Futures: AI Innovations in Water Quality Monitoring." The importance of water quality and the shortcomings of conventional monitoring techniques are emphasized in the opening of this chapter, along with the necessity for creative solutions. The article then explores India's current situation concerning water quality, looking at the difficulties encountered and their effects on the environment and human health. The next sections give an overview of artificial intelligence (AI), examine how it could transform the monitoring of water quality, and go over the essential elements of AI-based systems. This chapter presents successful applications of AI in water quality monitoring through real-world case studies, illustrating how useful it is for anomaly detection, trend prediction, and decision support. But it also tackles implementation difficulties like technical difficulties, legal roadblocks, and cost considerations, providing ways to get around these obstacles. This chapter seeks to further sustainable water management techniques and freshwater resource preservation by offering insights into the incorporation of AI into water quality monitoring.

References

American Water Works Association (AWWA). 2020. Water quality standards handbook. https://www.awwa.org/store/productdetail.aspx?productid=25054.

Bishop, C.M. 2016. *Pattern Recognition and Machine Learning*. Springer.

Central Pollution Control Board (CPCB). 2021. AI in Ganges River water quality monitoring. https://cpcb.nic.in/ai-ganges-monitoring.

Davenport, T.H. and Harris, J.G. 2007. *Competing on Analytics: The New Science of Winning*. Harvard Business Review Press.

Environmental Protection Agency (EPA). 2021. Water quality surveillance and monitoring. https://www.epa.gov/waterdata/water-quality-surveillance-and-monitoring.

Goodfellow, I., Bengio, Y. and Courville, A. 2016. *Deep Learning*. MIT Press.

Grand View Research, Inc. 2021. AI in environmental monitoring market analysis. https://www.grandviewresearch.com/industry-analysis/ai-in-environmental-monitoring-market.

Indian Institute of Technology (IIT). 2021. Groundwater quality management using predictive analytics. https://iit.ac.in/research/groundwater-predictive-analytics.

Indian Institute of Technology (IIT). 2021. Research on water quality monitoring and management. https://iit.ac.in/research/water-quality-monitoring-management.

Internation Water Association (IWA). 2020. Artificial intelligence and machine learning in water. https://iwa-network.org/learn/artificial-intelligence-machine-learning-water/.

LeCun, Y., Bengio, Y. and Hinton, G. 2015. Deep learning. *Nature*, 521(7553): 436–444.

Ministry of Electronics and Information Technology (MeitY). 2021. National AI strategy for India. https://www.meity.gov.in/press-release/launch-national-strategy-artificial-intelligence.

Ministry of Urban Development. 2021. AI in urban water supply monitoring. https://moud.gov.in/ai-urban-water-supply.

Murphy, K.P. 2012. *Machine Learning: A Probabilistic Perspective*. MIT Press.

National Academy of Sciences. 2017. Using AI to improve water quality monitoring and management. https://www.nationalacademies.org/our-work/using-ai-to-improve-water-quality-monitoring-and-management.

Provost, F. and Fawcett, T. 2013. Data science for business: What you need to know about data mining and data-analytic thinking. O'Reilly Media.

Russell, S. and Norvig, P. 2020. *Artificial Intelligence: A Modern Approach*. Pearson.

Twardowski, M.S. and Boss, E. 2011. Optical observations of the ocean. *Annual Review of Marine Science*, 4: 263–290.

United States Environmental Protection Agency (EPA). 2021. Water quality standards handbook. https://www.epa.gov/wqs-tech/water-quality-standards-handbook.

World Health Organization (WHO). 2020. Guidelines for drinking-water quality. https://www.who.int/water_sanitation_health/publications/drinking-water-quality-guidelines/en/.

Wurtsbaugh, W.A., Paerl, H.W. and Dodds, W.K. 2019. Nutrients, eutrophication, and harmful algal blooms along the freshwater to marine continuum. *Limnology and Oceanography*, 6: 1–27.

Chapter 15

Transforming Waste into Resources

AI's Impact on Wastewater Treatment

Ayushi Jha,[1,*] *Subharojyoti Bhattacharjee*,[1] *J Veeranna*[2] and *Narender Pal*[1]

Introduction

Water has been the source of life since prehistoric times. And it became the principal source of life, continuing the human life-line tendency. That's why we referred to it as a "Life". Inevitably, the vast majority of humanity has utilized it as a source of exploitation, gradually intoxicating the environment and turning it into a demise event. That is referred to as "water pollution".

What is water pollution? There are several ways to characterize pollution. When one or more compounds are released into the water that will negatively alter it, pollution of the water happens (Grégorio Crini and Eric Lichtfouse, 2018). These compounds have the potential to negatively impact the environment, humans, animals, and their ecosystems. Water contamination is categorized in many ways (Morin-Crini and Crini, 2017). "Point and non-point" can be used for the two main sources. On the other hand, the second refers to pollutants emitted from several sources. The first category includes pollutants that come from a single source, such as emissions from industry into the water. The causes of water pollution are classified in multiple ways, like: industrial wastes, mining activities, sewage and waste water, pesticides and

[1] Department of Soil and Water Engineering, Punjab Agricultural University, Ludhiana-141004, Punjab, India.

[2] Department of Soil and Water Conservation Engineering, College of Agricultural Engineering, Kandi, Sangareddy, India.

* Corresponding author: ayushijhagsm@gmail.com

chemical fertilizers, energy use, radioactive waste, urban development, etc. (Khalaf et al., 2016). The simple fact that water is used implies that it will become polluted: any activity, whether residential or agricultural, but also industrial, generates wastewater containing undesired pollutants, which can be poisonous. In this context, a constant effort must be made to protect water resources (Rathoure and Dhatwalia, 2016; Morin-Crini and Crini, 2017).

So "Wastewater treatment" has become crucial for reducing pollutants and enhancing environmental quality. In the process of treatment, the impact of the Tadeonal and Moder techniques is high. Traditional methods face significant challenges to the complex and variable nature of treated compounds which necessitates precise operational controls and introduces many uncertainties (Long et al., 2019; Loos et al., 2013; Mingzhi et al., 2009). Modern treatment plants must also comply with stricter emission restrictions and new energy efficiency and resource recycling regulations (Huang et al., 2015; Mamais et al., 2015; Wan et al., 2011).

Artificial intelligence (AI) offers powerful solutions to these challenges due to its high speed, ease of use, and acceptable accuracy without needing deep physical knowledge (Rajaee et al., 2019). AI can transform wastewater treatment by providing innovative approaches to convert waste into valuable resources. This chapter will provide an overview of the fundamentals of wastewater treatment and explore the potential of AI technologies in revolutionizing these processes. It will discuss AI techniques such as machine learning, neural networks, and data analytics, and their relevance to wastewater treatment.

Additionally, this chapter will provide a visual representation of the process and barriers to the adoption of artificial intelligence in wastewater treatment with regulatory, technical, and institutional issues, and potential solutions. It will explore the opportunities for further research, innovation, and collaboration in AI-enabled wastewater treatment. Finally, the chapter will summarize key findings and insights, reflecting on the potential impact of AI on transforming wastewater treatment into a resource recovery process and discussing future directions and emerging trends in this field.

Fundamentals of Wastewater Treatment

Water treatment is a process that improves water quality to make it appropriate for a specific end-use, such as drinking, industrial water supply, irrigation, river flow maintenance, water recreation, or many other uses. The primary objective of water treatment is to remove contaminants and undesirable components or reduce their concentration so that the water becomes fit for its desired end-use. Here are the fundamental steps and methods involved in typical water treatment processes: (a) **Screening** to remove large debris such as sticks, leaves, and rubbish; (b) **Coagulation and Flocculation** by aggregating smaller particles into larger ones so they can be more easily removed. (c) Sedimentation allows flocs to settle at the bottom of a sedimentation basin due to gravity. (d) **Filtration** by removing remaining particles from the water. (e) **Disinfection** by kill or inactivate harmful microorganisms. (f) **pH Adjustment** of the water to neutral or desired level. (g) Advanced Treatments by removing organic compounds, taste, and odors. (h) Distribution by delivering treated water to consumers.

From research it's clear from an analysis that 70% of the particles in sewage water are organic, making up just 0.1% of the total. 99.9% of sewage water is water. Over 70% of the solids in sewage effluents are made up of organic matter, which is mostly composed of proteins (65%), carbohydrates (25%), and lipids (10%). The high concentrations of suspended particles in organic effluents also hinder photosynthetic activity in aquatic organisms. Furthermore, organic waste from humans and animals may include harmful organisms (Nusrat et al., 2017).

The water pollution approach to wastewater is a broad approach to dissecting the concept's accountability into particular aspects resulting in excellent analysis. Chemical-induced contamination of water has grown to be a major worry and priority for the public sector, society, and, most crucially, the industrial world at large (Sonune and Ghate, 2004; Crini, 2005; Cox et al., 2007; Sharma, 2015; Rathoure and Dhatwalia, 2016). Wastewater has to be treated before being released into receiving water bodies because the organic matter and other impurities in it serve as a breeding environment for a variety of illnesses connected to water (Jegatheesan et al., 2008).

Nowadays, the most crucial task for mankind is to treat the contaminated polluted water. The most effective purification method should be used to meet the decontamination objectives (as set forth by law) when water gets contaminated and decontamination becomes essential. A purification procedure typically consists of the following five steps: The five stages of treatment are, physical and mechanical preliminary treatment; chemical and physicochemical primary treatment; chemical and biological secondary treatment or purification; physical and chemical tertiary or final treatment; and handling of the resulting sludge (supervised tipping, recycling, or incineration). Generally speaking, depending on the circumstances, the first two phases are grouped under the terms pre-treatment or preparatory step (Anjaneyulu et al., 2005; Crini and Badot, 2007, 2010). Pre-treatment consists of eliminating all suspended materials and (floating) solid particles from the wastewater. Before evaluating secondary treatment, this pre-treatment step—which can be completed mechanically or physically—is essential because particle pollution (such as SS,

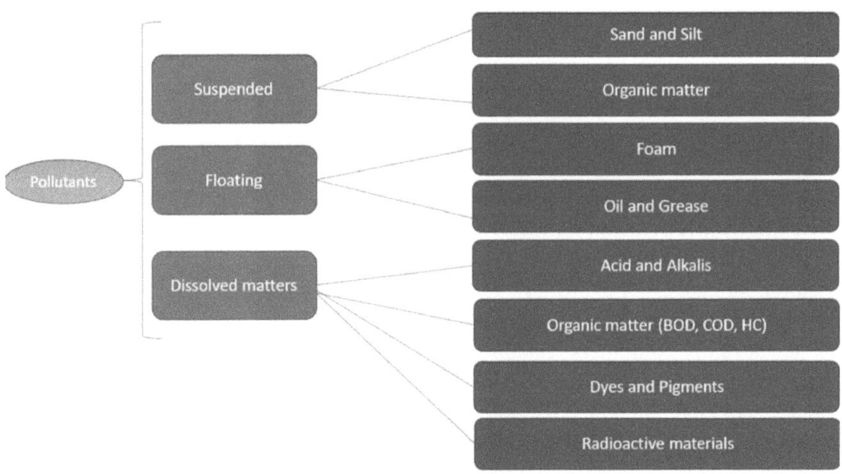

Fig. 1. Classification of wastewater streams containing pollutants.

colloids, lipids, etc.) will impede subsequent treatment, reduce its effectiveness, or harm the decontamination apparatus. It can also be necessary to perform primary chemical treatment, such as pH correction, pre-reduction of a large organic load, oxidation for cyanide destruction and Cr(VI) reduction. For example, a lot of SS, such as fibres, fillers, and other solids, are found in paper mill effluent (Pokhrel and Viraraghavan, 2004; Anjaneyulu et al., 2005; Sharma, 2015).

Despite frequently being alkaline and having a high organic content, textile mill wastewater has a very changeable pH. Pre-treating these effluents is therefore essential before thinking about further treatment. But frequently, these therapies by themselves are unable to satisfy the legal criteria. The pre-treated effluent must undergo secondary purification treatment using the most suitable biological, physical, or chemical technology available to eliminate the chemical pollutants before it is released into the environment or used again. To get rid of any leftover contaminants or molecules created during the secondary purification, a final or tertiary treatment may occasionally be necessary (e.g., the removal of salts formed by the mineralization of organic waste). Although tertiary therapy is not often used in Europe, it may be in the future if new regulations are put in place. Adsorption employing activated carbons (AC), ion-exchange, membrane filtration (ultrafiltration, reverse osmosis), advanced oxidation, and built wetlands (CW) are the primary tertiary treatments now used at a few industrial sites. The majority of CWs are used in Europe to treat municipal and household wastewater. On the other hand, the variety of CW designs makes them adaptable to the treatment of industrial effluents (such as wastewater from tanneries and post-treated pulp and paper).

Thus, the wastewater characteristics will determine which approach is chosen (Anjaneyulu et al., 2005; Crini, 2005; Crini and Badot, 2007; Cox et al., 2007). Every treatment has unique limitations concerning cost as well as practicality, efficiency, dependability, environmental effect, sludge generation, operation complexity, pre-treatment needs, and the creation of potentially hazardous byproducts. However, for technological and financial reasons, the industrial sector only often uses a small number of the several treatment procedures now recommended for wastewater treatment. Pollutants are typically removed from effluents using physicochemical and/or biological methods; research is focused on developing novel, more affordable alternatives or more efficient system combinations.

The Role of Artificial Intelligence in Wastewater Treatment

The area of computer science referred to as artificial intelligence (AI) aims to build machines with cognizant minds. It includes developing algorithms as well as systems that are capable of learning, thinking, solving problems, perceiving, and comprehending natural language. The primary goal of artificial intelligence (AI), a broad and multifaceted field which includes a wide range of technologies and methodologies, is to develop computers capable of accomplishing tasks that would typically require human intelligence. The following is an in-depth look at the fundamentals, approaches, and potential developments in AI. AI is a rapidly developing discipline of computer science that works to create mechanisms able to perform tasks which tend to need human intelligence. Artificial intelligence is a discipline with many different

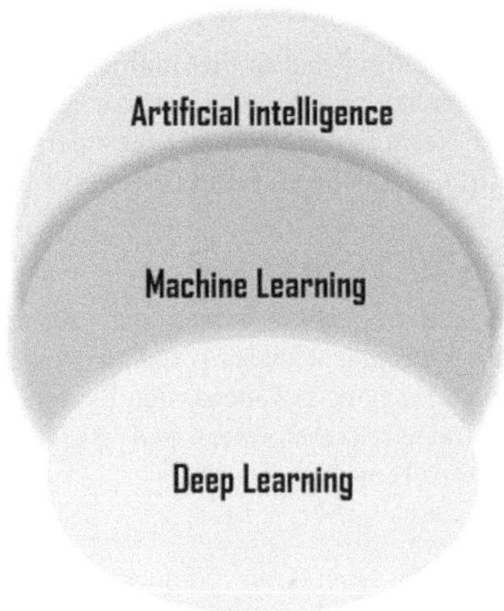

Fig. 2. Artificial intelligence methods.

approaches and technology, all of which add to its wide variety of uses. This whole field is classed based on the integrity of the internal working function, are (Russell and Norvig, 2016) as follows:

1) Weak AI/narrow AI: It is created and developed for a specific objective. Its approaches, image recognition software, and digital assistants like Siri are among the applications.
2) General AI (Strong AI): Artificial intelligence that is conceptual and has human-like comprehension, learning, and application skills for several kinds of activities.
3) Superintelligent AI: A level of intelligence surpassing human capabilities. This is currently theoretical and a subject of speculation and debate among experts.

As we have comprehensive knowledge about AI to understand its quality based on the working pattern, aspects of it have a great role in the selection of its particular algorithm to run the process explained below;

a. Machine Learning (ML) (Murphy, 2012):
   - An area of artificial intelligence that concentrates on creating algorithms which enable machines analyze data and make decisions or forecasts from it.
   - *Supervised Learning*: Involves using a labeled dataset to train a model, where the input gives the proper output. Regression (such as catching spam emails) and classification (for example, predicting housing values).
   - *Unsupervised Learning*: Involves training a model on data without labels. The model tries to find hidden patterns or intrinsic structures in the input data. Examples include clustering (e.g., customer segmentation) and dimensionality reduction (e.g., PCA).
   - *Reinforcement Learning*: Focuses on training agents to make a sequence of decisions by rewarding them for desirable actions. Robotic control and gaming (like AlphaGo) are two examples of applications.

b. Deep Learning (Murphy, 2012):
   - An area of machine learning that applies multilayered deep neural networks. It is quite good at jobs like speech and picture recognition. For sequential data, Recurrent Neural Networks (RNNs) and their derivatives (such as LSTMs) are utilized, whilst Convolutional Neural Networks (CNNs) are employed for picture data.

c. Natural Language Processing (NLP):
   - Involves communication between human (natural) languages and computers. Text summarization, sentiment analysis, and machine translation are important tasks. NLP techniques have advanced significantly with models like GPT (Generative Pre-trained Transformer) and BERT (Bidirectional Encoder Representations from Transformers).

d. Computer Vision:
   - Makes it possible for robots to comprehend and interpret the visual environment. Among the tasks are picture segmentation, facial recognition, and object identification. Computer vision relies heavily on technologies such as Generative Adversarial Networks (GANs) and CNNs.

e. Robotics:
   - Combines mechanical engineering and artificial intelligence to create robots that can operate either entirely or partially on their own. Robotics applications include surgical robots, personal assistants, and industrial automation.

f. Expert Systems:
   - Artificial intelligence (AI) systems which simulate human experts' decision-making processes. They control certain problems, frequently in industries like banking, medical, and customer service, by utilizing an inference engine and a knowledge base of human experiences.

Applications of AI in Various Fields

i. **Healthcare**:
   - **Diagnostics**: AI systems are capable of analyzing medical visual imagery, such as MRIs and X-rays, to identify diseases like cancer early on.
   - **Personalized Treatment**: AI can help tailor treatments based on individual patient data, leading to more effective care.
   - **Drug Discovery**: AI improves the search for new medications by forecasting the behavior of various components.

ii. **Finance**:
   - **Fraud Detection**: Artificial intelligence algorithms can identify odd trends and highlight possible fraudulent transactions.
   - **Algorithmic Trading**: Artificial Intelligence systems can examine a huge amount of data and make trading decisions faster than humans can.
   - **Personal Finance Management**: AI-driven apps can help individuals manage their finances by analyzing spending habits and suggesting savings strategies.

iii. **Retail**:
   - **Customer Service**: Artificial intelligence (AI)-powered chatbots may handle problems from customers, process requests, and provide recommendations for products.
   - **Inventory Management**: AI can forecast product demand, which helps retailers manage their inventory levels.
   - **Personalization**: AI uses consumer data analysis to provide customized marketing campaigns and individualized obtaining experiences.

iv. **Transportation**:
   - **Automated Vehicles**: AI is the fundamental technology that allows self-driving automobiles to navigate and react to the environment around them.
   - **Traffic Management**: AI systems can optimize city traffic flow, resulting in less traffic and improved safety.
   - **Logistics**: AI enhances the effectiveness of supply chain processes by forecasting demand and streamlining delivery routes.

v. **Education**:
   - **Personalized Learning**: AI provides the ability to modify course material depending on each student's needs, making for a more productive learning environment.
   - **Administrative Tasks**: AI may handle scheduling and grading, freeing up teachers to concentrate on instruction.
   - **Tutoring Systems**: Artificial intelligence-driven tutoring programs give learners extra assistance beyond the classroom.

vi. **Entertainment**:
   - **Content Creation**: AI can create writing, art, and music, resulting in new entertainment genres.
   - **Recommendation Systems**: Recommendation engines on websites like Netflix and Spotify are powered by AI algorithms, which aid users in finding new content.
   - **Gaming**: AI enhances the gaming experience by creating more realistic and adaptive non-player characters (NPCs).

vii. **Manufacturing**:
   - **Predictive Maintenance**: AI technologies identify when machines are likely to wear out, enabling rapid maintenance and minimising downtime.
   - **Quality Control**: AI is more effective than human inspection personnel for detecting flaws in items.
   - **Automation**: AI-driven robots and machinery can perform repetitive tasks, increasing efficiency and reducing human labor costs.

Potential of AI Technologies in Revolutionizing Wastewater Treatment Processes

Taking into account the increasing need for sustainable systems and clean water globally, the Internet of Things (IoT) can transform conventional methods and tackle pressing issues in the domains of wastewater treatment and water quality prediction. The disruptive uses of smart IoT technologies, such as machine learning (ML) and artificial intelligence (AI) models, are explored in detail in this extensive essay. The successful application of an Internet of Things (IoT)-based automated water quality monitoring system that makes use of cloud computing and machine learning techniques to efficiently handle the aforementioned problems is one example. The Internet of Things (IoT) has been used to simulate, automate, and optimize several processes, including wastewater treatment applications, water-related agriculture techniques like hydroponics and aquaponics, and the monitoring and management of natural systems (Alprol et al., 2024).

Water is the lifeblood of humans and the industrial process, but it is currently under severe threat from hazardous pollutants that result from both natural and human-caused processes. Globally, there is a major problem with access to clean and healthful drinking water. As such, it is imperative to classify and monitor water quality; the underlying issue is that, given existing technological capabilities, suitable parametric quality indicators are not available (Ray et al., 2023). The activities of aquatic systems (aquaculture, aquaponics, and hydroponics) have risen due to the growth in the human population. This has led to an increase in the nutrient load that drains into water bodies, mostly nitrogen and phosphorus, which has harmed many aquatic habitats. Thus, employing cutting-edge technology, aquaculture effluent wastewater treatment has to be developed and enhanced sustainably (Shen et al., 2023; Yadav et al., 2023).

While machine learning (ML) focuses on intelligent systems that can adjust their behaviour depending on new information presented during the training phase, artificial intelligence (AI) enables algorithms to execute activities and draw conclusions that would normally need human knowledge (Krittanawong et al., 2018). The idea behind artificial intelligence (AI) is to allow algorithms to carry out operations and draw conclusions that would normally need human intellect. Simopoulou et al. (2018) state that machine learning (ML) is focused on intelligent systems that can modify their behaviour in response to newly supplied information during the training phase. The capacity of AI and ML to improve and ease knowledge is the main factor driving its acceptance in academic groups across a range of areas and sectors (Fig. 3). This includes studying natural systems including river quality modelling, adsorption processes, membrane-filtration modelling, water treatment, including coagulation and chlorination dosing, and agricultural system health. According to earlier research, machine learning (ML) can be a useful strategy for resolving issues in various fields (Suquet et al., 2020).

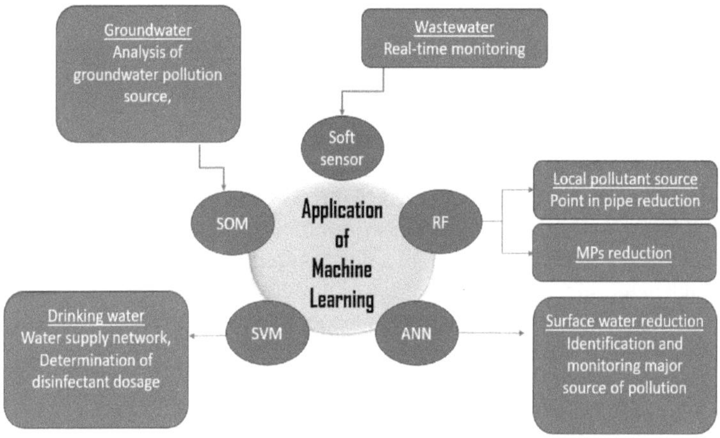

Fig. 3. Applications of various ML processes in water treatment.

According to the study, depicts the important AI models used in WWT. These models can be divided into three categories (Balabin et al., 2011) (Fig. 4):

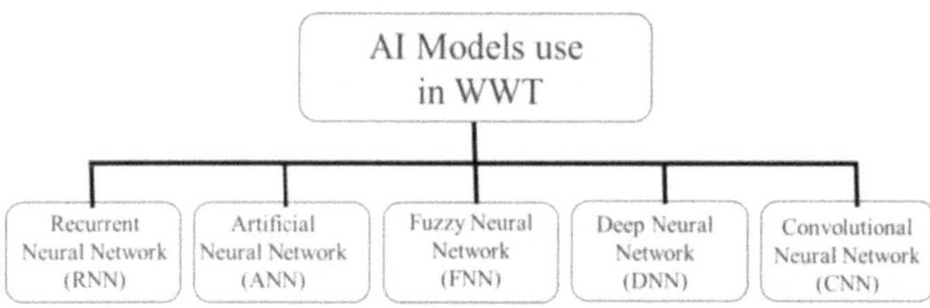

Fig. 4. AI models for wastewater treatment.

Using sequence data as input, an RNN, a kind of NN, iterates in the direction of sequence evolution. Learning nonlinear features in time-series issues is facilitated by RNN's memory, parameter sharing, and Turing completeness. According to Sherstinsky et al. (2020), the LSTM is the RNN that is most frequently utilized to solve the gradient disappearance issue. Numerous applications, including water, WWT, water quality management, and water-based agriculture, have seen noteworthy success using RNN.

To process data, artificial neural networks (ANNs) mimic the behaviour of biological neural networks (BNNs). With the use of unit nodes, artificial neural networks (ANNs) imitate neurons and process information by varying the connectivity weights between different nodes within the network. An ANN typically consists of an input layer, an output layer, and several hidden layers positioned in between. ANNs use numerous variable weights across neurons in addition to a variety of activation functions, including sigmoid, tanh, and ReLU functions, to execute complicated nonlinear calculations. An ANN's expressive skills may be improved and more complex nonlinear models can be created by increasing the number of hidden layers in the ANN (Basheer et al., 2000). By creating and refining loss functions, artificial neural networks (ANNs) may be trained to comprehend intricate nonlinear interactions between inputs and outputs. The most widely used artificial neural network (ANN) models include recurrent neural networks (RNNs), CNNs, FNNs, and DNNs (Ding et al., 2013).

An ANN and fuzzy logic are combined in the FNN, a hybrid neural network model, to handle unclear or uncertain scenarios. After processing input data using fuzzy logic reasoning, an ANN is used by the FNN to train and generate the output. Though fuzzy logic is used to establish the fuzzy relationship between inputs and outputs and the weights of connections among neurons (normalisation, fuzzy inference, and membership function), the topology of an FNN is similar to that of a traditional neural network. An input, membership function, fuzzy inference, normalisation, and output layers make up the FNN architecture. When handling difficult situations, FNN provides a great lot of advantages over ordinary NNs and is widely used in pattern recognition and control (Mahesh et al., 2020).

A DNN is an ANN type that has several hidden layers positioned in between the input and output layers. With the help of successive layers that combine lower-level characteristics to learn higher-level features, the deep architecture of DNN enables it to learn hierarchical data representations. Just like other neural networks, DNNs can include more neurons and hidden layers. They are frequently used to learn highly nonlinear mappings from inputs to outputs or to capture complicated patterns in data. However, because of their complex network design and the need for a substantial quantity of data for training, DNNs may be difficult and computationally costly to train (Yadav et al., 2020).

Convolutional computations and a deep structure feedforward neural network are two components of CNNs, a sort of deep learning technology. Natural language processing, computer vision, and other domains all make extensive use of CNNs. CNNs use fully connected layers for classification or regression tasks, pooling layers to lower the feature of dimensionality, and convolutional layers to extract complex features from input pictures (Alzubaidi et al., 2021). Input, output, convolutional, pooling, and fully connected layers are common components of a CNN design.

Optimization of Treatment Processes Using AI

The treatment of wastewater is important to know and understand for further enhancement of the process. Before wastewater from commercial, industrial, and municipal sources is released into natural water bodies, it must first be treated (WWT) to remove carbonaceous contaminants (such as organic carbon compounds) and nutrients (such as phosphorus and nitrogen species) (Russell, 2019). By lowering the dissolved oxygen content in surface water courses such as rivers and lakes, carbonaceous pollutants can harm the ecosystem (Jin et al., 2014). The aquatic life is at risk due to nutrient-induced eutrophication and algal bloom production in surface water (Elsayed et al., 2023a; Yaqub et al., 2020). Before being released into a river or another body of water for recycling (such as recycled water or consumption in agriculture), wastewater treatment facilities (WWTP) filter wastewater collected from cities, homes, companies, and other sources (Corominas et al., 2018). Water plants are intricate systems that use cutting-edge network technology to solve operational issues. Large linked tunnels are frequently used in WWTPs (particularly in populated regions) to store sanitary and precipitation-related flows for processing (Owolabi et al., 2022). Decisions about pumping stored wastewater from tunnels must be taken quickly at WWTPs because the wastewater cannot exceed the acceptable levels in the tunnel, which could result in an overflow of untreated water (Corominas et al., 2018; Schutze et al., 2002) or excessive chemical use. For this reason, computation time is a crucial concern (Schütze et al., 2002).

Overview of AI-driven Optimization Techniques for Wastewater Treatment Processes is most Important for Enhancement and Efficient Work Output

With the advent of artificial intelligence (AI) and other intelligent approaches, such as machine learning (ML), in membrane applications, environmental engineering challenges have significantly improved. Throughout the entire process, surface contamination is a particular concern. Current modeling approaches including genetic programming, artificial neural networks (ANNs), fuzzy logic (FL), and model trees are effectively employed in many parts of the world (Bagheri et al., 2019). Artificial intelligence (AI) research is always bringing new technical ideas and possibilities to life. High efficiency standards and a "greener" approach are two of the most common contemporary requests for any sort of new technology. This AI-based technology is a promising however acceptable technique to modernize and enhance existing membrane-based facilities—like desalination or water treatment units for better resource and energy management. It bears similar relevance when considering established technologies and appliances. However, although having no direct connection to any membrane-based processes, Bayes Model Averaging (BMA) offers a Bayesian probabilistic model that helps comprehend how uncertainty spreads from different input data sources. As opposed to the PCA, which alters the data by mostly using the primary components and occasionally disregarding the others.

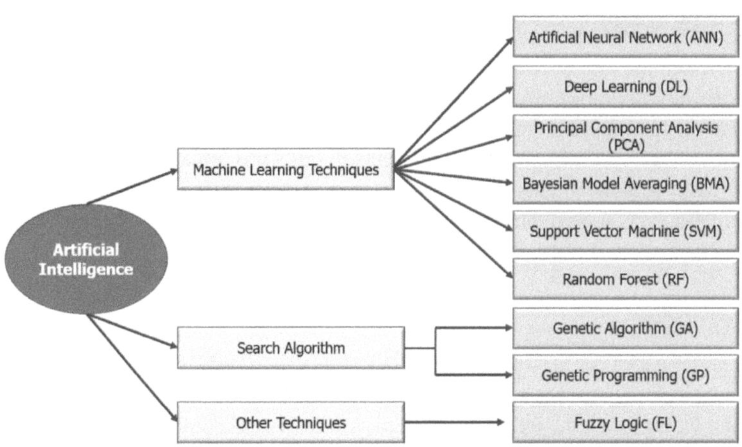

Fig. 5. A classification table of AI and ML techniques useful in water systems.

Research and Innovation

There is significant potential for research and innovation in the field of AI-enabled wastewater treatment. Advances in machine learning algorithms, particularly in handling complex and non-linear data, can lead to more accurate and efficient treatment processes. Innovations in sensor technologies and IoT can enhance data collection, providing the rich datasets needed for training AI models. Collaborative research initiatives between academic institutions, industry, and government agencies can drive the development of new AI applications tailored specifically for wastewater treatment.

Collaboration across sectors is crucial for overcoming the barriers to AI adoption. Partnerships between technology companies and wastewater treatment facilities can facilitate the transfer of knowledge and expertise. Collaborative efforts can also help in developing standardized protocols for data collection and processing, ensuring that AI models are trained on consistent and high-quality data. Additionally, collaboration with regulatory bodies can aid in creating a supportive regulatory environment that encourages innovation while ensuring safety and compliance.

Conclusion

AI technologies hold significant promise for revolutionizing wastewater treatment processes. Despite the challenges, the opportunities for innovation, research, and collaboration are vast. The integration of AI can lead to more efficient, reliable, and sustainable wastewater treatment systems, transforming waste into valuable resources and paving the way for a cleaner, healthier environment. As the field continues to evolve, ongoing research and cross-sector collaboration will be key to unlocking the full potential of AI in wastewater treatment.

Future Directions

Looking ahead, the future of AI in wastewater treatment is promising. AI has the potential to transform wastewater treatment into a resource recovery process, where waste materials are efficiently converted into valuable resources. Emerging trends such as the integration of AI with blockchain technology for secure and transparent data management, and the use of AI-driven predictive maintenance systems to optimize the performance of treatment plants, are areas ripe for exploration. Continuous advancements in AI and related technologies will likely lead to more sustainable and cost-effective wastewater treatment solutions, ultimately contributing to environmental conservation and public health.

References

Afonne, O.J. and Ifediba, E.C. 2020. Heavy metals risks in plant foods–need to step up precautionary measures. *Current Opinion in Toxicology*, 22: 1–6.

Alprol, A.E., Mansour, A.T., Ibrahim, M.E.E.D. and Ashour, M. 2024. Artificial Intelligence technologies revolutionizing wastewater treatment: Current trends and future prospective. *Water*, 16(2): 314.

Alzubaidi, L., Zhang, J., Humaidi, A.J., Al-Dujaili, A., Duan, Y., Al-Shamma, O., Santamaría, J., Fadhel, M.A. and Al-Amidie, M. 2021. Review of deep learning: Concepts, CNN architectures, challenges, applications, future directions. *Journal of Big Data*, 8: 1–74.

Anjaneyulu, Y., Sreedhara Chary, N. and Samuel Suman Raj, D. 2005. Decolourization of industrial effluents–available methods and emerging technologies—A review. *Reviews in Environmental Science and Bio/Technology*, 4: 245–273.

Bagheri, M., Akbari, A. and Mirbagheri, S.A. 2019. Advanced control of membrane fouling in filtration systems using artificial intelligence and machine learning techniques: A critical review. *Process Saf. Environ. Prot.*, 123: 229–252.

Baker, L.A. 1992. Introduction to nonpoint source pollution in the United States and prospects for wetland use. *Ecological Engineering*, 1(1-2): 1–26.

Balabin, R.M. and Lomakina, E.I. 2011. Support vector machine regression (SVR/LS-SVM)—An alternative to neural networks (ANN) for analytical chemistry? Comparison of nonlinear methods on near infrared (NIR) spectroscopy data. *Analyst*, 136: 1703–1712.

Basheer, I.A. and Hajmeer, M. 2000. Artificial neural networks: Fundamentals, computing, design, and application. *J. Microbiol. Methods*, 43: 3–31.

Corominas, L., Garrido-Baserba, M., Villez, K., Olsson, G., Cortés, U. and Poch, M. 2018. Transforming data into knowledge for improved wastewater treatment operation: A critical review of techniques. *Environ. Modell. Software*, 106(Aug): 89–103.

Cox, M., Négré, P. and Yurramendi, L. 2007. Industrial liquid effluents. *INASMET Tecnalia, San Sebastian*, 283.

Crini, G. and Badot, P.M. 2007. Traitement et épuration des eaux industrielles polluées. *PUFC, Besançon*, 353 p (in French).

Crini, G. and Badot, P.M. (eds.). 2010. Sorption processes and pollution. *PUFC, Besançon*, 489 p.

Crini, G. and Badot, P.M. (eds.). 2010. *Sorption Processes and Pollution: Conventional and Non-conventional Sorbents for Pollutant Removal from Wastewaters*. Presses Univ. Franche-Comté.

Crini, G. and Lichtfouse, E. 2018. Wastewater treatment: An overview. *Green Adsorbents for Pollutant Removal: Fundamentals and Design*, 1–21.

Davies, P.S. 2005. The biological basis of wastewater treatment. *Strathkelvin Instruments Ltd.*, 3.

Deblonde, T., Cossu-Leguille, C. and Hartemann, P. 2011. Emerging pollutants in wastewater: A review of the literature. *International Journal of Hygiene and Environmental Health*, 214(6): 442–448.

Ding, S., Li, H., Su, C., Yu, J. and Jin, F. 2013. Evolutionary artificial neural networks: A review. *Artif. Intell. Rev.*, 39: 251–260.

Druart, C., Morin-Crini, N., Crini, G. and Lichtfouse, E. 2018. Wastewater treatment: A overview. *Green Adsorbents for Pollutant Removal: Fundamentals and Design*, 1–21.

Ejder-Korucu, M., Gürses, A., Doğar, Ç., Sharma, S.K. and Açikyildiz, M. 2015. Removal of organic dyes from industrial effluents: An overview of physical and biotechnological applications. *Green Chemistry for Dyes Removal From Wastewater: Research Trends and Applications*, 1–34.

Elsayed, A., Hurdle, M. and Kim, Y. 2021. Comprehensive model applications for better understanding of pilot-scale membrane-aerated biofilm reactor performance. *J. Water Proc. Eng.*, 40(January): 101894.

Euvrard, E. and Crini, G. 2016. Chemical and ecotoxicological monitoring of discharge water from a metal-finishing factory. *Environ. Process.*, 3: 59–72.

Farhan, L. 2021. Review of deep learning: Concepts, CNN architectures, challenges, applications, future directions. *J. Big Data*, 8: 53.

Gaudio, M.T., Coppola, G., Zangari, L., Curcio, S., Greco, S. and Chakraborty, S. 2021. Artificial intelligence-based optimization of industrial membrane processes. *Earth Systems and Environment*, 5(2): 385–398.

Hou, H., Ren, A., Yu, L., Ma, Z., Zhang, Y. and Liu, Y. 2023. An environmental impact assessment of Largemouth Bass (Micropterus salmoides) aquaculture in Hangzhou, China. *Sustainability*, 15: 12368.

Huang, M., Ma, Y., Wan, J. and Chen, X. 2015. A sensor-software based on a genetic algorithm-based neural fuzzy system for modeling and simulating a wastewater treatment process. *Appl. Soft Comput. J.*, 27: 1–10.

Jegatheesan, V., Visvanathan, C. and Ben Aim, R. 2004. Advances in biological wastewater treatment. *Concise Encyclopaedia of Bioresource Technology. Haworth Press, Inc.*

Jin, L., Zhang, G. and Tian, H. 2014. Current state of sewage treatment in China. *Water Res.*, 66: 85–98.

Karri, R.R., Ravindran, G. and Dehghani, M.H. 2021. Wastewater—sources, toxicity, and their consequences to human health. In *Soft Computing Techniques in Solid Waste and Wastewater Management* (pp. 3–33). Elsevier.

Krittanawong, C. 2018. The rise of artificial intelligence and the uncertain future for physicians. *Eur. J. Intern. Med.*, 48: e13–e14.

Kulkarni, A., Yardimci, M., Kabir Sikder, M.N. and Batarseh, F.A. 2023. P2O: AI-driven framework for managing and securing wastewater treatment plants. *Journal of Environmental Engineering*, 149(9): 04023045.

Long, S., Zhao, L., Liu, H., Li, J., Zhou, X., Liu, Y. and Qiao, Z. 2019. Science of the total environment A Monte Carlo-based integrated model to optimize the cost and pollution reduction in wastewater treatment processes in a typical comprehensive industrial park in China. *Sci. Total Environ.*, 647: 1–10.

Loos, R., Carvalho, R., António, D.C., Comero, S., Locoro, G., Tavazzi, S., Paracchini, B., Ghiani, M., Lettieri, T., Blaha, L., Jarosova, B., Voorspoels, S., Servaes, K., Haglund, P., Fick, J., Lindberg, R.H., Schwesig, D. and Gawlik, B.M. 2013. EU-wide monitoring survey on emerging polar organic contaminants in wastewater treatment plant effluents. *Water Res.*, 7: 6475–6487.

Mahesh, B. 2020. Machine learning algorithms—A review. *Int. J. Sci. Res. (IJSR)*, 9: 381–386.

Mamais, D., Noutsopoulos, C., Dimopoulou, A., Stasinakis, A. and Lekkas, T.D. 2015. Wastewater treatment process impact on energy savings and greenhouse gas emissions. *Water Sci. Technol.*, 71: 303–308.

Manasa, R.L. and Mehta, A. 2020. Wastewater: Sources of pollutants and its remediation. *Environmental Biotechnology*, 2: 197–219.

Marcelino, K.R., Wongkiew, S., Shitanaka, T., Surendra, K., Song, B. and Khanal, S.K. 2023. Micronanobubble aeration enhances plant yield and nitrification in aquaponic systems. *ACS EST Eng.*, 3: 2081–2096.

Mingzhi, H., Jinquan, W., Yongwen, M., Yan, W., Weijiang, L. and Xiaofei, S. 2009. Control rules of aeration in a submerged biofilm wastewater treatment process using fuzzy neural networks. *Expert Syst. Appl.*, 36: 10428–10437.

Morin-Crini, N. and Crini, G. (eds.). 2017. Eaux industrielles contaminées (in French).

Oabi, T.A., Mohandes, S.R. and Zayed, T. 2022. Investigating the impact of sewer overflow on the environment: A comprehensive literature review paper. *J. Environ. Manage.*, 301(Jan): 113810.

Rajaee, T., Ebrahimi, H. and Nourani, V. 2019. A review of the artificial intelligence methods in groundwater level modeling. *J. Hydrol.*, 572: 336–351.

Rathoure, A.K. (ed.). 2015 . *Toxicity and Waste Management using Bioremediation.* IGI Global.

Ray, S.S., Verma, R.K., Singh, A., Ganesapillai, M. and Kwon, Y.-N. 2023. A holistic review on how artificial intelligence has redefined water treatment and seawater desalination processes. *Desalination*, 546: 116221.

Russell, D.L. 2019. *Practical Wastewater Treatment.* John Wiley & Sons.

Russell, S.J. and Norvig, P. 2016. Artificial Intelligence: A modern approach. Pearson.

Schütze, M., Campisano, A., Colas, H., Schilling, W. and Vanrolleghem, P.A. 2002. Real-time control of urban wastewater systems-where do we stand today? In *Global Solutions for Urban Drainage*, 1–17.

Shen, Y., Chen, C., Li, P., Huang, X. and Li, Y. 2023. Application of a smart pilot electrochemical system for recycling aquaculture seawater. *Aquac. Int.*, 2023: 1–17.

Sherstinsky, A. 2020. Fundamentals of recurrent neural network (RNN) and long short-term memory (LSTM) network. *Phys. D Nonlinear Phenom.*, 404: 132306.

Simopoulou, M., Sfakianoudis, K., Maziotis, E., Antoniou, N., Rapani, A., Anifandis, G., Bakas, P., Bolaris, S., Pantou, A. and Pantos, K. 2018. Are computational applications the "crystal ball" in the IVF laboratory? The evolution from mathematics to artificial intelligence. *J. Assist. Reprod. Genet.*, 35: 1545–1557.

Sonune, A. and Ghate, R. 2004. Developments in wastewater treatment methods. *Desalination*, 167: 55–63.

Suquet, J., Godo-Pla, L., Valentí, M., Verdaguer, M., Martin, M.J., Poch, M. and Monclús, H. 2020. Development of an environmental decision support system for enhanced coagulation in drinking water production. *Water*, 12: 2115.

Yadav, A., Noori, M.T., Biswas, A. and Min, B. 2023. A concise review on the recent developments in the Internet of Things (IoT)-based smart aquaculture practices. *Rev. Fish. Sci. Aquac.*, 31: 103–118.

Yadav, B., Gupta, P.K., Patidar, N. and Himanshu, S.K. 2020. Ensemble modelling framework for groundwater level prediction in urban areas of India. *Sci. Total Environ.*, 712: 135539.

Wan, J., Huang, M., Ma, Y., Guo, W., Wang, Y., Zhang, H., Li, W. and Sun, X. 2011. Prediction of effluent quality of a paper mill wastewater treatment using an adaptive network-based fuzzy inference system. *Appl. Soft Comput. J.*, 11: 3238–3246.

Chapter 16

Soil Moisture Evaluation by Artificial Intelligence and Computer Vision

Manimaran G,[1] ***Avinash K,***[1] ***Mageshen VR,***[2,*] ***Kamalesh Kanna S***[3] **and** ***Aswitha K***[4]

Introduction

Soil moisture (SM) is crucial for ecological processes and ecosystem services, impacting the hydrological cycle, climate regulation, nutrient availability, and primary productivity (Anagnostopoulos et al., 2017). It significantly influences land evapotranspiration, a vital process connecting water, energy, and carbon cycles (Green et al., 2019; Padrón et al., 2020). In agriculture, optimal soil moisture is essential for maximum crop productivity, making accurate SM data vital for effective water resource management to prevent drought and water stagnation (Li et al., 2021; Liu et al., 2019). SM is also critical for hydrological applications like flood prediction, as saturated soils increase flood risks (Kim et al., 2019). In forestry, it supports tree growth, forest health, and fire risk management (Krueger et al., 2022). Furthermore, SM is integral to climate models, influencing weather predictions, climate projections, and climate variability through its interaction with atmospheric conditions (Koster et al., 2010). Therefore, precise and feasible methods for soil moisture evaluation are essential given their diverse and significant roles.

[1] Department of Soil Science and Agricultural Chemistry, Tamil Nadu Agricultural University, Coimbatore 641003, Tamil Nadu, India.

[2] Amrita School of Agricultural Sciences, Arasampalayam, Coimbatore-642 109, Tamil Nadu, India.

[3] Department of Remote Sensing and GIS, Tamil Nadu Agricultural University, Coimbatore 641003, Tamil Nadu, India.

[4] Department of Soil Science, JKK Munirajah College of Agricultural Science, Erode-638 506, Tamil Nadu, India.

* Corresponding author: mageshsmart2@gmail.com

Traditional soil moisture evaluation methods like gravimetric methods, neutron probes, and tensiometers are accurate but have notable limitations. Gravimetric methods, though precise, are labor-intensive and time-consuming, providing only point-based measurements. Neutron probes are costly, involve radioactive sources, and need careful handling, while tensiometers require regular maintenance and are effective only at specific moisture levels. These limitations—high cost, laboriousness, and limited spatial coverage—hinder their use for large-scale or real-time assessments (Ochsner et al., 2013). Consequently, remote sensing systems have been developed, significantly improving soil moisture predictions (Wang and Qu, 2009). Advances in remote sensing technology now enable the monitoring of spatial time-series soil moisture using optical, thermal infrared (TIR), and microwave observations. New methods and algorithms combining satellite-based optical/TIR or optical/microwave data have enhanced soil moisture retrieval. However, TIR and passive microwave observations, while useful for regional and global scales, have limitations in spatial and temporal resolution for detailed agricultural applications.

The integration of deep learning and advanced machine learning techniques has revolutionized the retrieval of soil parameters, surpassing traditional models. These modern methods are adept at modeling the complex, nonlinear relationships inherent in soil moisture (SM) estimation. For example, deep learning approaches, like the Unsupervised Self-Training Method (USTM), has demonstrated higher precision and accuracy in estimating SM using various indices and ground truth data compared to traditional techniques (Sedighi et al., 2024). Machine learning algorithms are now pivotal in SM estimation studies, effectively simulating patterns and deriving insights from scientific data. They handle the nonlinear and complex nature of SM relationships adeptly and are used in scale transition processes such as gap filling, downscaling, and upscaling, enhancing the spatial completeness and resolution of SM products.

AI and computer vision techniques further advance SM assessment by analyzing visual data from various sensors, including RGB, multispectral, and thermal cameras, with high accuracy and non-invasively (Zhang et al., 2024). Integrating AI and computer vision into SM evaluation supports precision agriculture and daily management practices by enabling site-specific management and optimizing resource use for higher agricultural productivity. These techniques offer deeper insights into water movement within the soil, informing water resource management and ecosystem conservation strategies. AI and computer vision bridge the gap between point measurements and remote sensing, providing essential information on the spatial and temporal variability of SM, crucial for informed decision-making to sustain water resources, ensure food security, and implement ecosystem conservation strategies.

The benefits of integrating AI and computer vision are numerous. AI algorithms can process vast amounts of data from various sensors, improving the accuracy of SM measurements and enabling real-time monitoring. These techniques assess SM without disturbing the soil, preserving ecosystem integrity, and can be applied across large agricultural fields, offering comprehensive SM data that informs irrigation practices and crop management. By automating data collection and analysis, AI and computer vision reduce the need for expensive and labor-intensive traditional methods. AI-driven SM data can seamlessly integrate into precision agriculture systems, optimizing water use and improving crop yields. Moreover, accurate SM data supports climate models and weather predictions, aiding climate adaptation strategies for agriculture and forestry. These technologies provide valuable insights into SM dynamics, supporting sustainable agriculture, forestry, and ecosystem management. Leveraging AI and computer vision allows for informed decisions to optimize water resources, ensure food security, and conserve ecosystems amidst changing climatic conditions. With this background, we will delve into various machine learning techniques, including their classification, underlying relationships, advanced deep learning approaches used in soil moisture (SM) evaluation, and their further applications with SM products.

Machine Learning

Supervised Learning Techniques for Soil Moisture Prediction

Supervised learning is a central element in the methodology of soil moisture prediction that involves training predictive models on datasets labeled with noise-free data (Karthikeyan et al., 2020). As in the aforementioned supervised learning techniques, training is performed using input features—temperature, humidity, soil type, and remote sensing—separating the indices and furthering the test on the target variable, in other terms, the soil moisture for every input data point. The last step in this process is the training of the dataset on which they want the model to learn; that is, the soil moisture results. There are various key steps in the process of implementing supervised learning for soil moisture prediction (Fig. 1). Supervised learning techniques provide potent tools for soil moisture prediction based on labeled datasets, enabling the training of models to effectively predict soil moisture levels associated with a vast array of input features. Some commonly used algorithms (Fig. 2), such as decision trees, random forests, and SVMs, generate robust and highly interpretable predictions, which are very useful for proper management in terms of agriculture practices, water resources, and environmental conservation methods.

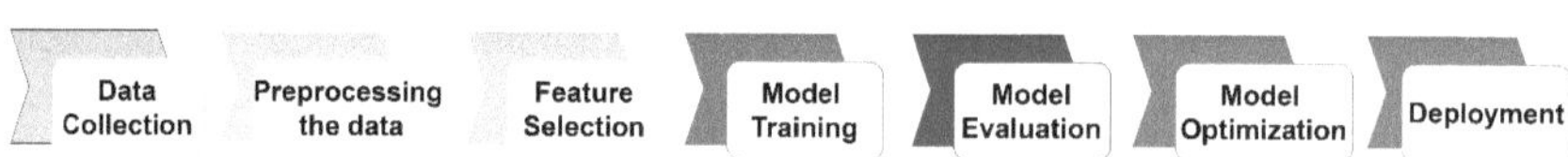

Fig. 1. Basic steps for implementation of machine learning techniques.

Classification

Classification models play a vital role in segregating soil moisture levels into different classes such as dry, moist, or wet. The models proved to be effective in classifying soil conditions for practical decision-making and were generally not suitable for predicting the exact numerical values of soil moisture. By categorizing continuous soil moisture data into meaningful classes, classification models provide more actionable information, especially in agriculture, where certain moisture thresholds must be closely monitored for irrigation and crop management.

Decision Trees

Decision trees are the simplest and most commonly used supervised learning algorithms for soil-moisture prediction. The basic architecture of decision trees relies on the segmentation of the input data into multiple groups based on the values of the input features, which results in a tree-like model with decision rules (Fig. 3). At every tree node, a decision related to a certain feature is made, and the next tree branch is considered as a result of this decision until we reach a leaf node in which the soil moisture level is to be predicted (Pekel, 2020). A key benefit of using decision trees is their simplicity and interpretability. Decision trees are a good illustrative way to gain insight into how a model arrives at its prediction. This makes them very

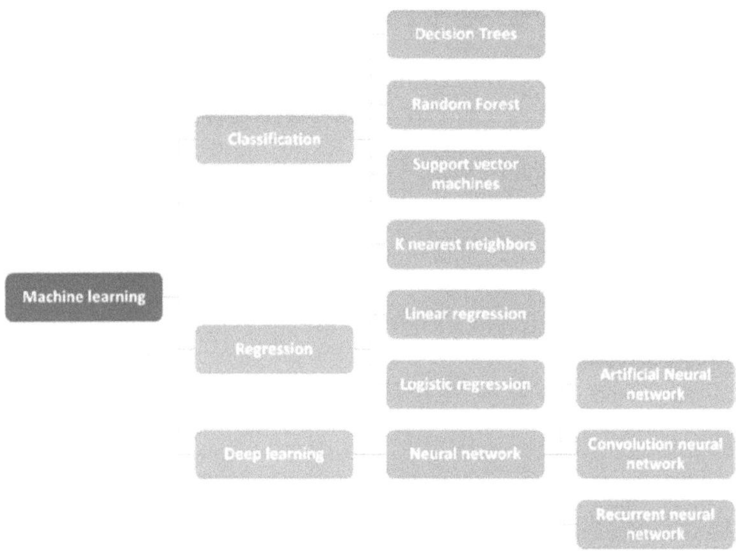

Fig. 2. Commonly used algorithms for Soil Moisture evaluation.

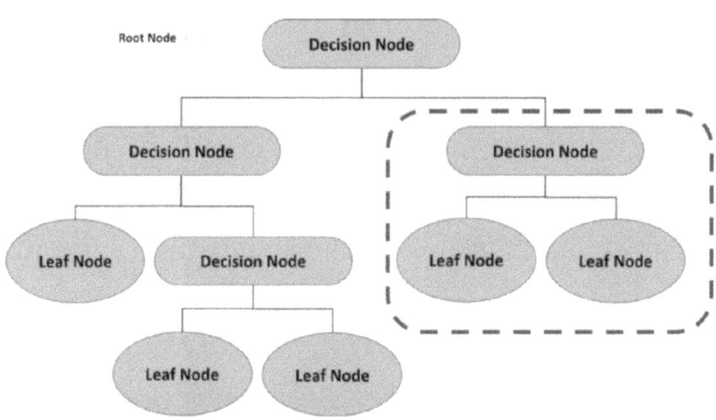

Fig. 3. Representative overview of decision tree algorithm.

attractive for situations in which we try to explain the rationale behind their decisions. However, one reason for using decision trees is their potential to overfit training data. That is when a model becomes too complex to generalize well to new, unseen data. Nevertheless, despite some drawbacks, decision trees are often used in soil moisture prediction in combination with approaches that help avoid overfitting the training set, such as pruning or ensemble approaches (Choi et al., 2024).

Random Forest

Random forest can be seen as a powerful ensemble learning technique that addresses the overfitting problem typically found in individual decision trees. By combining the wisdom of several decision trees, random forests build a solid and versatile prediction model to estimate soil moisture. The key concept of the random forest model is to build multiple uncorrelated trees by introducing randomness at different stages of tree generation (Breiman, 2001). During the training phase, every decision tree in the random forest is trained with a random subset of training data and a random subset of input features. This randomness allows each tree to capture different aspects of the underlying data distribution, thereby reducing the risk of overfitting to some subset of the training data that might be present. Bootstrapping (sampling with replacement) contributes to further diversity in individual trees and therefore strengthens the ensemble model.

To make the final prediction using a random forest, the different trees agglomerate the different predictions from individual trees. This ensemble method specifically solves the problem associated with individual high-variance predictions, thus enhancing the generalization of the model to new data (Fig. 4). Furthermore, random forests can handle a large number of input features. This makes them ideal for soil moisture prediction, which typically involves complex and high-dimensional data (Hengl et al., 2015). Random forests have several advantages over other

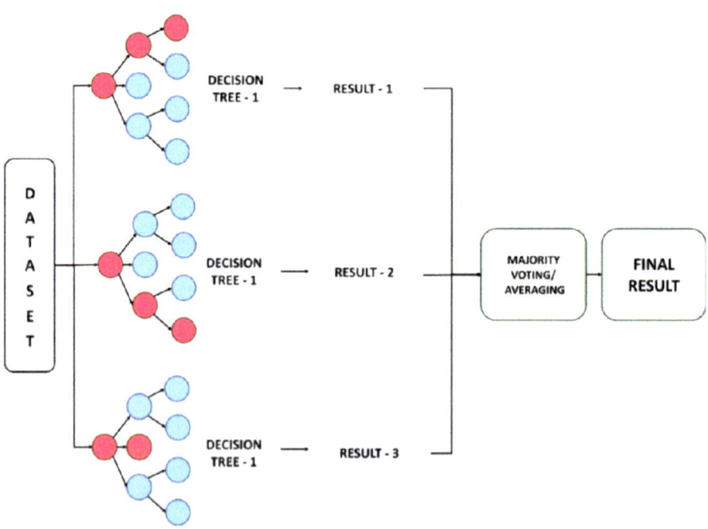

Fig. 4. Illustration of Random Forest.

techniques for soil-moisture prediction. They are naturally tolerant of noisy data and can capture nonlinear relationships between input features and soil moisture levels quite well. Increased predictability and reliability for soil moisture estimation in diverse environmental conditions and landscape configurations are other advantages of the ensemble nature of random forests, considering the collective knowledge learned from multiple trees. Overall, random forests are an important tool in the field of soil moisture mapping, which allows for higher accuracy, reliability, and generalization compared to single-tree models. Research done by Gumiere et al. (2020) demonstrated that the Random Forest (RF) method surpassed traditional physical models like HYDRUS 2D in predicting soil moisture, primarily because of its lower computational requirements and simpler implementation.

Support Vector Machines (SVMs)

Among the wide array of supervised learning algorithms, Support Vector Machines (SVMs) are a prominent candidate for soil moisture prediction. SVMs find the best hyperplane to separate data into classes, effectively distinguishing 'dry,' 'moist,' and 'wet' soil based on features like temperature, humidity, soil type, and remote sensing indices (Fig. 5). They excel in high-dimensional spaces by using kernel functions to transform input data into a feature space where linear separation is feasible, aiding both linear and nonlinear classification tasks (Adankon and Cheriet, 2009). Applying supervised learning for soil moisture prediction involves several key steps to achieve accurate and reliable models (Karthikeyan et al., 2020). Labelled datasets containing input features and corresponding soil moisture measurements are initially collected. Data preprocessing addresses issues like missing values, normalizing features, and partitioning the dataset into training and test sets. Feature selection methods, such as feature-importance scoring and correlation analysis, identify the most relevant features (Guyon and Elisseeff, 2003).

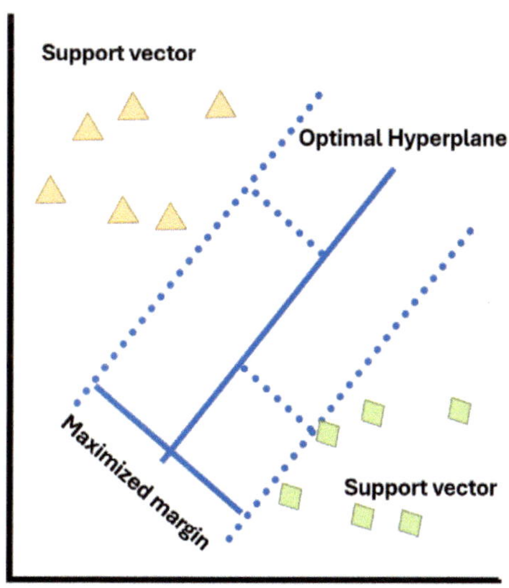

Fig. 5. Illustration of SVM.

Supervised learning algorithms, such as decision trees, random forests, or SVMs, are then trained on the training dataset to uncover relationships between features and soil moisture levels. Model evaluation involves estimating performance on the test set using metrics like MSE for regression or accuracy, precision, and recall for classification, ensuring generalization to new data (Kumar et al., 2024). Hyperparameter tuning and optimization techniques, such as cross-validation or grid search, further refine the model (Bergstra and Bengio, 2012). Once trained, validated, and optimized, the model can be applied in real-world scenarios through precision agriculture systems or environmental monitoring platforms to predict soil moisture and enhance management decisions.

K-Nearest Neighbors

K-nearest neighbors (KNN) is a simple yet powerful algorithm for classifying soil moisture levels (Gounaridis et al., 2019). It classifies data points based on the majority vote of their k-nearest neighbours in the feature space, using distance measures like Euclidean, Manhattan, or Minkowski (Nurdiawan et al., 2021; Ratih et al., 2021). The input features for KNN models typically include variables like temperature, humidity, soil type, and remote sensing indices such as NDVI (Valavi et al., 2019).

KNN is a non-parametric model, meaning it does not assume an underlying data distribution, making it flexible for various classification tasks (Fig. 6). This flexibility is advantageous for soil moisture classification, where relationships between input features and moisture levels can be complex and non-linear. Despite its simplicity and ease of implementation, KNN has high computational costs, especially with large datasets, due to the need to calculate distances for all data points. Moreover, KNN's performance is sensitive to the choice of k and the scaling of input features. Cross-validation can help determine the optimal k, and proper feature scaling is necessary to avoid biased classifications.

KNN helps classify fields into "dry," "moist," or "wet" conditions, aiding farmers in planning irrigation practices efficiently. An ensemble of Kernel K-nearest neighbors (KKNN) models has been used to predict daily soil surface moisture (SSM) at a 100-meter resolution, showing a 27% reduction in spatial error and improved temporal predictive performance compared to coarse resolution datasets (Warner

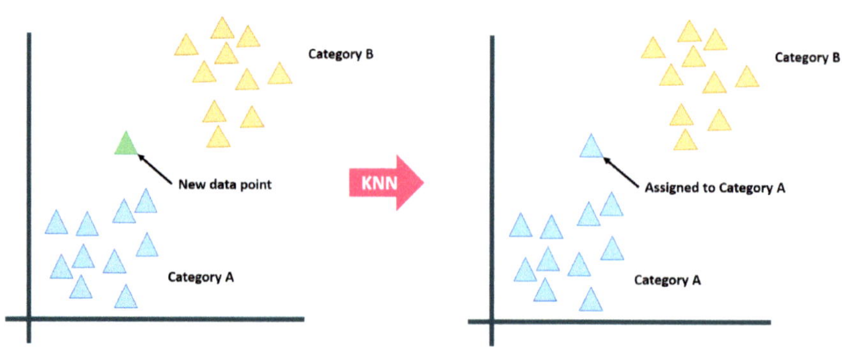

Fig. 6. Classification process of KNN.

et al., 2021). However, these models still struggle with accurately estimating SSM in wetland areas, highlighting the need for better in situ SSM monitoring across diverse landscapes.

Regression Models for Soil Moisture Prediction

Regression models are key to predicting continuous soil moisture levels based on different input characteristics. It was designed to determine any observed relationship between the independent variables, such as temperature, humidity, soil type, and remote sensing indices, and the dependent variable (soil moisture content). By modeling these relationships (Fig. 7), regression techniques permit the optimal quantification of soil moisture, which is important in applications that require forceful numerical predictions.

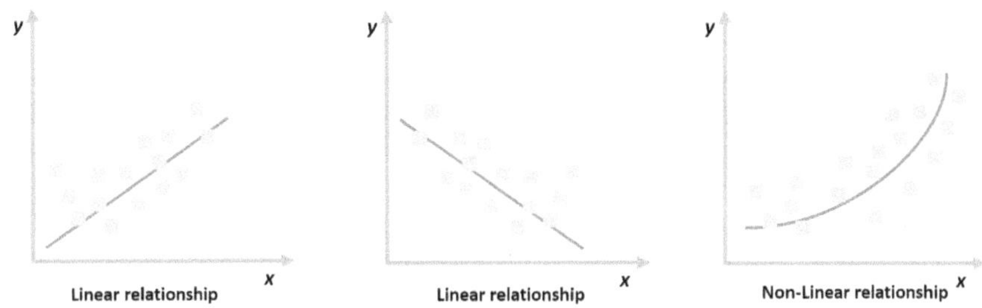

Fig. 7. Types of relationship in regression model.

Linear Regression

Linear regression is perhaps one of the most simple and common types of linear models that are used for regression operations. It builds relationships between the input features and soil moisture content using a linear equation. The underlying concept is to identify the best-fitting straight line—also referred to as the regression line—so that the residual sum of squares between the observed and those predicted by the line soil moisture values is minimized. This method is implemented when the assumption of a linear relationship between the input features and target variable holds. As linear regression is implemented and interpreted without difficulty, it constitutes an easy approach to soil moisture prediction. However, its simplicity could become a drawback when there is an occurrence of complicated, non-linear relationships observed in the data (Paul and Singh, 2021).

Polynomial regression extends the linear model to address the limitations of linear regression; it introduces the polynomial term of the input features. This enables a model to fit a polynomial equation to a dataset, thus enabling it to capture relations that might be more complex. For instance, a quadratic regression will have squared terms of input features, which will enable the model to handle the curvature of the data. Polynomial regression is particularly helpful when the relationship between the input features and soil moisture is non-linear. With higher-degree polynomials, the model can adjust sophisticated patterns and thus enhance the predictability of soil moisture. One needs to be careful not to overfit, as it would mean that the model becomes too complex and accounts for noise in the data rather than the underlying pattern (Jung et al., 2017). Linear regression provides the simplest and most interpretable solution. On the other hand, polynomial regression and ensemble methods, such as gradient boosting, enhance flexibility and the ability to consider higher-order relationships. Precise numerical predictions of the water content can be achieved, leading to better decision-making regarding environmental and agricultural applications (Palominos-Rizzo et al., 2022).

Logistic Regression

Logistic regression is generally used for classifying soil moisture due to its probabilistic approach. Unlike linear regression, logistic regression calculates the probability of an input fitting to a specific class using a logistic function (sigmoid function), transforming predictions into probabilities ranging from 0 to 1. The logistic function transforms the regression output into class probabilities, making them interpretable. It estimates the relationship between input features and the log odds of the target variable through an iterative optimization algorithm, fitting the model to the data. It can handle binary classification (e.g., "dry" vs. "not dry") and multi-class classification using extensions like multinomial logistic regression to predict probabilities for multiple categories (e.g., "dry," "moist," "wet"). One of the strengths of logistic regression is its interpretability. The model coefficients indicate the strength and direction of the relationships between input features and class probabilities, helping analysts understand which features are significant for classifying soil moisture levels (De Figueiredo et al., 2021).

However, logistic regression assumes a linear connection between input features and the log odds of the target variable, which might not be valid for complex datasets with nonlinear relationships. Thus, it might not perform well on complex data. Despite this, logistic regression is commonly used in agriculture for soil moisture classification, aiding land managers and farmers in irrigation scheduling and crop management by predicting the probability of different soil moisture levels. It helps interpret the factors influencing soil moisture, leading to better management strategies.

Deep Learning

Neural Networks

Deep learning, a subset of machine learning, includes neural networks (NNs) that process large, high-dimensional data for soil moisture prediction (Shen, 2018). Neural networks excel at uncovering hidden patterns and relationships that traditional machine learning models often miss (Rakshitkhajuria et al., 2018). For instance, feedforward neural networks can be trained on diverse input features to capture the processes generating soil moisture values (Zhou et al., 2023). Inspired by the human brain, neural networks are well-suited for deep learning tasks in remote sensing (Fig. 8). These architectures, comprising multiple layers, learn hierarchical representations of input features, such as temperature, humidity, soil type, and remote sensing indices, to predict soil moisture content. An important advantage of deep learning is its ability to automatically generate features from raw data, capturing subtle and nonlinear interactions. This is crucial in soil science, where environmental

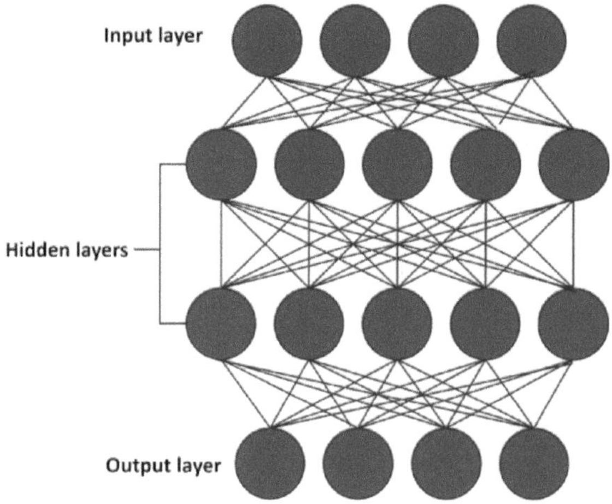

Fig. 8. Schematic diagram of an artificial neural network (ANN).

interactions are complex. However, deep learning requires large volumes of labelled data, which can be difficult and costly to obtain in soil science. Additionally, deep learning models are computationally intensive, requiring substantial resources for training and inference. Deep learning shows promise in integrating with remote sensing data. Convolutional neural networks (CNNs) have been applied to satellite imagery to extract spatial patterns relevant to soil moisture, enhancing the spatial resolution and coverage of predictions and enabling comprehensive monitoring across large areas.

Researchers are developing new architectures and hybrid models that combine deep learning with physical process-based models to leverage their strengths. Efforts to collect and annotate large-scale soil moisture datasets are underway, supporting the training of robust models to capture soil-vegetation-atmosphere dynamics. Overall, deep learning holds great potential for soil moisture prediction due to its ability to find hidden relationships in large datasets. Neural networks and advanced architectures enable accurate, high-resolution estimations of soil water dynamics, providing valuable input for environmental management effective agricultural practices, and further research and development.

Hachani et al. (2019) used Sentinel-1 SAR data and artificial neural networks (ANNs), incorporating auxiliary information like digital elevation models, local incidence angles, and NDVI. They combined satellite measurements with electromagnetic model simulation data (based on the IEM model) for robust, location-independent training, achieving a high correlation (R = 0.77, RMSE = 0.0184). Li et al. (2020) found that a basic feedforward three-layer ANN could simulate and predict soil moisture with lower computational costs and less reliance on prior conditions than traditional physical models. ANNs excel in predicting soil moisture due to their autonomous learning capabilities (Zhang, 2023). Hou et al. (2016) showed strong concordance between ANN predictions and real measurements using meteorological data. Li et al. (2018) used an adaptive genetic ANN model with atmospheric forcing data, significantly improving soil moisture predictions.

Table 1. Traditional machine learning models used for soil moisture evaluation.

Models	Input Variables	Spatial & Temporal	Feature Selection Method	Metric	Merits	References
OH model, SVR, GRNN	Sentinel-1, Sentinel-2, dual-polarization, altitude, LIA, MSAVI	Not specified	Not specified	RMSE: 0.015, R: 0.903	Improved accuracy by removing vegetation influence	Guo et al. (2022)
RF, ARIMA, StemGNN	IoT data	2 years	Not specified	MAPE: 0.052 (ARIMA), 0.127 (StemGNN)	ARIMA outperformed modern DL techniques	Custódio and Prati (2024)
SVM-EnKF	Precipitation, Temperature, Relative Humidity, Net Radiation, Wind Speed, PET	Xiang River Basin	Filter	Correlation, RMSE, ubRMSE, bias	–	Zhu et al. (2020)
PLSR, KNN, RF	RGB, Multispectral, Thermal	Ordos, Inner Mongolia Autonomous Region, China	Wrapper	R^2, RMSE, rRMSE	–	Zhang (2023)
RF, SVM, ELM	Vegetation Indices	Wugong County, Xianyang City, Shaanxi Province, NW China	Embedded	R^2, rRMSE, RMSE, MAE, GPI	–	Zhu et al. (2023)
Decision Tree-Based RF	Polarimetric features, vegetation coverage	Winter wheat areas	Optimal polarimetric feature combination	RMSE: 0.06	High-precision soil moisture retrieval for varying vegetation coverage	Li et al. (2020)
Extreme Random Tree	MODIS vegetation index, LST, soil moisture data	Shandianhe River Basin	Shortwave infrared transformed reflectance (STR)	RMSE: 0.054, R: 0.69	Higher accuracy than SVM and RF models	Cheng et al. (2021)
CIEM, Dubois Models	Synthetic aperture radar (SAR) data	Wheat growing season	Polarization decomposition	CIEM: R^2: 0.534, RMSE: 5.62 vol.%; Dubois: R^2: 0.634, RMSE: 5.16 vol.%	Effective SMC estimation, mitigating vegetation and surface roughness influence	Chen et al. (2022)

Convolutional Neural Networks (CNNs)

Convolutional Neural Networks (CNNs) are powerful tools for spatial data analysis, making them particularly effective for soil moisture evaluation through remote sensing images (Chouhan et al., 2024). CNNs automatically and adaptively learn spatial feature hierarchies from input images through layers of convolutions and pooling operations, capturing spatial patterns and relationships crucial for tasks like soil moisture estimation.

CNNs excel in identifying and extracting important features from remote sensing images, such as vegetation indices (e.g., NDVI) or soil reflectance properties correlated with soil moisture levels. By processing these images through multiple convolutional and pooling layers, CNNs detect minor changes in pixel intensities and spatial arrangements, capturing essential spatial information related to soil moisture dynamics.

A significant strength of CNNs is their ability to perform image segmentation, partitioning images into relevant sections. This capability allows CNNs to create detailed maps illustrating soil moisture distribution within landscapes, aiding researchers and land managers in decision-making for agriculture, hydrology, and environmental management. Li et al. (2020) proposed a Convolutional Neural Network (ICNN) method combining multi-polarization radar and multispectral data to improve soil moisture retrieval, achieving a correlation coefficient of 0.934 and an RMSE of 0.0145, without needing separate vegetation and soil roughness estimations. Similarly, Roberts et al. (2022) used CNNs with GNSS reflection measurements to complement soil moisture data from the SMAP Mission, improving spatial resolution and coverage, and showing a strong correlation with SMAP values.

Recurrent Neural Networks

Recurrent Neural Networks (RNNs), particularly Long Short-Term Memory (LSTM) networks, have emerged as powerful tools for handling time-series data, including soil moisture readings collected over time. Unlike classical feedforward neural networks that neglect the order of input features, RNNs are designed to exploit time dependency, making them ideal for sequences like soil moisture data. LSTMs, a type of RNN, excel at modeling long-range dependencies and retaining information for extended periods, which is crucial for predicting soil moisture levels based on historical data.

LSTM networks capture the temporal dynamics and seasonality inherent in soil moisture time series. They can identify patterns such as diurnal variations, seasonal trends, and long-term fluctuations, resulting in accurate future soil moisture predictions. Additionally, LSTMs are robust against irregular sampling intervals and missing values, common issues in environmental datasets, ensuring their performance in complex data scenarios. The LSTM networks involve several key processes. They handle sequential data, with each time step corresponding to a specific point in the data. LSTMs use memory cells to store information over long periods, maintaining historical context crucial for soil moisture prediction (Fig. 9). The network employs gates (input, forget, and output) to control information flow, which helps it manage important patterns and relationships over time. During training, LSTM networks

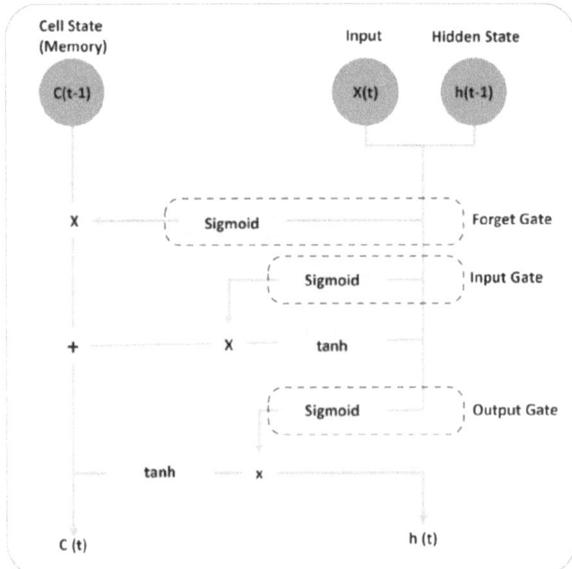

Fig. 9. Schematic representation of LSTM.

learn patterns and dependencies in soil moisture data, adjusting weights and biases to minimize prediction errors. This capability allows them to handle nonlinear relationships influenced by factors like rainfall, temperature, and soil properties, enhancing prediction accuracy.

LSTMs have demonstrated superior performance in soil moisture prediction compared to traditional models. Studies by Fang et al., have shown that LSTMs outperform conventional models when integrating meteorological data and soil characteristics. Li et al. (2023) introduced a Local-Global Dependency LSTM (LGD-LSTM) model for predicting soil textures at different depths, while Lal's team used spatial-temporal series clustering to analyze global vegetation area soil moisture patterns. Datta and Faroughi (2023) developed a multi-head LSTM model integrating soil moisture data from different scales for accurate monthly predictions. Li's team proposed an LSTM model with attention mechanisms to enhance prediction accuracy, followed by an attention-based convolutional LSTM model (AttConvLSTM) for multi-step surface soil moisture prediction. Wang et al. (2024) compared the predictive capabilities and computational costs of various models, noting the effectiveness of attention mechanisms in enhancing LSTM time series predictions.

However, LSTMs face challenges in predicting complex time series such as soil moisture due to their tendency to overfit or underfit nonlinear and multiscale features. Originally designed for predicting single future time points, LSTMs may struggle with long-term data analysis and impact prediction accuracy. Pan et al. noted that the memory characteristics of soil moisture vary at different scales and affect long-term predictions in data-driven models. The dynamic nature of environmental factors like temperature, rainfall, and soil structure, which exhibit high nonlinearity and temporal dependency, further complicates soil moisture prediction for traditional LSTM models.

Computer Visions

Computer vision technology plays a crucial role in soil moisture analysis by extracting meaningful information from remote sensing imagery. This involves a range of techniques for image data acquisition, preprocessing, and detailed analysis. Utilizing data from satellites and UAVs, computer vision leverages multispectral, hyperspectral, and thermal imaging to assess soil moisture accurately (Tang et al., 2022). Key computer vision techniques for soil moisture estimation include image preprocessing (radiometric and geometric corrections), feature extraction, and machine-learning-based classification. These methods improve image quality and identify relevant spatial and spectral soil patterns (Khan et al., 2022). Machine learning algorithms like support vector machines, random forests, and convolutional neural networks analyze complex relationships between features and soil moisture content, using extensive spectral, spatial, and temporal data from remote sensing imagery. By training these algorithms with ground-truth soil moisture measurements, researchers can develop predictive models for various landscapes and temporal scales. This capability enables accurate soil moisture estimation for agricultural management, hydrologic studies, and environmental monitoring (Brocca et al., 2017). Ultimately, computer vision technologies provide a comprehensive toolbox for extracting valuable information on soil moisture dynamics, aiding in informed decision-making for water resources, land-use planning, and ecosystem conservation.

Future Perspective

Soil moisture retrieval models have advanced significantly in recent decades, yet there is still potential for further enhancement. Future improvements should focus on achieving greater spatial and temporal resolution, extending time series monitoring, reducing transmission delays, enhancing accuracy, and improving model performance and interpretability.

Future Directions

Future advancements in soil moisture retrieval models should focus on multi-source data matching by integrating optical and SAR remote sensing to improve accuracy while addressing data format and resolution differences and exploring short-wave infrared data. Enhancing spatial and temporal resolutions is crucial for applications like irrigation and drought prediction, necessitating methods for root zone penetration and providing daily or hourly data. Establishing ground networks and launching more satellites will enable long-term data series with low latency, supported by improved satellite materials for extended operational life. Improving model accuracy involves refining empirical, semi-empirical, and physical models, enhancing machine learning interpretability, and integrating physical and AI methods to develop models requiring minimal auxiliary data. Technological advancements are needed to achieve better spatial and temporal resolution, longer monitoring, shorter delays, and higher accuracy through innovations in satellite materials, observation networks, and models.

Neural Network-Based Models

Table 2. Neural Network based models used for soil moisture evaluation.

Model/ Approach	**Input Variables**	**Spatial & Temporal Data**	**Metric**	**Merits**	**References**
cGBSAR, BP neural network, AIEM-OH	Backscattering coefficient, radar incidence angle	41° radar incidence angle	R: 0.8080, RMSE: 0.0385	Improved accuracy with radar incidence angle selection	Geng et al. (2021)
OH model, SVR, GRNN, CNNR	Sentinel-1, Sentinel-2, polarimetric decomposition features	Not specified	R^2: 0.8947, RMSE: 0.0208 cm^3/cm^3	Highest accuracy with CNNR model	Liu et al. (2021)
MLP	Canopy vegetation indices from UAV remote sensing	40 cm depth	R^2: 0.7447	Simplified model with robust accuracy	Deng et al. (2023)
CNN (SMNet)	Remote sensing, meteorological, topographic data	25 km to 1 km resolution	R: 0.82, RMSE: 0.052 m^3/m^3	Enhanced spatial resolution while maintaining accuracy	Zhang et al. (2024)
DL-MME, DL-CNN	Sentinel-1, Sentinel-2	Diverse agricultural areas	MAE: 0.01 m^3/m^3, RMSE: 0.036 m^3/m^3, R: 0.87	Higher accuracy and reliability	Sedighi et al. (2024)
REDF-LSTM	Time series data	Not specified	13.07% improvement in R^2, 20.98% in RMSE	Superior predictive capability	Li et al. (2024)
LSTM, BiLSTM, DNN, DBN	Net radiation, RH, air temperature, wind speed, SWC	Heterogeneous surfaces	MAPE: 16–50%, RMSE: 0.76 mm d^{-1}	Improved ET estimation with SWC	Feng et al. (2024)
CNN	Climate, Soil, Topography, Cropping Pattern	China	R^2: 0.91		Liu et al. (2021)

Table 3. Applications in Satellite-based Remote Sensing and UAV Imagery.

Aspect	Satellite-Based Remote Sensing	UAV (Drone) Imagery	Integration of Satellite and UAV Imagery	Applications
Multispectral Imagery	Analyz es reflectance across different spectral bands to determine soil moisture.	High-resolution cameras detect fine changes in soil moisture.	Combines broad-scale data with fine-scale UAV data for comprehensive understanding.	Agriculture: Irrigation scheduling, crop yield predictions, optimizing water use.
Hyperspectral Imagery	Offers finer spectral resolution, detecting minor variations in soil properties and moisture.	Flexible data collection supports targeted soil moisture mapping.	Supports informed decision-making in agriculture, hydrology, environmental management, and climate change adaptation.	Hydrology: Flood forecasting, drought monitoring, water resource management.
Thermal Imagery	Analyses surface temperature distributions to estimate soil moisture, combines with other data for accuracy.	UAVs offer dynamic monitoring at specific times and locations.	Allows comprehensive analysis by combining multispectral, hyperspectral, and thermal data.	Assessing ecosystem health, biodiversity conservation, land restoration.
Data Acquisition	Provides comprehensive data on soil moisture dynamics across large areas.	UAVs with high-resolution sensors facilitate fine-scale soil moisture mapping.	Enhances soil moisture estimation accuracy with a combined approach.	Climate resilience, vulnerability assessments, adaptation planning.
Flight Planning	Captures data at varying spatial and temporal resolutions.	UAVs offer greater flexibility in flight planning and data acquisition.	Supports adaptive management in various fields with detailed local-scale data.	Enhances capabilities to address food security, water management, environmental conservation, and climate adaptation.

References

Adankon, M.M. and Cheriet, M. 2009. Support Vector Machine. pp. 1303–1308. *In*: Li, S.Z. and Jain, A. (eds.). *Encyclopedia of Biometrics*. Springer US.

Anagnostopoulos, V., Petropoulos, G.P., Ireland, G. and Carlson, T.N. 2017. A modernized version of a 1D soil vegetation atmosphere transfer model for improving its future use in land surface interactions studies. *Environ. Modell. Software*, 90: 147–156. https://doi.org/10.1016/j.envsoft.2017.01.004.

Bergstra, J. and Bengio, Y. 2012. Random search for hyper-parameter optimization. *J. Mach. Learn. Res.*, 13: 281–305.

Breiman, L. 2001. Random forests. *Mach. Learn.*, 45(1): 5–32. https://doi.org/10.1023/A:1010933404324.

Brocca, L., Ciabatta, L., Massari, C., Camici, S. and Tarpanelli, A. 2017. Soil moisture for hydrological applications: Open questions and new opportunities. *Water*, 9(2): 140. https://doi.org/10.3390/w9020140.

Chen, H., Huang, J.J., Dash, S.S., Wei, Y. and Li, H. 2022. A hybrid deep learning framework with physical process description for simulation of evapotranspiration. *J. Hydrol.*, 606. https://doi.org/10.1016/j.jhydrol.2021.127422.

Cheng, Y., Li, Y., Li, F. and He, L. 2021. Soil moisture retrieval using extremely randomized trees over the Shandian river basin. *National Remote Sensing Bulletin*, 25(4): 941–951. https://doi.org/10.11834/jrs.20219396.

Choi, S.H., Lee, S.-H., Yang, U. and Kim, M.S. 2024. Data mining approach to predicting soil moisture based on meteorological factors and flow rates. *Horticultural Science and Technology*, 42(1): 1–14. https://doi.org/10.7235/HORT.20240001.

Chouhan, S.S., Singh, U.P. and Jain, S. 2024. Introduction to computer vision and drone technology. pp. 1–5. *In*: Chouhan, S.S., Singh, U.P. and Jain, S. (Eds.). *Applications of Computer Vision and Drone Technology in Agriculture 4.0*. Springer Nature.

Custódio, G. and Prati, R.C. 2024. Comparing modern and traditional modeling methods for predicting soil moisture in IoT-based irrigation systems. *Smart Agric. Technol.*, 7. https://doi.org/10.1016/j.atech.2024.100397.

Datta, P. and Faroughi, S.A. 2023. A multihead LSTM technique for prognostic prediction of soil moisture. *Geoderma*, 433. https://doi.org/10.1016/j.geoderma.2023.116452.

de Figueiredo, T., Royer, A.C., Fonseca, F., de Araújo Schütz, F.C. and Hernández, Z. 2021. Regression models for soil water storage estimation using the ESA CCI satellite soil moisture product: A case study in Northeast Portugal. *Water*, 13(1): 37. https://doi.org/10.3390/w13010037.

Deng, J., Pan, S., Zhou, M., Gao, W., Yan, Y., Niu, Z. and Han, W. 2023. Optimum sampling window size and vegetation index selection for low-altitude multispectral estimation of root soil moisture content for Xuxiang Kiwifruit. *Agric. Water Manage.*, 282: 108297. https://doi.org/10.1016/j.agwat.2023.108297.

Feng, J., Wang, W., Xu, F. and Wang, S. 2024. Evaluating the ability of deep learning on actual daily evapotranspiration estimation over the heterogeneous surfaces. *Agric. Water Manage.*, 291. https://doi.org/10.1016/j.agwat.2023.108627.

Geng, D., Zhao, T., Shi, J., Hu, L., Xu, H. and Hu, J. 2021. Surface microwave scattering model evaluation and soil moisture retrieval based on ground-based radar data. *National Remote Sensing Bulletin*, 25(4): 929–940. https://doi.org/10.11834/jrs.20219305.

Gounaridis, D., Chorianopoulos, I., Symeonakis, E. and Koukoulas, S. 2019. A random forest-cellular automata modelling approach to explore future land use/cover change in Attica (Greece), under different socio-economic realities and scales. *Science of the Total Environment*, 646: 320–335. https://doi.org/10.1016/j.scitotenv.2018.07.302.

Green, J.K., Seneviratne, S.I., Berg, A.M., Findell, K.L., Hagemann, S., Lawrence, D.M. and Gentine, P. 2019. Large influence of soil moisture on long-term terrestrial carbon uptake. *Nature*, 565(7740): 476–479. https://doi.org/10.1038/s41586-018-0848-x.

Gumiere, S.J., Camporese, M., Botto, A., Lafond, J.A., Paniconi, C., Gallichand, J. and Rousseau, A.N. 2020. Machine Learning vs. physics-based modeling for real-time irrigation management. *Front. Water*, 2. https://doi.org/10.3389/frwa.2020.00008.

Guo, M.-H., Xu, T.-X., Liu, J.-J., Liu, Z.-N., Jiang, P.-T., Mu, T.-J., Zhang, S.-H., Martin, R.R., Cheng, M.-M. and Hu, S.-M. 2022. Attention mechanisms in computer vision: A survey. *Comput. Visual Media*, 8(3): 331–368. https://doi.org/10.1007/s41095-022-0271-y.

Guyon, I. and Elisseeff, A. 2003. An introduction to variable and feature selection. *J. Mach. Learn. Res.*, 3: 1157–1182.

Hachani, A., Ouessar, M., Paloscia, S., Santi, E. and Pettinato, S. 2019. Soil moisture retrieval from Sentinel-1 acquisitions in an arid environment in Tunisia: Application of Artificial Neural Networks techniques. *Int. J. Remote Sens.*, 40(24): 9159–9180. https://doi.org/10.1080/01431161.2019.1629503.

Hengl, T., Heuvelink, G.B.M., Kempen, B., Leenaars, J.G.B., Walsh, M.G., Shepherd, K.D., Sila, A., MacMillan, R.A., Mendes de Jesus, J., Tamene, L. and Tondoh, J.E. 2015. Mapping soil properties of Africa at 250 m resolution: Random forests significantly improve current predictions. *PloS One*, 10(6): e0125814. https://doi.org/10.1371/journal.pone.0125814.

Hou, X., Feng, Y., Wu, G., He, Y., Chang, D. and Yang, H. 2016. Application research on artificial neural network dynamic prediction model of soil moisture. *Water Saving Irrigation*, 7(7): 70–72.

Jung, C., Lee, Y., Cho, Y. and Kim, S. 2017. A study of spatial soil moisture estimation using a multiple linear regression model and MODIS land surface temperature data corrected by conditional merging. *Remote Sens.*, 9(8). https://doi.org/10.3390/rs9080870.

Karthikeyan, L., Chawla, I. and Mishra, A. 2020. A review of remote sensing applications in agriculture for food security: Crop growth and yield, irrigation, and crop losses. *J. Hydrol.*, 586: 124905. https://doi.org/10.1016/j.jhydrol.2020.124905.

Khan, A., Vibhute, A.D., Mali, S. and Patil, C.H. 2022. A systematic review on hyperspectral imaging technology with a machine and deep learning methodology for agricultural applications. *Ecol. Inf.*, 69: 101678. https://doi.org/10.1016/j.ecoinf.2022.101678.

Kim, S., Zhang, R., Pham, H. and Sharma, A. 2019. A review of satellite-derived soil moisture and its usage for flood estimation. https://doi.org/10.1007/s41976-019-00025-7.

Koster, R.D., Mahanama, S.P.P., Yamada, T.J., Balsamo, G., Berg, A.A., Boisserie, M., Dirmeyer, P.A., Doblas-Reyes, F.J., Drewitt, G., Gordon, C.T., Guo, Z., Jeong, J.-H., Lawrence, D.M., Lee, W.-S., Li, Z., Luo, L., Malyshev, S., Merryfield, W.J., Seneviratne, S.I., Stanelle, T., van den Hurk, B.J.J.M., Vitart, F. and Wood, E.F. 2010. Contribution of land surface initialization to subseasonal forecast skill: First results from a multi-model experiment. *Geophys. Res. Lett.*, 37(2). https://doi.org/10.1029/2009GL041677.

Krueger, E.S., Levi, M.R., Achieng, K.O., Bolten, J.D., Carlson, J.D., Coops, N.C., Holden, Z.A., Magi, B.I., Rigden, A.J. and Ochsner, T.E. 2022. Using soil moisture information to better understand and predict wildfire danger: A review of recent developments and outstanding questions. *Int. J. Wildland Fire*, 32(2): 111–132. https://doi.org/10.1071/WF22056.

Kumar, R., Kumar, A., Bhatia, K., Nisar, K.S., Chouhan, S.S., Maratha, P. and Tiwari, A.K. 2024. Hybrid approach of cotton disease detection for enhanced crop health and yield. *IEEE Access*, 1–1. https://doi.org/10.1109/ACCESS.2024.3419906.

Li, K., Zhang, R., Duan, J. and Lyu, J. 2020. Wide-area soil moisture retrieval using SAR images and multispectral data. *Nongye Gongcheng Xuebao/Transactions of the Chinese Society of Agricultural Engineering*, 36(7): 134–140. https://doi.org/10.11975/j.issn.1002-6819.2020.07.015.

Li, N., Zhang, Q., Yang, F.X. and Deng, Z.L. 2018. Research of adaptive genetic neural network algorithm in soil moisture prediction. *Comput. Eng. Appl.*, 54(1): 54–69.

Li, P., Zha, Y., Shi, L., Tso, C.-H.M., Zhang, Y. and Zeng, W. 2020. Comparison of the use of a physical-based model with data assimilation and machine learning methods for simulating soil water dynamics. *J. Hydrol.*, 584. https://doi.org/10.1016/j.jhydrol.2020.124692.

Li, Q., Wang, Z., Shangguan, W., Li, L., Yao, Y. and Yu, F. 2021. Improved daily SMAP satellite soil moisture prediction over China using deep learning model with transfer learning. *J. Hydrol.*, 600. https://doi.org/10.1016/j.jhydrol.2021.126698.

Li, Q., Zhang, C., Shangguan, W., Li, L. and Dai, Y. 2023. A novel local-global dependency deep learning model for soil mapping. *Geoderma*, 438. https://doi.org/10.1016/j.geoderma.2023.116649.

Li, X., Zhang, Z., Li, Q. and Zhu, J. 2024. Enhancing soil moisture forecasting accuracy with REDF-LSTM: Integrating residual en-decoding and feature attention mechanisms. *Water*, 16(10): 1376. https://doi.org/10.3390/w16101376.

Liu, J., Xu, Y., Li, H. and Guo, J. 2021. Soil moisture retrieval in farmland areas with sentinel multi-source data based on regression convolutional neural networks. *Sensors*, 21(3): 877. https://doi.org/10.3390/s21030877.

Liu, Y., Liu, Y. and Wang, W. 2019. Inter-comparison of satellite-retrieved and Global Land Data Assimilation System-simulated soil moisture datasets for global drought analysis. *Remote Sens. Environ.*, 220: 1–18. https://doi.org/10.1016/j.rse.2018.10.026.

Nurdiawan, O., Kurnia, D.A., Solihudin, D., Hartati, T. and Suprapti, T. 2021. Comparison of the K-Nearest Neighbor algorithm and the decision tree on moisture classification. *IOP Conf. Ser.: Mater. Sci. Eng.*, 1088(1): 012031. https://doi.org/10.1088/1757-899X/1088/1/012031.

Ochsner, T.E., Cosh, M.H., Cuenca, R.H., Dorigo, W.A., Draper, C.S., Hagimoto, Y., Kerr, Y.H., Larson, K.M., Njoku, E.G., Small, E.E. and Zreda, M. 2013. State of the art in large-scale soil moisture monitoring. *Soil Sci. Soc. Am. J.*, 77(6): 1888–1919. https://doi.org/10.2136/sssaj2013.03.0093.

Padrón, R.S., Gudmundsson, L., Decharme, B., Ducharne, A., Lawrence, D.M., Mao, J., Peano, D., Krinner, G., Kim, H. and Seneviratne, S.I. 2020. Observed changes in dry-season water availability attributed to human-induced climate change. *Nat. Geosci.*, 13(7): 477–481. https://doi.org/10.1038/s41561-020-0594-1.

Palominos-Rizzo, T., Villatoro-Sánchez, M., Alvarado-Hernández, A., Cortés-Granados, V. and Paguada-Pérez, D. 2022. Estimation of soil moisture through multiple linear regressions in Llano Brenes, Costa Rica. *Agronomía Mesoamericana*, 33(2). https://doi.org/10.15517/am.v33i2.47872.

Paul, S. and Singh, S. (2021, February 15, 2021). Soil moisture prediction using machine learning techniques.*CIIS '20.*

Pekel, E. 2020. Estimation of soil moisture using decision tree regression. *Theor. Appl. Climatol.*, 139. https://doi.org/10.1007/s00704-019-03048-8.

RakshitKhajuria, D.S., Mr Siddharth, Dr, S. and Singh, S. 2018. A survey of existing leaf disease techniques using Artificial Neural Network. *Seventh Sense Research Group*, 59(1): 52–62.

Ratih, I.D., Retnaningsih, S.M. and Dewi, V.M. 2021. Classification of soil quality using K-Nearest Neighbors methods. *IOP Conf. Ser.: Earth Environ. Sci.*, 739(1): 012011. https://doi.org/10.1088/1755-1315/739/1/012011.

Roberts, T.M., Colwell, I., Chew, C., Lowe, S. and Shah, R. 2022. A Deep-Learning approach to soil moisture estimation with GNSS-R. *Remote Sens.*, 14(14): 3299. https://doi.org/10.3390/rs14143299.

Sedighi, A., Hamzeh, S., Alavipanah, S.K., Naseri, A.A. and Atkinson, P.M. 2024. Ensembles of multiple models for soil moisture retrieval from remote sensing data over agricultural areas: A deep learning-based framework. *Remote Sens. Appl.: Soc. Environ.*, 35: 101243. https://doi.org/10.1016/j.rsase.2024.101243.

Shen, C. 2018. A transdisciplinary review of deep learning research and its relevance for water resources scientists. *Water Resour. Res.*, 54(11): 8558–8593. https://doi.org/10.1029/2018WR022643.

Tang, Z., Jin, Y., Alsina, M.M., McElrone, A.J., Bambach, N. and Kustas, W.P. 2022. Vine water status mapping with multispectral UAV imagery and machine learning. *Irrig. Sci.*, 40(4): 715–730. https://doi.org/10.1007/s00271-022-00788-w.

Valavi, R., Elith, J., Lahoz-Monfort, J. and Guillera-Arroita, G. 2019. blockCV: An R package for generating spatially or environmentally separated folds for k-fold cross-validation of species distribution models. *Methods Ecol. Evol.*, 10(2): 225–232. https://doi.org/10.1111/2041-210X.13107.

Wang, L. and Qu, J.J. 2009. Satellite remote sensing applications for surface soil moisture monitoring: A review. *Front. Earth Sci. China*, 3(2): 237–247. https://doi.org/10.1007/s11707-009-0023-7.

Wang, Y., Shi, L., Hu, Y., Hu, X., Song, W. and Wang, L. 2024. A comprehensive study of deep learning for soil moisture prediction. *Hydrol. Earth Syst. Sci.*, 28(4): 917–943. https://doi.org/10.5194/hess-28-917-2024.

Warner, D.L., Guevara, M., Callahan, J. and Vargas, R. 2021. Downscaling satellite soil moisture for landscape applications: A case study in Delaware, USA. *J. Hydrol.: Reg. Stud.*, 38. https://doi.org/10.1016/j.ejrh.2021.100946.

Zhang, D., Lu, L., Li, X., Zhang, J., Zhang, S. and Yang, S. 2024. Spatial downscaling of ESA CCI soil moisture data based on deep learning with an attention mechanism. *Remote Sens.*, 16(8): 1394. https://doi.org/10.3390/rs16081394.

Zhang, G. 2023. Synergistic advantages of deep learning and reinforcement learning in economic forecasting. *Int. J. Glob. Econ. Manag.*, 1: 89–95.

Zhu, S., Chen, H., Dong, X. and Wei, J. 2020. Influence of persistence and oceanic forcing on global soil moisture predictability. *Clim. Dyn.*, 54(7-8): 3375–3385. https://doi.org/10.1007/s00382-020-05184-8.

Zhu, Z., Bo, Y. and Sun, T. 2023. Spatial downscaling of satellite soil moisture products based on apparent thermal inertia: Considering the effect of vegetation condition. *J. Hydrol.*, 616. https://doi.org/10.1016/j.jhydrol.2022.128824.

Chapter 17

Crop Yield Estimation

M Salomi,[1,*] *JK Kiruthika,*[2] *S Samsudeen*[3] and *G Maragatham*[4]

Introduction

India, a country with an extensive record of agriculture and a wide variety of farming landscapes, is always faced with the challenge of providing food security for its ever-expanding population. By 2030, the World Bank estimates that there will be 1.5 billion people living in India. In addition to boosting overall food production, feeding this expanding population will need effective management of the agricultural resources that are in existence. Navigating this difficulty requires accurate crop yield estimation (Yu, 2021). The amount of harvested produce, such as grains, per unit of land area is referred to as crop yield. Historically, India has assessed crop yields using self-reported data from farmers or manually conducted field surveys, as indicated in Fig. 1. Government organizations (Zhang, 2022). Although these approaches are useful, they frequently fall short of the timeliness and accuracy required to make good decisions.

Though valuable, traditional approaches are frequently imprecise and slow. For various reasons, a reliable estimation of agricultural production is crucial as depicted in Fig. 1.

[1] Assistant Professor, Department of Computational Intelligence, Faculty of Engineering and Technology, SRM Institute of Science and Technology, Kattankulathur, Tamil Nadu, India.

[2] Assistant Professor (Sr. G.), Department of Computer Science and Engineering, KPR Institute of Engineering and Technology, Coimbatore, India.

[3] Assistant Professor, Department of Computer Science and Engineering, GITAM University, Hyderabad, India.

[4] Professor, Department of Computational Intelligence, Faculty of Engineering and Technology, SRM, Institute of Science and Technology, Kattankulathur, Tamil Nadu, India.

* Corresponding author: salomis@srmist.edu.in

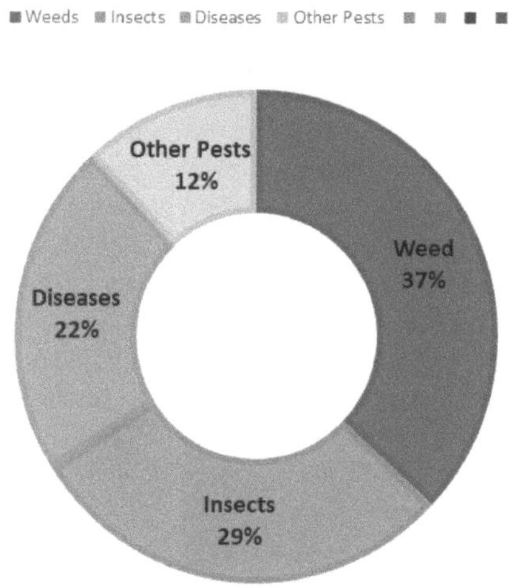

Fig. 1. Annual crop yield loss in India 2023.

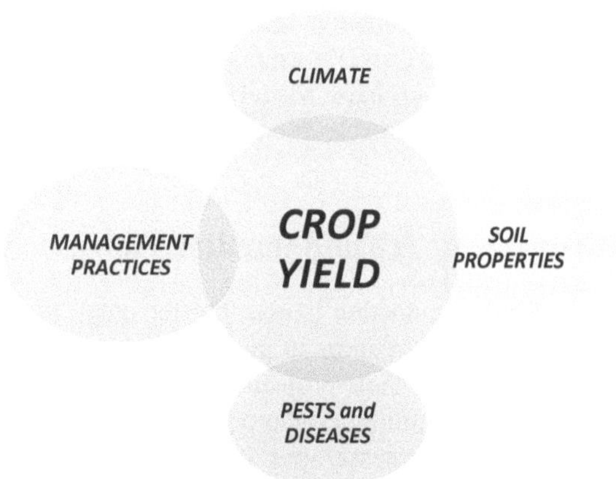

Fig. 2. Factors affecting crop cultivation.

Are farmers confident enough in using yield estimation methods to predict the accuracy of current yield crop percentage? This can be achieved using the recent unmanned ariel vehicle techniques and methodologies. First of all, it enables decision-makers to make well-informed choices. They can strategically arrange food storage, import/export plans, and farmer subsidies with the help of precise yield statistics (Benincasa, 2017). This guarantees a steady supply of food and reduces price swings that may affect customers, particularly those in vulnerable groups.

Secondly, accurate yield predictions give farmers insightful information during the growing season. They can spot possible yield gaps in certain locations and modify techniques like fertilizer or irrigation to make the most use of available resources and possibly boost total yield (Han, 2022).

Finally, accurate yield estimations can increase market effectiveness for growers and buyers equally. Based on reasonable yield forecasts, farmers can set pricing and establish marketing tactics. Furthermore, timely shortages or surpluses might avoid market disruptions through effective estimations and predictions (Javed, 2022). Establishing reliable techniques for estimating crop yields is essential for India's farming industry. It eventually contributes to food security and a more sustainable agricultural system by enabling decision-makers, producers, and market participants to make better decisions through timely and accurate data.

Both satellite photography and labour-intensive field scouting can be costly, time-consuming, and lacking in the information necessary for precise decision-making. Unmanned aerial vehicles (UAVs), sometimes known as drones, are revolutionizing the agricultural landscape of India at present. When compared to conventional techniques, UAV data collection is substantially faster. Farmers can rapidly and effectively cover large fields, enabling them to collect vital data on time. This quick data collection is especially helpful for tasks like tracking the efficiency of irrigation or highlighting locations that need to be cared for immediately.

Hence, farmers should be provided with access to an effective tool for precision farming (Arroyo, 2017). Their flexibility, speed, and high-resolution imaging provide an abundance of advantages over conventional techniques, enabling farmers to make decisions based on information, allocate resources optimally, and eventually enhance crop yields and general farm management techniques.

This chapter reveals the crop yield estimation using UAV integrated with the Normalized Difference Vegetation Index to elevate agriculture ahead to the next level which reduces the risk factor of decreased crop cultivation for the increasing human population in India.

Estimating Crop Yield: A Fundamental View

Precise assessment of crop production is essential for many farming techniques. It determines the optimal amount of seeds to plant, how much fertilizer to apply, and how much watering is required. Yield projections have an impact on market pricing tactics as well, which helps to minimize shortage risks. It also makes proactive risk management strategies against illnesses, pest infestations, and unfavourable weather possible (Zheng, 2018). Traditional ground-based surveys require a lot of time and effort, and can involve human error especially when dealing with big fields but satellite imagery has drawbacks such as spectral resolution limitations, reliance on weather, and is expensive for frequent monitoring.

Increasing Use of UAVs in Farming

Drones, or unmanned aerial vehicles, or UAVs, have changed the precision agriculture industry as depicted in Fig. 3. An airframe, a motor system, a flight control system, and a payload that can hold sensors or cameras make up these flying machines (Li, 2022). Main benefits include scalability, cost-effectiveness, Flexibility and Data Richness.

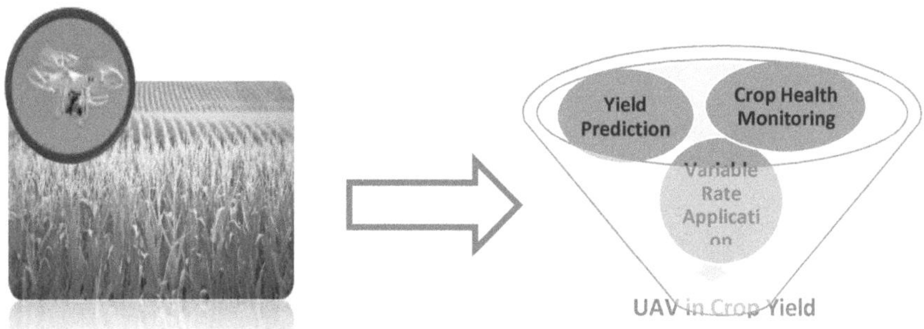

Fig. 3. Contribution of UAV in agriculture.

Scalability

Compared to ground-based techniques, UAVs efficiently cover huge regions in a short amount of time, which makes them perfect for expansive fields.

Cost-Effectiveness

UAVs are a more economical option for data acquisition when compared to manned aircraft.

Flexibility

Their capacity to function at low altitudes makes it possible to gather comprehensive data that offers insightful information about crop health.

Richness of Data

UAVs may obtain multispectral and high-resolution photos, providing a complete picture of crop conditions.

NDVI Need for Crop Yield Estimation

Several important aspects make fast and accurate crop output estimation necessary as depicted in Fig. 4.

Food Security: Increasing agricultural output is necessary to feed the world's expanding population (Nasi, 2018). Planning and resource allocation can be improved by estimating agricultural yields, providing food security for present and future generations.

Predicting the Market: Market pricing techniques are informed by precise yield projections. Farmers can minimize losses and maximize profits by anticipating possible shortages or surpluses and modifying their selling strategy accordingly.

Agricultural Administration: Accurate yield predictions enable informed decision making about the use of fertilizer, irrigation techniques, and pest management

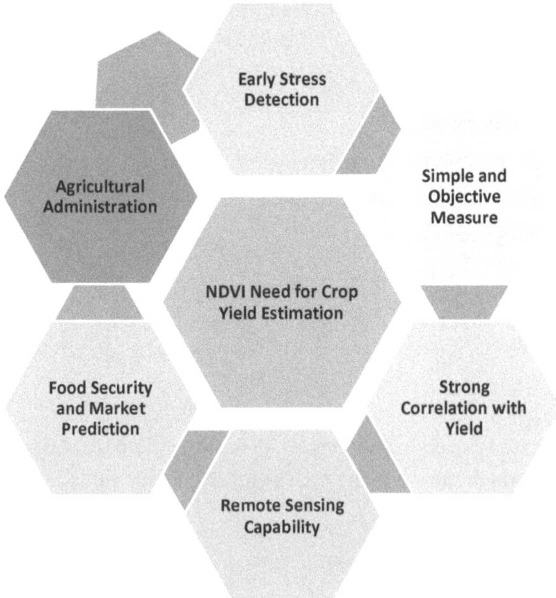

Fig. 4. NDVI needs for crop yield estimation.

tactics. In the end, this results in improved farm productivity and resource utilization (Chouhan, 2024a and b).

These needs cannot be fully satisfied by traditional yield estimation techniques. Although ground-based surveys yield precise information, they are labor-intensive, time-consuming, and unfeasible for large fields (Kumar, 2024). On the other hand, spectral resolution, weather sensitivity, and expenses for regular use are often the reasons behind the limitations of satellite photography. NDVI is a useful technique to deal with these issues.

The NDVI's Power in Crop Yield Estimation

For farmers as well as other agricultural stakeholders, accurate crop yield estimation is essential. Finally, it fosters market stability, enhances the ability to make well-informed decisions, and improves global food security (Chouhan, 2021). Conventional techniques frequently fall short because they are not precise, timely, or scalable. This is where the usefulness of the Normalized Difference Vegetation Index (NDVI) becomes apparent.

One significant benefit of NDVI is that it's an objective, straightforward metric. The NDVI offers a consistent method for evaluating crop growth and health since it is based on the reflectance characteristics of vegetation at particular wavelengths. Rich in chlorophyll, healthy plant reflects more near-infrared (NIR) light and absorbs more visible light. As a result, the NDVI value rises, signifying a robust association with multiple metrics gauging crop health and productivity, such as biomass, leaf area index (LAI), and ultimately, the farmer's harvest.

NDVI has a significant advantage over other metrics, namely, remote sensing capability. Data from a variety of platforms, including handheld sensors, multispectral satellites, and unmanned aerial vehicles, can be used to compute the NDVI. Large agricultural areas can now be effectively and non-intrusively monitored thanks to this technology. Throughout the growing season, farmers can use NDVI to collect data at regular intervals (Wang, 2021). This temporal monitoring enables the tracking of crop growth patterns and the identification of places with potential changes in production. Equipped with this data, agriculturalists can adjust management strategies, such as watering or applying fertilizer, to maximize crop productivity throughout the whole area.

Moreover, NDVI works flawlessly with contemporary farming technologies. It provides useful input for AI models that estimate agricultural yield. In addition to other variables like weather patterns, soil conditions, and past yield data, AI systems are capable of analyzing (Khajuria, 2018) NDVI data. Compared to conventional techniques, our multi-modal approach greatly increases accuracy and refines yield estimates. NDVI also promotes a standardized approach. NDVI makes it easier to compare crops across various fields, farms, and even geographical areas by offering a standardized indicator of crop health. In addition to empowering individual farmers, this encourages cooperation and knowledge exchange among agricultural researchers, which eventually advances the field.

NDVI's Limitations on Crop Yield Estimation

Even though NDVI has many advantages, it's critical to recognize its drawbacks. A significant obstacle is its susceptibility to outside influences as depicted in Fig. 5. NDVI values can be influenced by soil moisture, background soil colour, and even cloud cover, which could result in yield projections that are off. For example, an area with dry soil may have a lower NDVI value, which could be an indication of poor crop health when there may be a water shortage.

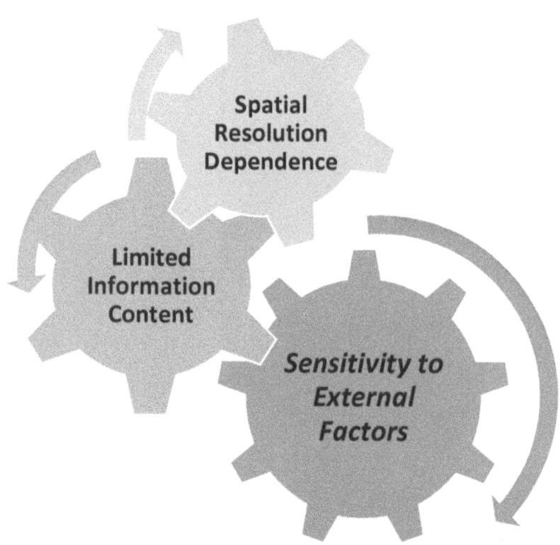

Fig. 5. NDVI's limitations on crop yield estimation.

Further constraint pertains to the data that NDVI offers. It provides a general overview of the vegetation health, but it is not detailed enough to identify certain crop types or stressors. A low NDVI value may be a sign of a pest infestation, nutrient shortage, or even a crop type that grows slowly by nature. It's challenging to identify the precise issue and implement focused remedial measures in the absence of more information.

Finally, the spatial resolution of the data source affects how accurate NDVI-based yield estimates are. Data with lower resolution, such as that gathered from some satellites, may overlook significant differences within a field. Consider a scenario in which pest damage is present in a particular region of the field. If the data resolution is coarse, the localized issue may be obscured by the field's overall NDVI value, leading to an overestimation of potential yield loss.

Overcoming the Challenges of NDVI

NDVI has certain limitations even if it's a helpful tool for estimating crop productivity. Figure 6 elaborates on how we may improve its efficacy and overcome the mentioned obstacles.

Multi-modal Data Integration: A more complete picture is produced when NDVI is combined with information from other data sources, including weather patterns, soil type, and past yield statistics. This makes it possible to comprehend NDVI reading-influencing elements more deeply and produces more accurate yield projections.

Advanced Data Processing Techniques: You can isolate particular regions of a field for examination by using techniques like picture segmentation. By further separating the reflectance signals from the soil, plant, and other background components, spectral unmixing helps reduce extraneous noise in NDVI data.

These techniques have the potential to greatly increase the precision and dependability of NDVI-based crop output estimation, giving farmers the flexibility to make data-driven choices and maximize their farming methods.

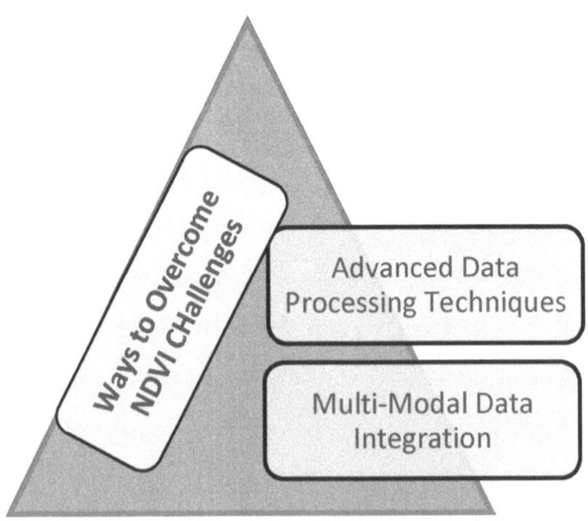

Fig. 6. Overcoming challenges of NDVI.

Data Acquisition

The following data is captured in the year 2023 in Chengalpattu district of the state Tamil Nadu, India (12°40′54.8″N 79°59′19.7″E). The boundaries of the field area are identified and then extracted the shape files. A DJI Mini2 drone is used to capture the field images. The video capture duration has an interval of 20 minutes between each flightt to capture the field. The data captured through UAV drones in the field and the data set consists of 10,000 images. The image resolution is of 18000 by 15000 pixels. The field consists of Papaya, Mango and Guava trees as depicted in Fig. 7. The field image is annotated and pre-processed and can be used as inputs to perform crop yield estimation using the Normalized Difference Vegetation Index.

Fig. 7. DJI Mini2 drone field images.

Calculation of Vegetation Index

Initially, a combination of visible and near-infrared wavelengths was used to calculate various ratios and normalized indices as depicted in Equation 1.1. NDVI values span from +1 (strongest vegetative growth) to −1 (typically water). The NR ($\lambda = 700 - 1300$ nm) and VS ($\lambda = 550 - 700$ nm). The optical properties of the cellular structure and air-cell wall of leaf tissues determine their reflectance levels. Air cell wall contacts between the chloroplasts and the protoplasm.

$$NDVI = (NR - VS) \div (NR + VS) \tag{1.1}$$

whereas,

NDVI stands for Normalized Difference Vegetation Index

NR - Near Infrared Radiation

VS -Visible Red Spectrum

This formula states that the difference in the red and infrared range's reflected light intensities divided by the sum of these intensities equals the density of vegetation (NDVI) at a specific place in the image.

Results and Discussions

The comparison results as depicted in Table 1 and Fig. 8 reveal that the Normalized Difference Vegetation Index provides higher accuracy in estimating crop health compared to other existing models.

Table 1. Comparison results analysis of various techniques for crop yield estimation.

Models	Accuracy
Linear Regression	0.64
Support vector Machine	0.559
Convolution Neural Network	0.673
Normalized Difference Vegetation Index	0.789

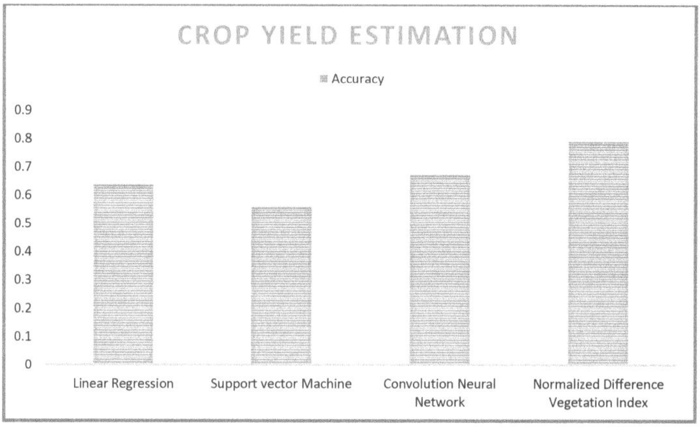

Fig. 8. Comparison results analysis of various techniques for crop yield estimation.

Conclusion

In conclusion, the combination of computer vision, artificial intelligence (AI), and the power of NDVI is revolutionizing the conventional approaches to crop yield estimation. Due to its ease of use, positive association with yield, and capacity to utilize remote sensing data, NDVI offers a strong foundation for data-driven yield estimation. By evaluating enormous datasets and extracting complex information, AI algorithms in conjunction with computer vision techniques enable NDVI to reach its maximum potential. This cooperative strategy improves yield forecasts, facilitates early stress identification, and gives farmers the ability to make data-driven decisions. Future agricultural practices offer a more productive and sustainable environment, providing food security for future generations.

Declaration

Ethical Responsibilities of Authors

"All authors have read, understood, and have complied as applicable with the statement on "Ethical responsibilities of Authors" as found in the Instructions for Authors".

Ethical Approval

Not applicable.

Use of AI Tool

The corresponding author declares that no AI tool has been used to prepare this manuscript.

Conflict of Interest

On behalf of all the authors the corresponding author states that there is no conflict of interest to disclose.

Competing Interests

Not applicable.

Authors' Contributions

Author[1]: Conceptualization, Methodology, Writing - Original Draft, Writing – Ideology - Review & Editing, **Author**[2]: Data Collection - Writing - Ideology, Supervision, **Author**[3]: Writing - Formal analysis, Ideology, **Author**[4]: Formal Writing, Methodology.

Funding

No funding has been received.

References

Arroyo, J.A. 2017. Assessing nitrogen nutrition in corn crops with airborne multispectral sensors. *Lect. Notes Comput. Sci. (including Subser. Lect. Notes Artif. Intell. Lect. Notes Bioinformatics)* 10351 LNCS, 259–267.

Benincasa, P. 2017. Reliability of NDVI derived by high resolution satellite and UAV compared to in-field methods for the evaluation of early crop N status and grain yield in Wheat. *Exp. Agric.*, 54: 604–622.

Chouhan, S.S., Singh, U.P. and Jain, S. 2024b. Introduction to computer vision and drone technology. *In*: Chouhan, S.S., Singh, U.P. and Jain, S. (eds.). *Applications of Computer Vision and Drone Technology in Agriculture 4.0.* Springer, Singapore. https://doi.org/10.1007/978-981-99-8684-2_1.

Chouhan, S.S., Singh, U.P. and Sharma, U. 2024a. Classification of different plant species using deep learning and machine learning algorithms. *Wireless Pers. Commun.*, 136: 2275–2298. https://doi.org/10.1007/s11277-024-11374-y.

Chouhan, S.S., Uday Pratap Singh, U.P., Sharma, U. and Jain, S. 2021. Leaf disease segmentation and classification of Jatropha Curcas L. and Pongamia Pinnata L. biofuel plants using computer vision based approaches. *Measurement*, 171: 108796, ISSN 0263-2241. https://doi.org/10.1016/j.measurement.2020.108796.

Han, S. 2022. Monitoring key wheat growth variables by integrating phenology and UAV multispectral imagery data into random forest model. *Remote Sens.*, 14.

Javed, T. 2022. Recent advances in agronomic and physio-molecular approaches for improving nitrogen use efficiency in crop plants. *Front. Plant Sci.*, 13: 1–21.

Khajuria, R., Sunanda and Singh, S. 2018. A survey of existing leaf disease techniques using Artificial Neural Network. *International Journal of Computer Trends and Technology (IJCTT)*, 59(1): 52–62, May 2018. ISSN:2231-2803, doi:10.14445/22312803/IJCTT-V59P109 (https://ijcttjournal.org/archives/ijctt-v59p109).

Kumar, R. 2014. Hybrid approach of cotton disease detection for enhanced crop health and yield. In *IEEE Access*, https://ieeexplore.ieee.org/document/10589357.

Li, M. 2022. Retrieval of nitrogen content in apple canopy based on unmanned aerial vehicle hyperspectral images using a modified correlation coefficient method. *Sustain.*, 14.

Nasi, R. 2018. Estimating biomass and nitrogen amount of barley and grass using UAV and aircraft based spectral and photogrammetric 3D features. *Remote Sens.*, 10: 1–32.

Wang, L. 2021. Estimation of paddy rice nitrogen content and accumulation both at leaf and plant levels from uav hyperspectral imagery. *Remote Sens.*, 13: 1–21.

Yu, J., Wang, J. and Leblon, B. 2021. Evaluation of soil properties, topographic metrics, plant height, and unmanned aerial vehicle multispectral imagery using machine learning methods to estimate canopy nitrogen weight in corn. *Remote Sens.*

Zhang, J. 2022. In-season variable rate nitrogen recommendation for wheat precision production supported by fixed-wing UAV imagery. *Precis. Agric.*, 23: 830–853.

Zheng, H. 2018. Combining unmanned aerial vehicle (UAV)-based multispectral imagery and ground-based hyperspectral data for plant nitrogen concentration estimation in rice. *Front. Plant Sci.*, 9: 1–13.

Chapter 18

Soil Fertility Evaluation

***VR Mageshen,*[1,*] *G Manimaran,*[2] *K Aswitha*[3] **and** *S Thirumeninathan*[4]**

Introduction

Soil fertility evaluation is the assessment of the soil's capacity to provide essential nutrients to plants. Assessing the fertility of the soil is a crucial component of an efficient agricultural system. Through comprehending the soil's nutrient condition, farmers and agronomists may make well-informed choices to boost soil health, optimize crop yields, and assure the implementation of sustainable farming techniques (Salomon and Cavagnaro, 2022). Conducting regular soil testing and implementing suitable management actions are crucial for preserving productive and healthy soils. The basis for assessing soil fertility is on a thorough comprehension of soil characteristics and their impact on plant development (Guo et al., 2021). Soil fertility evaluation involves five primary procedures. Firstly, it examines the indications of nutrient deficiency in plants. Additionally, the second way entails conducting plant and tissue tests, while the third method focuses on cultivating higher plants and microbes. The fourth approach involves soil chemical analysis, while the last technique utilizes isotopic dilution techniques. Through a methodical examination of various soil characteristics, we can ascertain its ability to provide necessary nutrients, preserve ideal circumstances for

[1] Department of Soil Science and Agricultural Chemistry, Amrita School of Agricultural Sciences, Amrita Vishwa Vidyapeetham, Coimbatore - 642 109.

[2] Department of Soil Science and Agricultural Chemistry, Tamil Nadu Agricultural University, Coimbatore - 641 003.

[3] Department of Soil Science and Agricultural Chemistry, JKK Munirajah College of Agricultural Science, Erode - 638506.

[4] Department of Agronomy, Pandit Jawaharlal Nehru College of Agriculture and Research Institute, Karaikal - 609693.

* Corresponding author: mageshsmart2@gmail.com

Orcid Id: https://orcid.org/0000-0001-6611-6223

growth, and facilitate environmentally-friendly farming techniques (Cakmakcı et al., 2023). This comprehensive approach guarantees that soil management practices are both effective and efficient, while also being ecologically sustainable.

Artificial intelligence is a discipline within the field of computer science which utilizes various methods, including machine learning and deep learning, to analyze and interpret data to imitate human intellect (Chouhan et al., 2024b). These kinds of networks make recommendations by continuously linking input and output elements, allowing for the development of several solutions to both simple and intricate scenarios (Tyagi and Chahal, 2022). AI's interdisciplinary nature, incorporating computer science, mathematics, statistics, and cognitive science, empowers it to address intricate difficulties in agriculture. AI technology is advancing the productivity of agriculture, maintaining food security, and promoting sustainable farming practices by imitating human intellect in robots and other automated systems (Shaikh et al., 2022). Machine learning allows computers to acquire knowledge and make decisions autonomously, without the need for explicit programming, by recognizing and analyzing patterns in data (Javaid et al., 2023). This requires training algorithms on large datasets, allowing them to steadily improve their performance as they acquire information.

Scenario of Soil Testing

Assessing soil fertility may be a tedious and time-consuming procedure, but its comprehensiveness is crucial for obtaining precise and significant outcomes. The complexity occurs due to the necessity of comprehending and incorporating diverse soil variables and their interactions with plant development (Alaoui et al., 2022). Conventional methods for measuring soil nutrients involve using the random grid sample approach to collect soil cores. Subsequently, the cores are carefully enclosed and sent to a laboratory (Lawrence et al., 2020). Furthermore, this procedure will require costly laboratory equipment and supplementary solvents for further chemical examination. While the laboratory approaches exhibit great accuracy, they do not possess the capability to provide continuous real-time examination of agricultural soil. Large scale farmers have two significant problems when trying to map their terrain: duration per sample and expense per sample. Farm terrain exhibits variability, indicating that the characteristics represented on one crop land may differ from those found on another over the same terrain.

Rise of AI in Soil Fertility Management

Machine learning models in soil nutrient analysis can accurately forecast nutrient levels, optimize fertilizer utilization, and improve crop yields by assessing various data sources such as soil samples, remote s atellite images, and meteorological events (Folorunso et al., 2023). Regression algorithms and neural networks are widely used for soil nutrient monitoring due to their ability to accurately predict continuous values using input data. Both machine learning (ML) and deep learning (DL) methods are used to predict soil nutrients. ML applications utilize both supervised and unsupervised learning methods to enhance data analysis and produce ample data for statistical solutions (Trontel and Chambers, 2021). Deep learning techniques are utilized to efficiently analyze a vast quantity of data for the purpose of soil nutrient testing. These approaches provide precise forecasts of soil nutrient levels by using extensive datasets that encompass diverse soil features and environmental circumstances. This predictive capability aids in optimizing the use of fertilizer and enhancing strategies for managing crops.

AI Techniques for Soil Fertility Management Decision Tree

Decision tree is another popular machine learning method used for classifying and predicting data in problems involving regression as well as classification. It represents decisions and their potential outcomes using a hierarchical structure like a tree. The decision tree operates by constantly dividing the dataset into smaller groups, or subsets, according to the value of a certain characteristic (Priyanka and Kumar, 2020). The objective is to enhance the uniformity of the desired outcome within each subset. This technique generates a hierarchical arrangement of choices that culminates in a definitive forecast.

Decision trees are useful in categorizing soil samples according to their nutrient content and other characteristics during the assessment of soil fertility. The nodes of the tree symbolize determination of nutrient thresholds, while the leaves symbolize the anticipated fertility status. This assists farmers and agronomists in making informed decisions based on information.

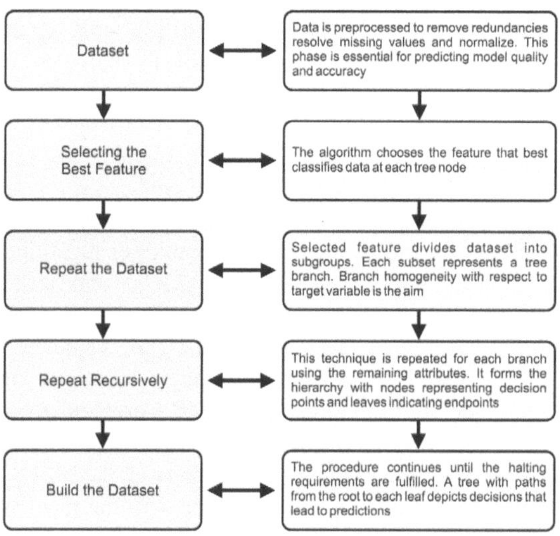

Fig. 1. Working of decision tree algorithm.

Artificial Neural Networks

Artificial Neural Networks (ANNs) are computer models which draw inspiration from neural networks seen in the brains of individuals. These algorithms are used for diverse machine learning endeavors, including classification, regression, and pattern identification (Khajuria and Sunanda, 2018). Artificial neural networks (ANNs) are particularly efficient in assessing soil fertility by acquiring intricate connections

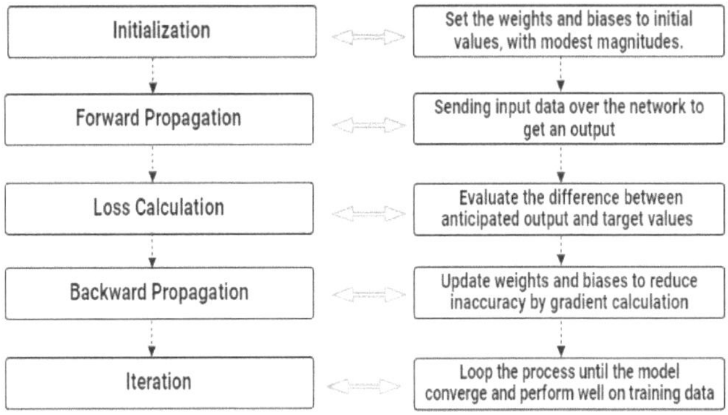

Fig. 2. Working of ANN algorithm.

between soil parameters and fertility levels (Sujatha and Jaidar, 2023). This can result in enhanced prognostic accuracy and improved choices for soil management and agricultural activities. Artificial Neural Networks function by emulating the cognitive processes of the human brain in processing data. Neural networks consist of interconnected neurons arranged in layers, with each neuron applying a behavior perspective against the cumulative weight of its inputs. The network uses forward propagation and back propagation to optimize its weights and biases, therefore reducing the discrepancy between anticipated and actual values (Gulcu, 2022). Artificial Neural Networks (ANNs) are very effective tools for analyzing intricate patterns and correlations in data, render them suitable for a diverse range of operations, including the assessment of soil fertility.

K-Nearest Neighbor

KNN is a non-parametric and robust technique that is utilized to address classification and regression difficulties. It does not rely on hypotheses regarding the fundamental data dissemination. The K-Nearest Neighbor (KNN) approach operates on the principle of similarity. Through the examination of the numerical values of its K nearest neighbors in the training dataset, it predicts the value for a new data point (Uddin et al., 2023). There is no need to make any conjectures about the allocation of data. The K value denotes the total quantity of nearest neighbors. KNN aims to ascertain the precise category for the test data by computing the proximity that exists between the test data and each training point (Zhang et al., 2021). There are several techniques available for computing this distance, with the most widely recognized ways being Euclidean distance (continuous data), Manhattan distance (continuous data), and Hamming distance (categorical data). The computational cost of updating distance measures with every loop is high, which is why KNN is considered a lazy learning algorithm.

The KNN algorithm is both time-consuming and resource-intensive. The algorithm encounters difficulty in accurately categorizing the data points when the number of dimensions is very large. It is also susceptible to the issue of excessive fitting.

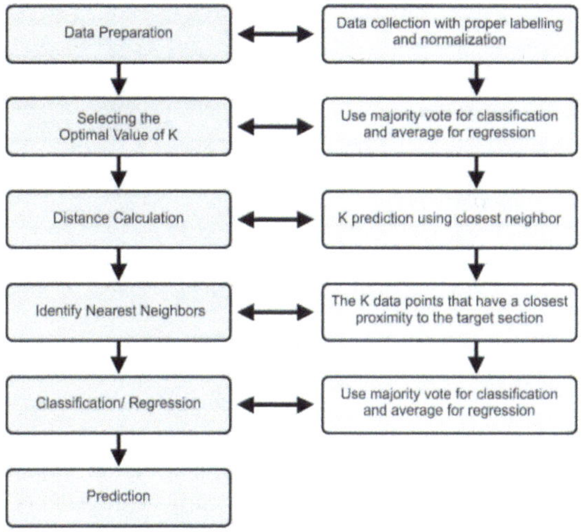

Fig. 3. Working of KNN algorithm.

Support Vector Machine

The Support Vector Machine (SVM) represents a resilient and adaptable trained machine learning method frequently employed for classification problems, such as assessing soil fertility. The core principle underlying Support Vector Machines (SVM) was to ascertain an ideal hyperplane which efficiently distinguishes among data points owing towards different groups (Cervanates et al., 2020).

SVM can effectively forecast the fertility status of soil samples by identifying the best hyperplane that splits various classes, using information about their nutrient content and other features (Sujatha and Jaidhar, 2023). This facilitates the process

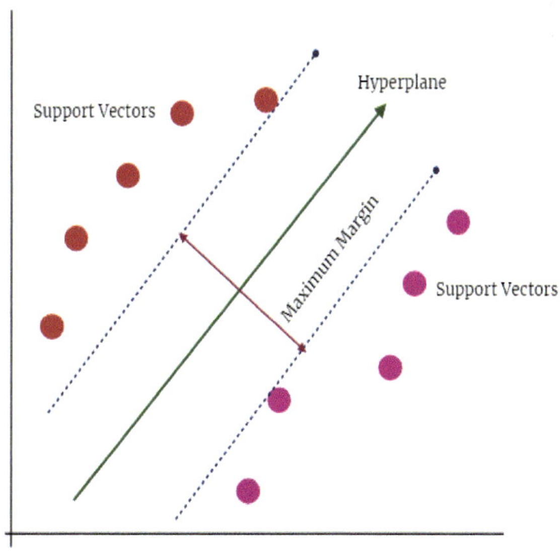

Fig. 4. Concept of Support Vector Machine.

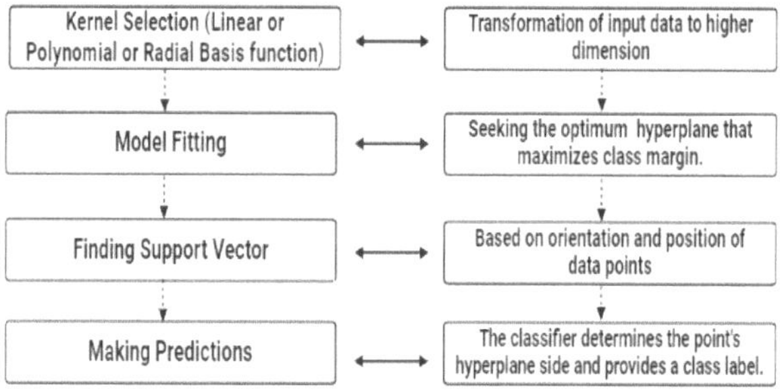

Fig. 5. Working of SVM algorithm.

of making well-informed judgments on soil management and agricultural practices. Support Vector Machines (SVM) employ a kernel mechanism to convert the input characteristics into high dimensional space with a higher number of dimensions, particularly when the data points cannot be linearly segregated (Gaye et al., 2021). This enables the utilization of a hyperplane concerning effectively separating given data. This phenomenon is commonly referred to as the "kernel trick". The goal is to reduce misclassification errors while simultaneously maximizing the margin.

Computer Vision Technologies in Soil Fertility Evaluation

Drones in Soil Fertility Evaluation

The adoption of unmanned aerial vehicles (UAVs) for assessing the quality and productivity of the soil is a cutting-edge method that exploits sophisticated technology to improve agricultural techniques. Drones are employed for to monitor soil fertility by gathering precise and up-to-date information from the atmosphere (Chouhan et al., 2024a). Drones utilize modern imaging technology and spatial analysis to offer comprehensive information on soil health, facilitating accurate and effective soil management techniques (Raza et al., 2023). The introduction of drone technology, coupled with sophisticated sensors, has completely transformed the process of soil sampling and analysis in the field of agriculture. Their objective is to provide farmers with precise and timely information that may be used to improve soil management techniques, increase crop yield, and promote sustainable agriculture practices worldwide (Kumar et al., 2024). Drones outfitted with multispectral cameras are transforming soil analysis by delivering prompt, thorough, and practical data on a previously unachievable scale using conventional techniques (Mushtaq et al., 2024). This method signifies a notable advancement in sustainable agriculture since it maximizes resource use and enhances crop productivity while reducing negative effects on the environment.

Drones offer accurate and up-to-date information that aids the agricultural community in efficiently managing soil health. Drones may combine their obtained data with other sources, such as satellite imaging, meteorological data, and historical crop production data, to create thorough models for assessing soil fertility (Weiss et al., 2020). The utilization of sophisticated data, processing and machine learning approaches significantly enhances understanding of this data (Chouhan et al., 2021).

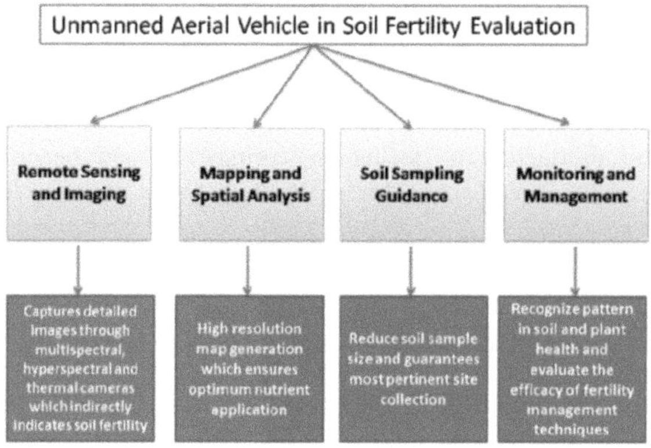

Fig. 6. Role of Drone in soil fertility evaluation.

Farmers may optimize their time, obtain a precise assessment of every square inch of land, and, most significantly, reduce costs compared to doing physical soil analysis (Khan et al., 2021). As technology continues to progress, drones are anticipated to have a progressively crucial function in addressing the issues of providing food for a rapidly expanding global population while also preserving natural resources.

Internet of Things

Internet of Things (IoT) has transformed soil fertility assessment through the provision of uninterrupted, continuous surveillance and data acquisition (Narasegouda et al., 2020). This method facilitates the more accurate and effective control of soil health, thereby improving agricultural yield. The concept of IoT in soil fertility evaluation involves establishing a networked system that consistently gathers, transmits, analyzes, and utilizes real-time data from diverse sensors to enhance and automate soil management techniques (Parween et al., 2021). This technique results in agricultural operations that are more accurate, effective, and environmentally friendly.

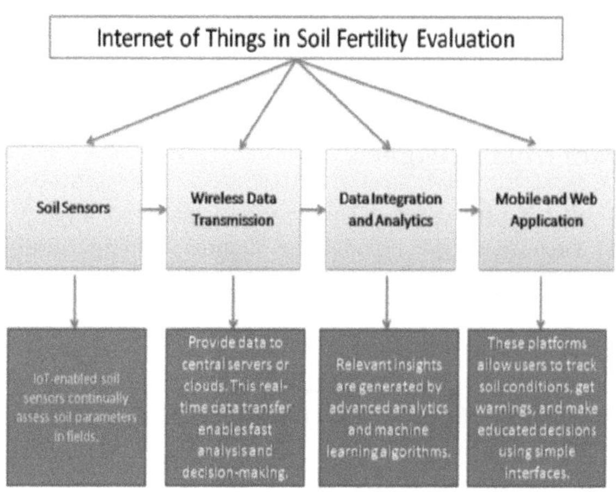

Fig. 7. Role of IoT in soil fertility evaluation.

IoT soil nutrient sensors are a very successful tool for contemporary agriculture, providing farmers with an unparalleled ability to efficiently regulate and track soil health (Khan et al., 2021). The insertion of sensor technology into agricultural techniques is anticipated to boost productivity, efficiency, as well as sustainability in global food production as technology progresses and becomes increasingly accessible.

Practical Implication and Challenges

Variations in Soil Composition across Different Regions

Developing accurate and reliable AI models for identifying soil nutrient levels is inherently difficult due to the substantial variability in soil characteristics across diverse geographical regions.

Effect of Data on Performance

The technological expense of deep learning algorithms has considerable effects on the speed and effectiveness of the training process. Analyzing and processing high-dimensional data, such as hyperspectral photographs with a large number of spectral bands, necessitates significant computer resources (Ang and Seng, 2021). The complexity of the task is determined by various aspects, including the size of the dataset, the complexity and structure of neural networks, and the need for meticulous data preparation and feature extraction. To overcome these obstacles and enhance the effectiveness of deep learning in soil fertility evaluation, it is crucial to optimize algorithms and utilize efficient computer resources.

Ultra-High-Resolution Hyperspectral Images

The substantial dimension of hyperspectral data is a result of its extensive range of spectral bands or channels. Hyperspectral imaging is a technique that collects data over several closely spaced spectral bands, allowing for precise spectral information to be obtained for every pixel in an image (Lu et al., 2020). The abundance of data enables precise detection and characterization of materials, but it also presents obstacles in terms of data processing, storage, and analysis due to its high complexity.

Interference from Sunlight

Obtaining soil samples using a smartphone camera to forecast soil pH might be challenging because of the inconsistent natural lighting conditions. Variations in lighting conditions can impact the fidelity of colors in photographs, resulting in inconsistent and untrustworthy results (Alhakamy and Tuceryan, 2020). To address this issue, it is essential to either maintain regulated lighting settings or utilize AI algorithms that can compensate for differences in illumination. This would enhance the accuracy and dependability of pH predictions made from smartphone photos.

Conclusion

The cost associated with implementing deep learning algorithms has a significant impact on the speed and efficiency of the training process. The use of artificial intelligence (AI), remote sensing and the Internet of Things (IoT) are transforming the evaluation of soil health, enhancing its accuracy, efficiency, and comprehensiveness. Although there are challenges in making information more available and combining data, these advancements have significant potential for enhancing agricultural production and sustainability. More research is needed to fully utilize the capabilities of AI in the field of soil science. Future research should prioritize the development of durable and scalable artificial intelligence algorithms capable of managing large amounts of complex soil nutrient data. Additionally, there should be an emphasis on integrating AI with other technologies like sensor networks and remote sensing.

Declaration

Ethical Responsibilities of Authors

"All authors have read, understood, and have complied as applicable with the statement on "Ethical responsibilities of Authors" as found in the Instructions for Authors".

Ethical Approval

Not applicable.

Conflict of Interest

On behalf of all the authors the corresponding author states that there is no conflict of interest to disclose.

Competing Interests

Not applicable.

Authors' Contributions

Mageshen: Conceptualization, Methodology, Writing - Original Draft, Writing - Review & Editing. **Manimaran**: Investigation, Formal analysis, Visualization, Software, Data Curation, Supervision **Ashwitha**: Writing - Original Draft, Formal analysis, Methodology, **Thirumeninathan:** Formal analysis, Visualization.

Funding

No funding has been received.

References

Alaoui, A., Barao, L., Ferreira, C.S. and Hessel, R. 2022. An overview of sustainability assessment frameworks in agriculture. *Land*, 11(4): 537. https://doi.org/10.3390/land11040537.

Alhakamy, A.A. and Tuceryan, M. 2020. Real-time illumination and visual coherence for photorealistic augmented/mixed reality. *ACM Computing Surveys (CSUR)*, 53(3): 1–34. https://doi.org/10.1145/3386496.

Ang, K.L.M. and Seng, J.K.P. 2021. Big data and machine learning with hyperspectral information in agriculture. *IEEE Access*, 9: 36699–36718. 10.1109/ACCESS.2021.3051196.

Cakmakcı, R., Salık, M.A. and Çakmakçı, S. 2023. Assessment and principles of environmentally sustainable food and agriculture systems. *Agriculture*, 13(5): 1073. https://doi.org/10.3390/agriculture13051073.

Cervantes, J., Garcia-Lamont, F., Rodríguez-Mazahua, L. and Lopez, A. 2020. A comprehensive survey on support vector machine classification: Applications, challenges and trends. *Neurocomputing*, 408: 189–215. https://doi.org/10.1016/j.neucom.2019.10.118.

Chouhan, S.S., Singh, U.P. and Jain, S. 2024a. Introduction to computer vision and drone technology. In *Applications of Computer Vision and Drone Technology in Agriculture 4.0*. Springer, Singapore. https://doi.org/10.1007/978-981-99-8684-2_1.

Chouhan, S.S., Singh, U.P., Sharma, U. and Jain, S. 2024b. Classification of different plant species using deep learning and machine learning algorithms. *Wireless Pers. Commun.*, 136: 2275–2298. https://doi.org/10.1007/s11277-024-11374-y.

Chouhan, S.S., Singh, U.P., Sharma, U. and Jain, S. 2021. Leaf disease segmentation and classification of Jatropha Curcas L. and Pongamia Pinnata L. biofuel plants using computer vision based approaches. *Measurement*, 171: 108796. https://doi.org/10.1016/j.measurement.2020.108796.

Folorunso, O., Ojo, O., Busari, M., Adebayo, M., Joshua, A., Folorunso, D. and Olabanjo, O. 2023. Exploring machine learning models for soil nutrient properties prediction: A systematic review. *Big Data and Cognitive Computing*, 7(2): 113. https://doi.org/10.3390/bdcc7020113.

Gaye, B., Zhang, D. and Wulamu, A. 2021. Improvement of support vector machine algorithm in big data background. *Mathematical Problems in Engineering*, 5594899. https://doi.org/10.1155/2021/5594899.

Guo, M. 2021. Soil health assessment and management: Recent development in science and practices. *Soil Systems*, 5(4): 61. https://doi.org/10.3390/soilsystems5040061.

Gülcü, Ş. 2022. Training of the feed forward artificial neural networks using dragonfly algorithm. *Applied Soft Computing*, 124: 109023. https://doi.org/10.1016/j.asoc.2022.109023.

Javaid, M., Haleem, A., Khan, I.H. and Suman, R. 2023. Understanding the potential applications of Artificial Intelligence in agriculture sector. *Advanced Agrochem.*, 2(1): 15–30. https://doi.org/10.1016/j.aac.2022.10.001.

Khan, N., Ray, R.L., Sargani, G.R., Ihtisham, M., Khayyam, M. and Ismail, S. 2021. Current progress and future prospects of agriculture technology: Gateway to to sustainable agriculture. *Sustainability*, 13(9): 488. https://doi.org/10.3390/su13094883.

Khajuria, R. and Sunanda, S.S. 2018. A survey of existing leaf disease techniques using Artificial Neural Network. *International Journal of Computer Trends and Technology (IJCTT)*, 59(1): 52-62: 2231–2803. doi:10.14445/22312803/IJCTT-V59P109.

Kumar, R., Kumar, A., Bhatia, K., Nisar, K.S., Chouhan, S.S., Maratha, P. and Tiwari, A.K. 2024. Hybrid approach of cotton disease detection for enhanced crop health and yield. *IEEE Access*. 10.1109/ACCESS.2024.3419906.

Lawrence, P.G., Roper, W., Morris, T.F. and Guillard, K. 2020. Guiding soil sampling strategies using classical and spatial statistics: A review. *Agronomy Journal*, 112(1): 493–510. https://doi.org/10.1002/agj2.20048.

Lu, B., Dao, P.D., Liu, J., He, Y. and Shang, J. 2020. Recent advances of hyperspectral imaging technology and applications in agriculture. *Remote Sensing*, 12(16): 2659. https://doi.org/10.3390/rs12162659.

Mushtaq, Z., Saeed, S., Demir, C., Fırat, M., Baran, M.I. and Siddiqui, A.R. 2024. Role of sprayers drone in sustainable agriculture. *Water-Air-Soil for Sustainable Agriculture and People Well-being*, 21. 10.5281/zenodo.10841204.

Narasegouda, S. 2020. A decade survey on internet of things in agriculture. Internet of Things (IoT) *Concepts and Applications*, 351–370. 10.1007/978-3-030-37468-6_19.

Parween, S., Hameed, R.S. and Sinha, K. 2021. Iot and its real-time application in agriculture. In *Handbook of Research on Knowledge and Organization Systems in Library and Information Science*. IGI Global, 103–123. 10.4018/978-1-7998-7258-0.ch006.

Priyanka and Kumar, D. 2020. Decision tree classifier: A detailed survey. *International Journal of Information and Decision Sciences*, 12(3): 246–269. https://doi.org/10.1504/IJIDS.2020.108141.

Raza, I., Zubair, M., Zaib, M., Khalil, M.H., Haidar, A., Sikandar, A. and Ashfaq, M.A. 2023. Precision nutrient application techniques to improve soil fertility and crop yield: A review with future prospect. *International Research Journal of Educational and Technology*, 05(8): 109–123.

Shaikh, T.A., Rasool, T. and Lone, F.R. 2022. Towards leveraging the role of machine learning and artificial intelligence in precision agriculture and smart farming. *Computers and Electronics in Agriculture*, 198: 107119. 10.1016/j.compag.2022.107119.

Sujatha, M. and Jaidhar, C.D. 2023. Machine learning-based approaches to enhance the soil fertility—A review. *Expert Systems with Applications*, 122557. https://doi.org/10.1016/j.eswa.2023.122557.

Trontelj, M.L., Janez and Chambers, O. 2021. Machine learning strategy for soil nutrients prediction using spectroscopic method. *Sensors*, 21(12): 4208. https://doi.org/10.3390/s21124208.

Tyagi, A.K. and Chahal, P. 2022. Artificial intelligence and machine learning algorithms. In *Research Anthology on Machine Learning Techniques, Methods, and Applications*. IGI Global, 421–446. 10.4018/978-1-6684-6291-1.ch024.

Uddin, S., Haque, I., Lu, H., Moni, M.A. and Gide, E. 2022. Comparative performance analysis of K-nearest neighbour (KNN) algorithm and its different variants for disease prediction. *Scientific Reports*, 12(1): 6256. https://doi.org/10.1038/s41598-022-10358-x.

Weiss, M., Jacob, F. and Duveiller, G. 2020. Remote sensing for agricultural applications: A meta-review. *Remote Sensing of Environment*, 236: 111402. https://doi.org/10.1016/j.rse.2019.111402.

Zhang, S. 2021. Challenges in KNN classification. *IEEE Transactions on Knowledge and Data Engineering*, 34(10): 4663–4675. 10.1109/TKDE.2021.3049250.

Chapter 19

Advancements in Soil Moisture Evaluation

Sensors, Remote Sensing and Artificial Intelligence

Ayushi Jha,[1,*] *Ankita Shinde*,[2] *Jatinder Kaur*,[1] *Isha Swati*,[1] *Rajat Mishra*[3] and *Sanskriti Sundar*[1]

Introduction

Soil is an essential component of life, serving as a reservoir for the nutrients and water that support us. Air, water, and life can all be found in the complex combination of minerals and organic components that makes up healthy soil. Soil moisture is an important component of the climate system. It is also known as soil water content and refers to the total amount of water, including water vapour, in the soil's active layer. Depending on the kind of soil and vegetation, the total thickness of this active layer varies, although it is usually in the first 1–2 meters (Robock et al., 2000). In addition to being an important factor in agricultural water security, soil moisture is a reliable indicator of crop output in the future (Tian et al., 2019). The total amount of water storage within the soil is the most important factor in soil moisture. The main factor influencing soil moisture is the extent of the root zone, which is the area where roots

[1] Department of Soil and Water Engineering, Punjab Agricultural University, Ludhiana-141004, Punjab, India.
[2] Indian Agricultural Research Institute, Pusa, New Delhi-110012, India.
[3] Department of Soil and Water Conservation Engineering, GBPUAT, Pantnagar, Uttarakhand.
* Corresponding author: ayushijhagsm@gmail.com

can absorb water. Soil density, structure, texture, organisms, soil–plant–atmosphere interactions, and meteorological and climatic conditions all affect the accessibility of soil moisture (Liu et al., 2020). Soil moisture storage serves as the main source of water for plants in between precipitation events, which affects the production of crops. Soil moisture levels can influence on-farm water management decisions such as planting timing, pesticide and herbicide use, fertilizer application, and irrigation schedule. Such data can be extremely beneficial to many governmental bodies and commercial firms interested in weather and climate, crop yield estimation, water quality, and so on. The amount of moisture in the soil is a crucial state variable that determines the temperature of the surface because a wetter surface will be less warm and more energy will be used for evaporation and transpiration than for surface heating. The amount of rainfall or snowmelt that instantly enters into rivers, streams, or in the worst situations, causes flooding, is determined by runoff, which is also influenced by soil moisture present in the soil. Drought is frequently associated with low soil moisture, and the interplay between soil moisture and the environment may be critical to the duration of droughts.

Even though soil moisture makes up only 0.05% of the water in the entire global hydrological cycle (Robinson et al., 2008) and 0.001% of all freshwater that is available. Its significance in the hydrological cycle has led to it being recognized as one of the Essential Climate Variables (Dorigo et al., 2011). Precise soil moisture data helps farmers to make the best irrigation decisions possible, which lowers the expenses associated with over-irrigation, including higher fertilizer and pumping costs, undesired nitrate leaching, and greenhouse gas emissions (Doorenbos and Kassam, 1979). However, there is significant evidence which indicates that over and under-irrigation both lead to yield penalties in numerous kinds of crops like groundnut, beans, soybean, cotton, spring wheat, maize, sunflower, potato, sugarcane, and winter wheat (Doorenbos and Kassam, 1979; Irmak, 2014). On the other hand, inadequate or inaccurate soil moisture information increases the likelihood of under-irrigation. Considering these difficulties, using measurements of soil moisture in agricultural production choices could result in a 40% increase in the effectiveness of water consumption (Irmak, 2012). Therefore, a key indicator of monthly to seasonal climate fluctuations is the moisture content of the soil.

Several numerical methods have been proposed and applied for studying the dynamics as well as distribution of soil moisture at different scales. Soil moisture can be measured using two different methods: (I) direct and (II) indirect techniques. Direct methods include the gravimetric method, volumetric method and feel and appearance method. Indirect methods includes infrared method, tensiometer, time-domain reflectometry (TDR), gamma attenuation, frequency-domain reflectometry (FDR), infrared method, global positioning systems (GPS), remote sensing, electromagnetic (EM) technology, resistive sensors, capacitance type devices, and cosmic ray neutron method (Irmak, 2019a, 2019b; Romano, 2014; Schmugge et al., 1980; Zazueta and Xin, 1994). When it comes to direct methods of estimating soil moisture content, gravimetric measurement is the most precise method and the most accurate indirect techniques are frequency domain reflectometry (FDR) and time domain reflectometry (TDR) (Chen et al., 2020; Peng et al., 2017). However, they have poor spatial representation. This issue is partially resolved by new technologies known as global positioning systems (GPS) and cosmic-ray neutron sensing (CRNS). According to Rahimzadeh-Bajgiran et al. (2013), there are several benefits to using soil moisture estimating equipment, including mobility, ease of installation, operation, and maintenance, relative maturity, and the capacity to make direct measurements at different depths with high temporal resolution. However, according to Zhang et al. (2014), these methods are labour-intensive, costly, advanced and some may also be destructive (e.g., gravimetric sampling).

For such uses, further contemporary techniques are required, such as electronic sensors that provide an instantaneous measurement of soil moisture. The most extensive range of soil moisture sensors available for purchase nowadays is based on the soil medium's electromagnetic sensing. These sensors fall into three categories: capacitance-type devices, TDR, and electrical resistivity (ER) (Heng et al., 2002; Jones et al., 2002). To achieve accurate data, these sensors need to be calibrated properly. Although the scientific literature provides a variety of methods for calibrating dielectric soil moisture sensors, their usage in field investigations is limited due to factors impacting calibration, the technical difficulties of calibration, and a shortage of confidence in their efficacy.

Compared to capacitance-based sensors, Time Domain Reflectometry sensors have been demonstrated to have greater accuracy and to work at higher frequencies (Walker et al., 2004; Robinson et al., 2008). On the other hand, capacitance sensors are far less expensive despite having inferior accuracy. This makes it possible to employ more instruments and, thus, denser networks. Considering the significant temporal and spatial variability of soil moisture at comparatively small scales (Teuling and Troch, 2005), it has been recommended that some applications would benefit from selecting less expensive but less accurate sensors to reduce the sampling error caused by spatial variability (Teuling et al., 2006; Bogena et al., 2007; Robinson et al., 2008).

Remote sensing technologies comprises radar (microwaves), satellite, and other non-contact approaches. The assessment of electromagnetic energy reflected or emitted from the soil surfaces is necessary to determine the soil moisture by using remote sensing. Depending on the temperature of the soil, the dielectric characteristics, or a mix of the two, the intensity of this radiation with soil moisture may change. As moisture content affects the dielectric constant, it is possible to use the attenuation of microwave energy in active radar to determine the moisture content of porous material. The visible/infrared hyperspectral (VIR/NIR) and thermal infrared (TIR) remote sensing techniques are used to measure the surface soil moisture content. The soil water content of surface soils, which can be categorized into warm and cool soils, with temperatures ranging from 2°C to 50°C and –2°C to –50°C, respectively, is measured using thermal infrared (TIR) sensors. Only warm soils allow for the measurement of soil moisture using thermal infrared (TIR) sensors. On the other hand, cool soil moisture is measured by visible/infrared hyperspectral (VIR/NIR) sensors (Gerhards et al., 2019).

Artificial intelligence in the form of Deep Learning/Machine Learning Technique utilizes algorithms to address problems without explicit programming. This procedure is frequently rapid and more efficient than conventional ones. New algorithms that can precisely forecast the moisture content of soil have been created using machine learning. These algorithms can subsequently be utilized for irrigation or other applications (Ahmad et al. 2010). Using remotely sensed data, machine learning and deep learning algorithms have been successfully deployed to predict soil moisture (Prasad et al., 2018). Using deep learning and machine learning techniques, several vegetation indices and spectral bands along with meteorological, soil, and crop factors could be used as input variables for determining soil moisture (Li et al., 2022; Xie et al., 2021). Artificial neural networks and convolutional neural networks are two examples of machine learning techniques that have been effectively applied for this purpose in recent years (Yinglan et al., 2022). It has been shown that these algorithms outperform conventional rule-based methods (Song et al., 2016). High-performance soil moisture sensors that are resistant to weather fluctuations can be created using this method. Farmers will be able to save money on irrigation water, which is a valuable resource in dry areas (Cai et al., 2019; Lee et al., 2019; Sarwar et al., 2019; Zhao et al., 2022).

These days, technology is a huge part in so many facets of our life. Agriculture might seem to be exempt from that, as all it involves is planting seeds and watching them grow, but that is definitely not the case. Farming is not an easy task. The population's exponential increase presents extremely dynamic challenges to maintaining food supplies. To feed more people, food must be produced in greater quantities. However, an increase in population also means that more buildings and water use will occur, depleting the precious land and water resources required for food production. With fewer resources, the agriculture sector must rely on technical advancements to feed this expanding population. Technological advancements like Genetic engineering, GPS, aerial imaging, herbicide tolerance, robotics, soil sensors, and precision agriculture are just a few examples that have transformed farming to a high-level industry. Both irrigation fields and gardens for plants depend heavily on the soil's moisture content. Over the past ten years, many farmers have found that implementing remote sensors to monitor a field's soil moisture content is an essential management tool. By using precision approaches, it helps to minimize fertilizer runoff from over-saturation, irrigate crops precisely when they need it, and lower input costs.

Soil Moisture Measurement Methods

The procedure of measuring soil moisture is not easy, yet it is crucial for much of the research on water management, including scheduling irrigation. Numerous approaches have been put out, and each has pros and cons of its own. Some of the methods most commonly used are discussed below:

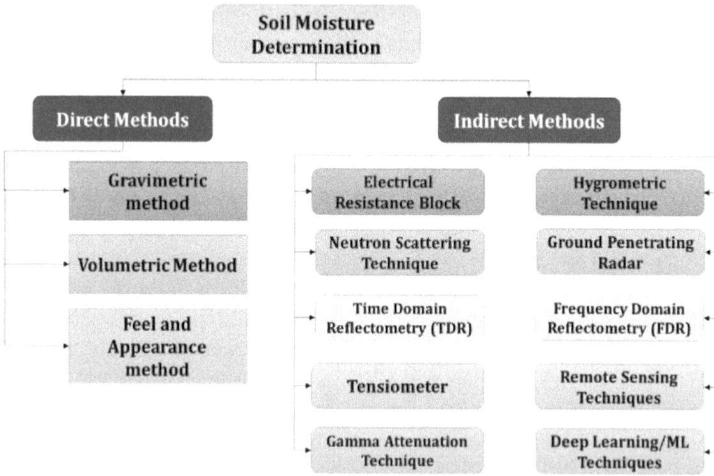

Fig. 1. Techniques for soil moisture measurement.

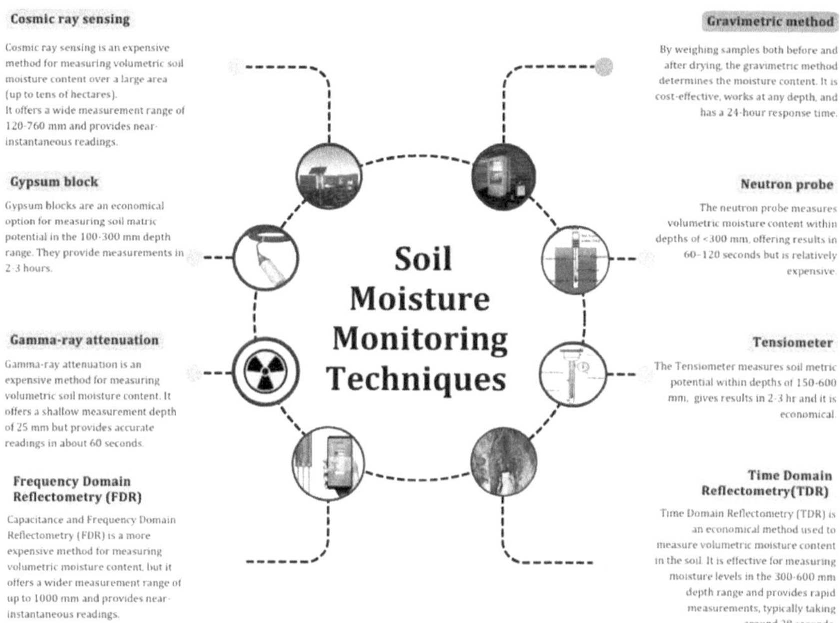

Fig. 2. Evaluation of popular soil moisture monitoring techniques in various classes.

Traditional Soil Moisture Measurement Methods

Gravimetric or Oven Drying Technique

Currently, the most common practical approach for determining soil moisture content (θm) is the gravimetric method (Kashyap et al., 2021). All indirect methods are calibrated using this most natural and traditional approach. Using an auger, gather a sample of soil from the field and weigh it to determine its wet and dry contents. Weighing the soil sample (M_s) at the moment of sampling yields the moist weight, and drying the soil sample in an oven at 105°C for 24 hours yields the dry weight. The soil water (M_w) is represented by the weight loss.

$$\theta m = \frac{Mw}{Ms}$$

This technique guarantees an accurate assessment of the moisture content regardless of the type and salinity of the soil. Nevertheless, a repeated measurement is not possible due to the method's destructive character (Zazueta et al., 1994). However, because of the labour and time-intensive techniques, it is not appropriate for large-scale or real-time field studies. Stated differently, the gravimetric approach-based measurements of soil moisture are not very practical for agricultural or hydrological applications where a large number of *in situ* observations are required on a regular basis, nor for validating remote sensing-based products that necessitate sampling within a limited time window during satellite passes (Fig. 2).

Fig. 3. Schematic illustration of moisture measurements using the gravimetric method.

Volumetric Method

The volume of water in a particular amount of dry soil, typically in 1 m^3 is known as the volumetric water content. VWC can be computed using volume (cm^3/cm^3) or mass (g/g). This moisture content percentage is displayed for output. With this method, a soil sample is taken from the field using a core sampler of known volume at appropriate root zone depth. Its wet and dry weights are then determined using a procedure similarly to the gravimetric water content method. The amount of water in the sample is shown by the difference between the wet and dry masses of soil.

Benefits and Challenges of Traditional Soil Moisture Evaluation Techniques

Traditional soil moisture evaluation approaches have been around for an extended period and provide a few unique advantages while also posing certain challenges.

Benefits

- **Accuracy**: Many tried and tested approaches, like the gravimetric method, yield very accurate findings when applied appropriately. They offer an accurate standard for calibrating more recent sensor technologies.
- **Simplicity**: Certain methods, like the feel and appearance of the soil or using a tensiometer, are quite simple to perform and do not require a lot of specialist tools.
- **Standardization**: Since these techniques frequently follow well defined steps, the data gathered can be compared between other studies and fields.

Challenges

- **Labor intensive**: It can take a lot of time and labour to gather and analyse soil samples for gravimetric procedures.
- **Destructive**: Certain techniques, such as the gravimetric procedure, obliterate the soil sample under study. This can be problematic if you need to monitor the same spot over time.
- **Limited scope**: Most traditional methods provide data from a single point. This can be misleading if soil moisture varies significantly across a field.
- **Time-consuming**: Techniques like the gravimetric method can take a while to get results, which is not ideal for real-time irrigation management.

Conventional techniques yield useful information, but they might not always be the best or most sensible course of action. Newer sensor-based systems that provide continuous monitoring and data collection across bigger areas are emerging as technology progresses.

Case Studies Highlighting the Application of Sensors in Soil Moisture Evaluation

Then et al., 2016 studied the microstrip ring resonator sensor in Peat and sandy soils which works on dielectric principle. This prediction model used an analytical ring resonant model with polynomial interpolation approximation via lumped element model. The sensing area and the measurement accuracy were also improved. This prediction model was found to agree well with the commercial dielectric probe in the dielectric prediction of peat (P 28% m.c.) and sandy (P 10% m.c.) soils. Kojima et al., 2016 developed a low-cost sensor which works on the basis of capacitance principle and is used in the greenhouse of a Grape field having sandy loam type of soil. The sensor consists of copper film substrate and it was prepared by etching copper on a polyethylene terephthalate (PET) film that costs less than $300. The sensor recorded dynamic variations in soil moisture at depths of 10 cm, 20 cm, and 30 cm. After sensor placement, 1014 days were needed for the soil-capacitor contact to stabilize.

Iwata et al., 2017 assessed the impact of three different sensor installation techniques which affects the capacitive soil moisture sensor's output and concluded that the installation technique had a major impact on the sensor's output value. Simultaneously, the sensor's output is boosted by compacting the dirt. Because this is more important in the soil. The study's findings revealed that the installation depth may have an impact on sensor stability. Chakraborty et al., 2019 carried out a study to examine the impact of the polymethyl methacrylate-coated sensor's form and coating on its accuracy and distribution of the fringe electric field. A total of four distinct shapes—a triangle, rectangle, circle, and point edge—were used to investigate the distribution of the electric field and calculate the standard root mean square error value to assess performance. The circular soil sample performs well in measuring water content, according to both simulation and real data.

In a greenhouse of CAU grassland with Yunnan soils, Saeed et al. (2019) examined the performance of a completely autonomous high-resolution sensor that operates on the dielectric principle. Three linear calibration models were developed for various soil types. a cheap sensor that can measure soil moisture three different vertical depths. Acceptable validation statistics were displayed for every depth of the sensor. The coefficients of linear fitting (R^2) varied between 0.95 and 0.99. A TDM sensor was developed by Saeed et al., in 2019 to detect soil moisture simultaneously at four different soil levels like 10 cm, 20 cm, 40 cm, and 60 cm. The sensor's sensitivity was evaluated in a micro-control watering system. The sensor's ability to measure soil moisture continuously was demonstrated by the results, which will be helpful for arranging irrigation techniques in irrigated and desert locations. Lloret et al. (2021) studied a wireless sensor network comprising inexpensive soil moisture sensor nodes to optimize irrigation procedures for precision agriculture. Four soil moisture sensors, each capable of measuring moisture at various depths, made up each wireless node. Every sensor consisted of a pair of coils twisted around a plastic pipe. The functionality of sensors relied on mutual induction between coils to track the soil's water content percentage. Tests were conducted on multiple prototypes with various features. The more successful prototype has a 1:2 winding ratio and operates at 93 kHz with 15 and 30 spires.

Soil Moisture Sensor

Soil moisture is essential for determining optimal irrigation management strategies (Singh et al., 2019). It is important to soil health and plant growth, representing the amount of water held within the soil matrix (Usharani et al., 2019). This optimal availability of moisture in the soil is crucial for various biophysical processes, like seed germination, plant growth, nutrient cycling, and maintaining natural soil biodiversity (Kashyap et al., 2021). The soil moisture regulates atmospheric interactions, maintaining water balance, determining infiltration and percolation rates, influencing groundwater recharge, and affecting stream water fluxes (Altdorff et al., 2023). It substantially impacts the hydrological cycle, plant growth processes, and the structural integrity and stability of soil.

Agriculture, consuming two-thirds of global water usage, holds a critical role in water conservation potential (Datta et al., 2023). Approximately 70% of the world's water is allocated for irrigation, with a significant portion of this use considered unsustainable (Water Facts and Trends, 2020). Traditionally, agriculture requires substantial water volumes, often leading to inefficiencies and potential nutrient leaching in the root zone, which can compromise crop yields over time Thus, optimizing soil moisture content in agriculture is imperative for mitigating the strain on water systems, ensuring their sustainability and resilience in supporting life and ecosystem stability.

Soil moisture content is a critical variable across multiple disciplines, including geotechnical engineering, geological engineering, and soil science (SU et al., 2014). It substantially impacts the hydrological cycle, plant growth processes, and the structural integrity and stability of soil. The precision and resolution of soil moisture predictions are dependent upon the specific application and the spatial scale of the interest (Lunt et al., 2005). Accurate measurement of soil moisture distribution is key for assessing the hydro-geological conditions of in-situ ground and optimizing irrigation schemes for agricultural production (Cuomo et al., 2013; Gui et al., 2014). An in-depth understanding of soil moisture dynamics not only facilitates efficient water usage but also enhances crop yield and soil health. The moisture content in the soil also governs the dissolution and movement of nutrients significantly impacting microbial activities and soil health (Taghvaeian et al., 2020). Excessive moisture levels can disrupt crop root respiration, while inadequate moisture levels can lead to nutrient and fertilizer leaching, resulting in groundwater pollution and hindered fertilizer uptake by crops, exacerbating soil salinization. Therefore, integrating advanced soil moisture monitoring technologies with precision agriculture practices is critical for sustainable agricultural development, ensuring optimal plant growth, efficient resource use, and minimal environmental impact.

Implementing advanced irrigation management techniques can significantly monitor and assess the soil moisture content, thereby augmenting crop yields, improving grain quality, and concurrently reducing nutrient leaching. Pramanik et al. (2022) reported on the automation of soil moisture sensor-based basin irrigation systems. Water scarcity in arid regions is a critical challenge, and an automatic check gate connected to IoT through wireless soil moisture sensors has been developed to improve irrigation efficiency. The system showed significant improvement in irrigation application efficiency, with up to 86.6% efficiency. The assessment of soil moisture content, quantified as volumetric soil water content (VSWC), is needed to understand the complex hydrological processes in soil profiles. VSWC in Fig. 3 represents the proportion of water volume to total soil volume, providing insights into the spatial distribution of soil water near the surface (Aniley et al., 2018). Volumetric soil water content-based soil moisture sensors are used to determine the amount of water present in the soil. VSWC can be calculated by mass (g/g) or volume (m^3/m^3). It gives output in per cent content.

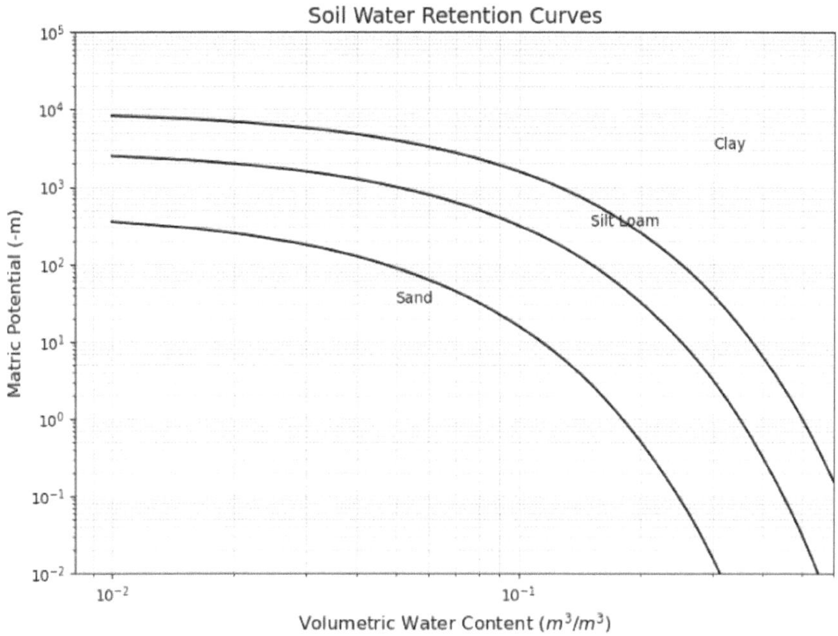

Fig. 4. Typical soil water characteristic curves for different soil textures.

Sensor-based methods such as time domain reflectometry (TDR), neutron probes, and capacitance probes offer non-invasive alternatives, each operating on distinct principles to measure VSWC efficiently. TDR estimates water content by measuring soil dielectric characteristics; neutron probes use neutron scattering to assess moisture; and capacitance probes use soil capacitance changes to calculate VSWC. Using the TDR approach, the research investigated the effect of various vegetation coverings on volumetric water content in higher soil layers. The bare ground had the greatest VWC values, whereas plots with vegetation covers had lower values. The presence of vegetation cover on fallow land decreased VWC values relative to the bare ground but stabilized daily fluctuations (Brant et al., 2006). These techniques enhance spatial and temporal resolution in soil moisture studies through point measurements like gravimetry (Cao et al., 2020). An effective approach to enhancing irrigation efficiency involves integrating soil moisture sensor technology into irrigation scheduling.

Factors Influencing Soil Moisture Variability

The variability of soil moisture is influenced by multiple factors, including soil texture, topography, crop cover, and irrigation practices. Soil texture, defined by the proportions of sand, silt, and clay, governs water retention and infiltration capacities. This classification is based on the composition of mineral particles within the soil, evaluated using diverse retrieval systems. Topography plays a crucial role in water

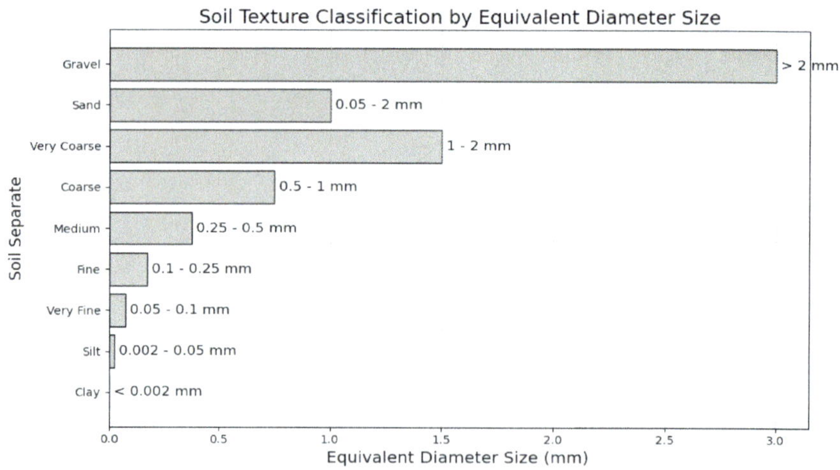

Fig. 5. Soil texture classification.

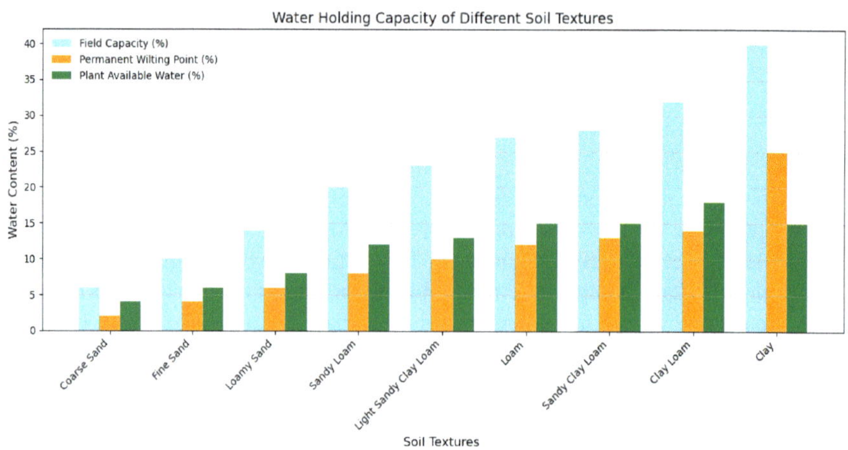

Fig. 6. Water holding capacity of different soil textures.

distribution and movement: slopes facilitate runoff, while low-lying areas encourage water accumulation. The interaction of soil texture, topography, crop cover, and irrigation practices results in significant spatial and temporal variations in soil moisture (Hendrickx et al., 2024). The soil textural triangle provides a standardized method for classifying soils based on the heterogeneous sizes of soil particles.

Vegetation cover intercepts precipitation, reduces surface evaporation, and enhances soil structure through root systems. Meanwhile, irrigation practices, in terms of frequency and application method, directly regulate soil moisture levels by adding water to the soil profile. Collectively, these factors dictate the soil's capacity for water retention and transmission, thereby critically influencing both agricultural productivity and ecosystem functioning (Farouki et al., 1981). Soil texture classification, detailed in provides a framework for quantifying sand, silt, and clay proportions in soil samples, directly influencing key properties such as water holding capacity and infiltration rate. These properties are critical for the precise assessment of soil moisture content, essential for the effective management of agricultural productivity and ecosystem health.

Soil Moisture Sensors and their Application in Agriculture

Soil water sensors (SWS) are among the most effective tools for implementing scientific irrigation scheduling (Brant et al., 2004). Accurate monitoring of soil water content is critical for precision agriculture, enabling optimized productivity, resource conservation, and environmental protection (Saleem et al., 2024). SWS estimates water availability through various methods, providing essential data for irrigation management, drought monitoring, and soil health assessment. In recent decades, a plethora of techniques have emerged to estimate volumetric soil water content (θ_v). These include direct methods like gravimetry and volumetry, while in indirect methods such as neutron scattering, tensiometers, frequency domain reflectometry (FDR) sensors, resistive sensors, time-domain reflectometry (TDR), and capacitance-type devices (Umarfarooque et al., 2011). These devices offer precise measurements of soil moisture levels, thereby supporting effective water management strategies. Soil moisture sensor placement and the optimum number of sensors are also important actors that significantly affect sensor-based surface irrigation automation, performance and cost-effectiveness. This section focuses on different types of soil moisture sensors, their principles of working and their respective strengths and limitations.

Gaikwad (2015) describes a variety of soil moisture sensors, each with its own set of characteristics and applications. The study introduced a potentiometric sensor that is affordable, lightweight, and probe size independent, making it ideal for on-site investigation. Kojima et al. (2016), developed a low-cost soil moisture profile probe utilizing thin-film capacitors and a capacitive touch sensor. Soil moisture content was measured through the capacitance changes in the capacitors, constructed from electrode probes serving as electrode plates. Yamamoto et al. (2002), presented sensors that utilize the thermal characteristics of soil and water, offering high sensitivity in low moisture concentrations and minimal ion effects. You et al. (2013) developed a handheld microwave reflectometer-sensor system for precise moisture monitoring in various soil types. Placidi (2020) focused on the experimental characterization of a low-cost capacitive soil moisture sensor, demonstrating its reliability in determining soil moisture for IoT applications. These sensors collectively provide a range of options for soil moisture analysis, each with its own strengths and potential applications. Different types of soil moisture sensors are discussed below.

Tensiometers

Tensiometers measure the matric potential (also known as soil water tension or soil suction) of water in the soil. The matric potential indicates how tightly water is held in the soil, which is an indirect measure of the soil moisture content. The device simulates the function of a plant's root system by assessing the ease with which a plant can absorb water from the soil. It comprises a stem filled with distilled water, a porous ceramic tip at the base, and a pressure/vacuum gauge at the top, allowing precise measurement of soil water potential. Manjhi and Sarkar (2019) reported in their study on soil moisture variations that tensiometers and soil moisture meters showed deviations of 8.51% and 5.45%, respectively, compared to the gravimetric method. These findings underscore the importance of accurate soil moisture measurement for effective irrigation scheduling and crop yield improvement. While these sensors are invaluable for irrigation management due to their direct measurement of soil water potential, they do have limitations, such as a restricted measuring range and the need for regular maintenance to prevent air entry.

Capacitive Soil Moisture Sensors

Capacitive sensors measure the dielectric permittivity of soil by leveraging the higher dielectric constant of water compared to dry soil. These sensors typically consist of two metallic plates forming a capacitor (Eller and Denoth, 1996). Variations in soil moisture alter the dielectric permittivity of the soil, which in turn changes the capacitance between these plates (Mander et al., 2014). This principle enables capacitive sensors to gauge soil moisture levels accurately. The durability, power efficiency, and affordability of capacitive sensors make them suitable for widespread application. Their ease of use, reliable measurement results, and cost-effectiveness make capacitive sensors ideal for various purposes, including agricultural and environmental monitoring.

The underlying mechanism is based on the proportionality of capacitance to the relative permittivity (ε_r) of the medium between the capacitor's electrodes. Air has a relative permittivity (ε_r, air) of approximately 1, whereas water has a relative permittivity (ε_r, H_2O) of about 81. Consequently, the theoretical maximum capacitance change between air and water is 81-fold, highlighting the sensor's sensitivity to changes in soil moisture content Soil and sensor form a capacitor where the capacitance varies according to the water content present in the soil. The capacitance is converted into voltage level basically from 1.2V to 3.0V maximum.

Despite their relatively high cost, capacitive sensors are extensively employed in Decision Support Systems (DSS) due to their compatibility with small processor units such as Arduino and Raspberry Pi. Their cost-effectiveness and user-friendliness make capacitive sensors a preferred choice over other sensing technologies, particularly in small projects and primary agricultural practices. They provide a reliable, low-power, and economical solution for monitoring soil moisture, facilitating effective soil management in small-scale gardening and farming activities.

Soil Moisture Resistive Sensors

Resistive soil moisture sensors measure electrical resistance between two metal probes inserted into the soil. As water is a superior conductor compared to dry soil, an increase in soil moisture results in decreased resistance. This variation in resistance can be accurately measured and correlated with soil moisture levels. Although these sensors are cost-effective and easy to deploy, they are susceptible to corrosion over time, which can compromise their accuracy and reliability. Consequently, due to their lower precision and vulnerability to environmental factors, resistive sensors are more suitable for basic applications and educational purposes rather than high-precision scientific measurements.

Time Domain Reflectometry (TDR) Sensors

Time Domain Reflectometry sensors measure the travel time of an electromagnetic pulse along a waveguide inserted into the soil. Soil moisture levels influence the travel time, as the presence of water slows the pulse. This delay is calibrated and converted into volumetric soil moisture content using appropriate conversion functions. TDR sensors offer high-precision measurements at various soil depths but are expensive and require specialized equipment and expertise for data interpretation. Despite these challenges, they are extensively utilized in research, precision farming, and hydrological studies, providing a deeper understanding of soil moisture dynamics essential for scientific investigations (Kashyap et al., 2021).

Frequency Domain Reflectometry (FDR) Sensors

The FDR sensors measure the dielectric constant of soil by analyzing the frequency response of the soil's capacitive properties. Operating at higher frequencies, these sensors provide greater accuracy across varying depths. However, FDR sensors are expensive and require calibration for different soil types. Like TDR sensors, FDR sensors are utilized in research and precision agricultural applications, particularly in areas where detailed moisture profiling is essential. Zhu et al. (2019) conducted experiments comparing TDR and FDR sensors under varying temperature conditions and found that TDR sensors are more sensitive to temperature changes than FDR sensors.

Gypsum Block Sensors

Gypsum block sensors measure the electrical resistance of a gypsum block in equilibrium with soil moisture. In this case, electrical resistance changes along with the degree of wetness surrounding the block. These sensors are cheaply made and require little skill to use but they have lower accuracy rates and become less effective over time. The use of these sensors in agriculture, particularly when cost is a major factor, is limited to basic soil moisture monitoring. Other methods are usually employed together with them for a comprehensive understanding of the amount and distribution of soil water. Using experimental results, Ezekiel et al. (2021) demonstrated that the mean difference between the Gravimetric method, compared to gypsum, Tensiometer and Sensor, was 6.30, 3.95 and 3.02, respectively. It could be concluded that the Sensor used with the Arduino board measures moisture content more accurately than the Gypsum and Tensiometer. Veeranna et al. (2016) developed the automated Wireless Gypsum Block Sensor Network System which performed extremely well with results highly correlated concerning actual soil moisture as 0.98 r-square value between them.

Neutrons Probing Sensors

Measuring hydrogen atom density in the soil using a neutron source at a point is how neutron probe sensors work. The slowdown of neutrons by hydrogen atoms and this change are measured to determine the moisture content of the soil. Neutron probe sensors require special handling because they have radioactive sources that need safety precautions but they give accurate readings as well as can measure various depth levels of soil moisture. Neutron probes find applications in research activities and critical monitoring of soil moisture. They are highly accurate and hence suitable for detailed hydrological investigations and large-scale agricultural schemes. Evett et al. (2009) determined spatiotemporal variability of electromagnetic (EM) and Neutron Probe Sensors to determine the soil profile water content. They stated that the EM sensor had low accuracy and high variability than the Neutrons Probing Sensors with a standard comparison with gravimetric samples. Similarly, a study conducted by Kodikara et al. (2014) reported the low variation in the lower subsurface measurements than the upper subsurface using a Neutron Probe sensor and gave a prediction equation to determine the subsurface soil moisture (Fig. 7).

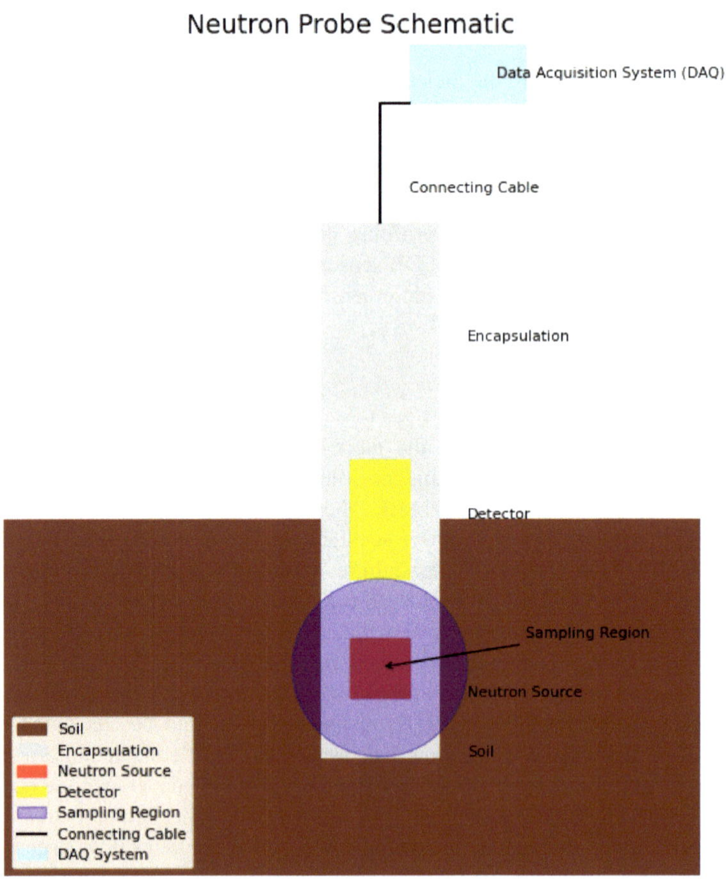

Fig. 7. Neutron probe for soil moisture sensing.

Dielectric Soil Moisture Sensors

Using high-frequency electromagnetic waves, dielectric soil moisture sensors check the dielectric constant in the soils which provides real-time data regarding changing patterns in water wetness over time making it a continuous measurement by wave interaction with the soils. These sensors offer continuous measurements of soil moisture but are costly and sensitive to salinity levels in the ground. Plauborg et al. (2005) experimented with TDR and dielectric soil moisture sensors in drip-irrigated sandy soils and discovered that TDR produces better results with higher precision.

Remote Sensing Techniques for Estimation of Soil Moisture

An essential factor in soil and water study is soil moisture (SM). SM is arguably the most significant link that interlinks water cycle and life—plant—and its significant effects on farming, biology and nature.

A thoroughly researched aspect of the terrestrial water cycle is land surface hydrology. In land-surface hydrology, the primary factors are streamflow, vegetation, precipitation, SM, and surface temperature. Currently, satellites are used to monitor surface temperature, vegetation, and precipitation. The only variable that is still either not observed is SM. For precisely this reason, satellite SM has seen an increase in use over the last ten years (Lakshmi et al., 2013).

Overview of Remote Sensing Techniques for Estimation of SM

Determination of SM from remote sensing has been studied for many years and a growth has been observed in this field of study. The methods used during Thermal Infrared (TIR) remote sensing uses the relationship between SM and surface temperature. This results in complicated derivative processes and many approximations and hence valuable information of SM is extracted by combining optical and TIR remotely sensed data (Zhang et al., 2014).

The advancement in satellite technology offers quantitative estimations and a potent way to track the spatial-temporal variations of regional SM. Many specialists have developed different methods to obtain SM by examining the correlations between SM and land surface parameters obtained from satellites. Through the usage of Visible Infrared (VIR) band, deriving SM through the relation between different types of SM data or vegetative traits and soil reflectance that occur during water stress has been studied. These methods are categorized into 2 divisions as shown in the diagram below. Figures 8 and 9 show the methods of determination of SM when Thermal Infrared bands are used (Peng et al., 2017).

Merging of remotely sensed data from Optical and TIR sensors yields accurate data. Techniques based on spatial and temporal information each have benefits and drawbacks. Therefore, it's critical to figure out how to combine these techniques in a sensible way to produce SM that is accurate. Remotely Sensed techniques are used in estimation of SM (Petropoulos et al., 2015), providing a thorough summarization

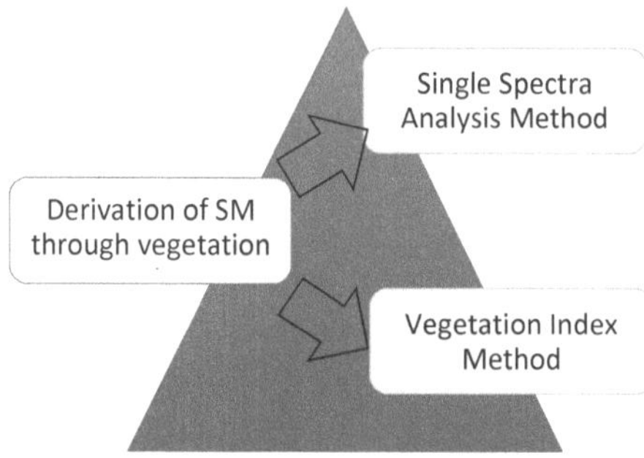

Fig. 8. Methods to derive SM through vegetation traits.

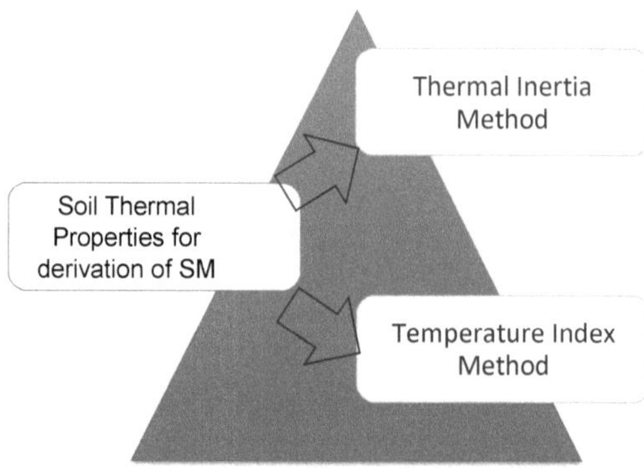

Fig. 9. Methods to derive SM through soil thermal properties.

of various remote sensing methods employed to determine SM, highlighting their respective benefits and limitations.

Microwave remote sensing, particularly Synthetic Aperture RADAR (SAR) data, is useful data because it can penetrate cloud coverage and provide data irrespective of the weather condition. SAR systems emit microwave signals and measure the backscatter, which is influenced by soil's moisture content. These measurements are utilized to derive accurate SM levels across massive zones. The high resolution of SAR data allows for detailed spatial analysis, making it a valuable resource in precision agriculture and hydrology. Passive microwave sensors, employed from 1978 to 2015, have significantly advanced soil moisture (SM) estimation by providing a detailed understanding of sensor technologies and their applications. Over this period, sensors like SMMR, SSM/I, MSMR, SMOS, AMSR-E, TRMM, and SMAP have progressively enhanced SM monitoring through improvements in frequency range, swath width, polarization capabilities, and temporal resolution. These advancements have revolutionized the ability to capture accurate, large-scale SM data, contributing to critical applications in hydrology, agriculture, and environmental monitoring. The evolution of these technologies underscores the growing precision and reliability of remote sensing methods for tracking SM variations over time. Mentioned below are some microwave remote sensing sensors used between 1978–2015:

1. **Scanning Multichannel Microwave Radiometer (SMMR):**
    - Launch Date: 1978–1987.
    - Frequency: Operates on multiple bands, including 6.6 GHz, 10.7 GHz, 18 GHz, 21 GHz, and 37 GHz.
    - Swath Width: 822 km.
    - Polarization: Horizontal and Vertical (H & V).
    - Temporal Resolution: Not specified.
2. **Special Sensor Microwave/Imager (SSM/I):**
    - Launch Date: 1987/1992/1995.
    - Frequency: Operates on 19 GHz, 22 GHz (Vertical polarization), 38 GHz, and 85.5 GHz.
    - Swath Width: 1,400 km.
    - Polarization: H & V.
    - Temporal Resolution: 1 day.
3. **Multifrequency Scanning Microwave Radiometer (MSMR):**
    - Launch Date: 1999–2001.
    - Frequency: Bands include 6.6 GHz, 10.65 GHz, and 18.21 GHz.
    - Swath Width: 1,360 km.
    - Polarization: H & V (excluding 22.23 GHz).
    - Temporal Resolution: 2 days.
4. **Soil Moisture and Ocean Salinity (SMOS):**
    - Launch Date: 2009.
    - Frequency: 1.4 GHz.
    - Swath Width: Not specified.
    - Polarization: H & V.
    - Temporal Resolution: 3 days.
5. **Advanced Microwave Scanning Radiometer for EOS (AMSR-E):**
    - Launch Date: 2002–2011.
    - Frequency: Multiple bands, including 6.6 GHz, 10.65 GHz, 18.7 GHz, 23.8 GHz, 36.5 GHz, and 89 GHz.
    - Swath Width: 1,445 km.
    - Polarization: H & V.
    - Temporal Resolution: 2 days.
6. **Tropical Rainfall Measuring Mission (TRMM):**
    - Launch Date: 1997–2001.
    - Frequency: 10.65 GHz, 19.35 GHz, 22.24 GHz, 37 GHz, and 85.50 GHz.

  - Swath Width: 880 km.
  - Polarization: H & V.
  - Temporal Resolution: 3 days.

7. **Soil Moisture Active Passive (SMAP):**
  - Launch Date: 2015.
  - Frequency: 1.41 GHz.
  - Swath Width: 1,000 km.
  - Polarization: H & V.
  - Temporal Resolution: 4 days.

The evolution of these sensors highlights advancements in satellite remote sensing technology for SM monitoring. The increased precision in frequency, wider swath widths, and improved temporal resolutions have enhanced the accuracy of SM data, supporting large-scale hydrological and agricultural applications (Das et al., 2015).

Airborne Sensing Platforms for Assessment of SM

Airborne remote sensing platforms play a crucial role in SM assessment, offering the capability to collect data over large areas quickly and efficiently. These platforms utilize various sensors and instruments capable of measuring electromagnetic radiation across different wavelengths, enabling them to capture detailed information about SM content. Mentioned below is description of various remote sensing instruments used for Earth observation, highlighting their operational frequencies and primary usage in environmental monitoring and research.

(a) The Advanced Microwave Scanning Radiometer 2 (AMSR-2) is a passive remote sensing instrument on the GCOM-W1 satellite that drives at frequencies ranging between 6.9–89 GHz. It is primarily used to measure precipitation, concentration of sea ice, and sea surface temperatures.

(b) The Advanced Microwave Scanning Radiometer–EO System (AMSR-E), a microwave radiometer mounted on the Aqua satellite that operates at frequencies between 6.9–89 GHz and is used to measure SM, sea surface temperatures, and precipitation.

(c) Aquarius is a joint active and passive L-band microwave radiometer on the Aquarius satellite that operates at frequencies of 1.26 and 1.41 GHz. Aquarius primarily measures ocean surface salinity as well as sea surface temperature and wind speed.

(d) Advanced Synthetic Aperture Radar (ASAR) is an active radar instrument on the ENVISAT satellite that operates at a frequency of 5.33 GHz. ASAR is used to measure sea and land surface features, such as sea ice cover and land deformation.

(e) Advanced Scatterometer (ASCAT) is an active radar scatterometer on the MetOp satellite that operates at a frequency of 5.25 GHz. ASCAT is used to measure wind speed and direction over oceans and land surfaces.

(f) Microwave Imaging Radiometer using Aperture Synthesis (MIRAS) is a passive microwave radiometer on the SMOS satellite that operates at a frequency of 1.4 GHz. MIRAS measures SM and sea surface salinity.

(g) The NASA-ISRO Synthetic Aperture Radar (NISAR) is an active radar instrument that will operate at L and S bands when launched. NISAR is designed to measure changes in the Earth's surface, including surface elevation and deformation.

(h) Phased Array L-band Synthetic Aperture Radar (PALSAR) is an active radar instrument on the ALOS satellite that operates at a frequency of 1.27 GHz. PALSAR is used for forest mapping, disaster monitoring, and terrain analysis.

(i) RADARSAT-1 and 2 are active radar instruments that operate at a frequency of 5.40 GH and are primarily used for ice mapping, maritime surveillance, and environmental monitoring.

(j) Tandem-1 is an active radar instrument on the Tandem-1 satellite that operates at a frequency of 1.2 GHz. It's used for terrain analysis, environmental monitoring, and disaster response.

(k) Sentinel-1A and 1B are active radar instruments that operate at C-band. These are mainly used to monitor sea ice, surveillance of marine environments, and disaster response.

(l) SM Active Passive (SMAP) is a combined active-passive radiometer on the SMAP satellite that operates at frequencies of 1.41 and 1.26 GHz. It's used to measure the amount of vegetation water, and freeze/thaw state and SM.

(m) Special Sensor Microwave Imager (SSM/I) is a passive radiometer that operates at a frequency of 19.35 GHz. SSM/I is used for measuring sea ice extent, atmospheric moisture content, and precipitation.

(n) WindSAT is a passive radiometer on the Coriolis satellite which operates at a frequency of 6.8–37 GHz. WindSAT measures sea surface winds, temperature & water vapour (Mohanty et al., 2017).

Case Studies Highlighting the Application of Remote Sensing in SM Monitoring

Zhou and Zhang (2016) demonstrated recent advancements in SM determination through RS (Remote Sensing). Their article addresses the evolution of retrieval techniques, the validation of SM products produced from satellites, and their applications in hydrology and agriculture. The study suggests that there has been a major increase in remote sensing to study SM, with the launch of dedicated satellite missions such as SMOS and SMAP. Validation studies are being conducted to assess the precision of geospatial SM products. New methods are now being explored to enhance resolution of these products. For example, unification of SAR data and Optical or TIR data, Single Spectral Analysis Method, Vegetation Index Method (VCI, AMTNDVI, NDWI, NMDI, PDI, MPDI). The study emphasizes the importance of developing comparable SM indices that can be normalized and standardized to allow for accurate comparisons across different data sources and time periods, which is crucial for monitoring true SM conditions. Das and Paul (2016) discussed types of Microwave Remote Sensing and the factors that influence microwave remote sensing such as Dielectric constant (ε), Backscattering Coefficient (σ°), Surface Roughness, Bulk Density (ρb), Soil texture, Vegetation cover, Incident angle (□), Bands, Polarization. The study mentions algorithms and models such as empirical models, physical models, and data assimilation techniques that convert microwave signals into SM estimates and integrate data from multiple sensors and incorporate ancillary information (land cover, surface temperature) to enhance accuracy. The study emphasizes that SAR data is very beneficial to determine SM although both active and passive SAR sensors have advantages and disadvantages. Ongoing and future satellite missions, along with improved algorithms can enhance accuracy and reliability of SM data.

Zhang et al. (2014) addresses the estimation of surface soil water content (SSWC) using thermal remote sensing as traditional SM estimation methods have limitations, prompting the need for innovative approaches. The study area chosen was the central area of Spain and discusses the computation of SM index TRRVDI (Temperature Rising Rate Vegetation Dryness Index), and Theoretical Dry Edge is determined.

$$TRRVDI = \frac{RT_{(i)} - RT_{wet}}{RT_{dry} - RT_{wet}}$$

Ground based findings of SM were used to calibrate and verify the model. The study concludes with emphasizing on the potential of using temporal variations in LST, significance of Thermal Remote Sensing for monitoring and determination of SM content.

O'Neill et al. (1996) utilized a vegetation scattering model to estimate microwave backscatter comparable to radar measurements, employing the Peake (1959) formulation to express emissivity via bistatic scattering coefficients. This approach was applied to obtain SM using SAR data, validated through comparisons of modelled and measured radar backscatter. By inputting model-generated values for vegetation transmissivity (γ) and single scattering albedo (α) into equations along with field measurements, soil reflectivity was calculated, enabling SM estimation. The study practiced MACHYDRO '90 test in Pennsylvania, demonstrating the model's effectiveness in accounting for vegetation effects on SM predictions. The estimated and measured SM in corn fields showed promising results on comparison, validating the model's accuracy in practical field conditions (O'Neill et al., 1996).

Introduction to AI Techniques used in Soil Moisture Evaluation

Artificial Intelligence (AI) is a technology that enables machines to imitate various complex human skills (Sheikh et al., 2023). Artificial Intelligence (AI)'s role in soil moisture assessment is to apply computational models and techniques to accurately gauge, track, and forecast the soil's water content. This process utilizes diverse data inputs, including ground sensors, satellite imagery, climatic data, and geographical inputs. AI encompasses two primary methodologies, viz., machine learning (ML) and deep learning (DL). ML is defined as the development of algorithms that enable programs to autonomously enhance their performance with experience (Mitchell, 1997), while DL applies intricate algorithms and neural networks with multiple layers to train models (LeCun et al., 2015). ML algorithms are categorized into supervised and unsupervised learning:

Supervised learning involves training models with datasets that have known outcomes. The goal is for the model to learn to predict outputs based on input data. It is primarily employed for classification and regression tasks. Soil moisture assessment frequently uses supervised ML algorithms like Support Vector Machines (SVMs), linear regression, logistic regression, neural networks, and Random Forest (RF).

Unsupervised learning deals with unlabeled data, allowing algorithms to identify patterns and infer insights autonomously. It is used for pattern discovery, data grouping, exploratory analysis, and reducing data dimensions. Clustering and association are key techniques in unsupervised learning. Algorithms such as k-means, hierarchical clustering, Apriori algorithm, expectation maximization algorithm, and principal component analysis are examples of unsupervised ML methods.

Recent studies have demonstrated the efficacy of standard cameras paired with AI algorithms for monitoring soil moisture, presenting a cost-effective and straightforward method for smart farming practices. The integration of machine learning algorithms has revolutionized soil testing, providing more precise and efficient analyzes. Deep learning, in particular, has shown promise in simulating soil moisture dynamics, with research exploring various network structures to enhance prediction capabilities (Wang et al., 2024). These advancements in AI-driven soil analysis contribute to sustainable water conservation and soil texture analysis (Awais et al., 2023), while addressing the limitations of deep learning methods by incorporating physical mechanisms into data-driven models (Geng et al., 2024).

Machine Learning Algorithms Applied in Soil Moisture Assessment

In the field of soil moisture study, machine learning (ML) algorithms are predominantly utilized. This chapter provides a concise overview of the most frequently employed ML algorithms in soil moisture evaluation, which are well-supported by extensive literature. The general methodology for soil moisture assessment through remote sensing using ML algorithms (Rani et al., 2022).

Linear Regression (LR)

It is a statistical approach used to model the relationship between soil moisture levels (the dependent variable) and one or more environmental and soil characteristics (the independent variables). By fitting a linear equation to the observed data, linear regression can predict soil moisture levels based on input features such as temperature, humidity, rainfall, and soil properties.

K-nearest Neighbors (KNN)

It is a machine learning approach used to predict soil moisture levels by analyzing the similarities between observed environmental and soil characteristics. It leverages historical data to find the most similar instances (neighbors) and uses them to make predictions about new data points.

Support Vector Machine (SVM)

It is a supervised machine learning approach that can be used to predict soil moisture levels by finding the optimal hyperplane that separates the data into different classes or regression outputs. SVM is effective in high-dimensional spaces and can be used for both classification and regression tasks, making it versatile for soil moisture analysis.

Decision Tree (DT)

It is a supervised machine learning technique used to predict soil moisture levels by dividing the data into smaller subsets based on the most important features. This method creates a model in the form of a tree, where each internal node represents a decision based on a specific feature, branches represent the outcomes of these decisions, and leaf nodes represent the predicted soil moisture levels.

Random Forest (RF)

It is an ensemble learning method that combines multiple decision trees to improve the accuracy and robustness of predicting soil moisture levels based on environmental and soil characteristics. It operates by constructing a multitude of decision trees during training and outputs the average prediction (for regression) or the mode of predictions (for classification) of the individual trees.

Artificial Neural Network (ANN)

It is a computational model inspired by biological neural networks in the human brain. It consists of interconnected nodes (neurons) organized in layers: an input layer where environmental and soil characteristics are fed, one or more hidden layers that process this information, and an output layer that predicts soil moisture levels.

Convolutional Neural Network (CNN)

It is a specialized type of artificial neural network designed to analyze spatial relationships in data, particularly suited for processing grid-like data such as images or spatial datasets. In the context of soil moisture assessment, CNNs can be adapted to analyze spatial patterns and relationships in environmental and soil characteristics to predict soil moisture levels.

Applications of AI in Processing and Analyzing Soil Moisture Data

Artificial Intelligence (AI) technologies, including various machine learning and data analysis techniques, are increasingly being applied to process and analyze soil moisture data. Here are some key applications of AI in this domain:

- **Predictive Modeling and Forecasting**: AI algorithms like linear regression, decision trees, random forests, support vector machines (SVM), and neural networks can predict soil moisture levels based on historical data and current environmental conditions (e.g., temperature, humidity, rainfall). AI techniques can analyze temporal patterns in soil moisture data to forecast future trends, helping in agricultural planning and water management.
- **Precision Agriculture**: AI can optimize irrigation schedules by integrating soil moisture data with weather forecasts and crop water requirements. This ensures efficient water usage, reduces water waste, and enhances crop yield. AI can recommend variable irrigation rates based on spatial variability in soil moisture, soil type, and crop type, maximizing irrigation efficiency.
- **Environmental Monitoring and Management**: AI models can detect anomalies or deviations in soil moisture levels that may indicate drought conditions or potential flooding risks, enabling timely interventions. By analyzing soil moisture data alongside other environmental parameters, AI can assess ecosystem health, biodiversity, and habitat suitability.
- **Data Integration and Fusion**: AI algorithms integrate data from various sources such as satellite imagery, remote sensors, and ground-based sensors to provide comprehensive soil moisture maps with high spatial and temporal resolution. AI techniques handle large volumes of heterogeneous data to extract meaningful insights, correlating soil moisture with factors like land use, topography, and climate change impacts.
- **Decision Support Systems**: AI-driven decision support systems provide farmers and land managers with actionable insights and recommendations for optimal land use practices, crop selection, and resource allocation based on soil moisture data.AI can assess risks associated with soil moisture variability, supporting adaptive management strategies to mitigate adverse impacts on agriculture and ecosystems.
- **Research and Innovation**: AI enables the development of advanced models, such as deep learning models for image-based soil moisture estimation using aerial or satellite imagery. Continuous AI research enhances algorithms for better accuracy, robustness, and scalability in processing and analyzing soil moisture data.

AI's application in processing and analyzing soil moisture data represents a significant advancement in agricultural and environmental sciences. By leveraging AI techniques, stakeholders can make informed decisions, enhance resource efficiency, and mitigate risks associated with soil moisture variability, thereby contributing to sustainable land management practices and food security.

Benefits and Challenges of Integrating AI into Soil Moisture Evaluation Systems

Integrating AI into soil moisture evaluation systems brings significant advancements but also introduces a set of challenges that need to be carefully managed.

Benefits

- **Improved Accuracy and Precision**: AI algorithms, such as machine learning and deep learning, can analyze complex and large datasets to accurately predict soil moisture levels, considering various environmental factors. AI provides high-resolution insights into soil moisture variability across different parts of a field, enabling precise irrigation management.
- **Increased Efficiency**: AI-driven systems adjust irrigation schedules based on real-time soil moisture data, reducing water waste and ensuring crops receive the right amount of water. Continuous, automated monitoring reduces the need for manual soil moisture checks, saving time and labor.
- **Cost Savings**: By optimizing the use of water and nutrients, AI systems help reduce operational costs. Automation of data collection and analysis minimizes labour costs associated with manual monitoring.
- **Environmental Sustainability**: Efficient water use reduces runoff and leaching, decreasing the environmental impact and promoting sustainable farming. AI systems help farmers adapt to climate variability by providing accurate soil moisture forecasts and irrigation recommendations.
- **Risk Management**: AI can detect abnormal soil moisture levels that may indicate drought or flood risks, allowing for timely preventive measures. AI-driven insights support informed decision-making, enhancing resilience against environmental and climate-related risks.
- **Comprehensive Data Integration**: AI integrates data from various sources like satellite imagery, IoT sensors, and weather stations to provide a holistic view of soil moisture conditions. AI systems can handle extensive datasets and large geographic areas, making them suitable for both small-scale and large-scale agricultural operations.

Challenges

- **Data Quality and Availability**: Inconsistent or incomplete data can affect the accuracy of AI models. Reliable and continuous data collection is crucial. Soil moisture sensors may have limitations in accuracy and durability, which can affect data quality.

- **Technical Complexity**: Building and maintaining AI models requires expertise in AI and soil science, which may not be readily available. Incorporating AI systems into existing agricultural infrastructure can be complex and may require significant changes.
- **Cost of Implementation**: The cost of deploying AI systems, including sensors, software, and training, can be high, which may be a barrier for small farmers. Regular maintenance, sensor calibration, and model updates incur additional costs.
- **Scalability Issues**: AI models trained on data from one region may not perform well in other regions due to differences in soil types, climate, and farming practices. Ensuring AI models are adaptable to various conditions and scales is a significant challenge.
- **User Adoption**: Farmers and land managers need adequate training to effectively use AI systems and interpret results. There may be reluctance to adopt new technologies among those accustomed to traditional practices.

The integration of AI into soil moisture evaluation systems offers substantial benefits, including improved accuracy, efficiency, cost savings, sustainability, and enhanced risk management. However, addressing challenges related to data quality, technical complexity, costs, scalability, user adoption, and ethical concerns is crucial. Collaborative efforts among researchers, technologists, policymakers, and the agricultural community are essential to ensure AI technologies are accessible, reliable, and beneficial for all stakeholders involved.

Case Studies Highlighting the Application of AI in Soil Moisture Evaluation

Jung et al., 2017 studied the spatial soil moisture estimation using a multiple linear regression model in South Korea. The study used Terra MODIS land surface temperature (LST) and normalized difference vegetation index (NDVI) data from January 2013 to December 2015. For three chosen LST stations, the original LST and observed LST had a coefficient of determination (R^2) of 0.71, whereas the corrected LST and observed LST had an R^2 of 0.95. The MLR model can be an effective tool for estimating soil moisture, especially in regions where *in situ* data is scarce. Anusha et al., 2020 emphasized soil moisture as a key variable in environmental processes, including evaporation and plant transpiration. The researchers used a sensor connected to a microcontroller to measure soil moisture. This sensor was inserted into the soil to retrieve values that are stored in a database. The KNN algorithm is applied to analyze the collected soil moisture data, helping to make informed decisions regarding soil management and agricultural practices. Hegazi et al., 2023 develop a CNN model that can accurately predict soil moisture using Sentinel-2 satellite imagery. The model consists of six convolutional layers, one pooling layer, and two fully connected layers. The study found that using combinations of Sentinel-2 bands as input data for the CNN architecture is more effective than using single bands or all bands together. The best results were achieved using a combination of Red, Red Edge 1, Red Edge 2, Red Edge 3, NIR, and Red Edge 4 bands ($R^2 = 0.7094$, MAE = 0.0277, RMSE = 0.0418).

Imanpour et al., 2023 discussed the downscaling of Soil Moisture Active Passive (SMAP) soil moisture data from 3 km to 1 km resolution using remote sensing. The study evaluates two methods, regression and artificial neural network (ANN), across different climates and land covers. Land Surface Temperature (LST), Normalized Difference Vegetation Index (NDVI), and brightness temperatures were used as the input features. DEM and SWIR data do not significantly improve downscaling. The performance was best in homogeneous climatic conditions, with high accuracy in bare soil and flat regions. Liu et al., 2023 employed random forest (RF), Extra Tree (ET_r), and linear regression (LR) algorithms to estimate soil moisture at different depths using optical satellite data sources in Farming Land of Northern China. The random forest model was particularly noted for its ability to handle large datasets and complex relationships between variables, making it well-suited for predicting soil moisture levels from remotely sensed data.

Challenges, Future Directions, and Conclusion

Challenges in integrating remote sensing and AI techniques for soil moisture evaluation include addressing data quality and availability issues, ensuring consistent and reliable data for accurate modeling. Technical complexities arise in developing and maintaining AI models that effectively capture the variability of soil moisture across different environmental conditions, requiring specialized expertise in both AI and soil science. The high initial implementation costs and ongoing maintenance expenses pose financial barriers, particularly for small-scale farmers. Scalability remains a challenge, as AI models may not generalize well across diverse geographic regions without robust calibration and adaptation. Moreover, achieving widespread user adoption necessitates comprehensive training and support to overcome reluctance toward adopting new technologies in agriculture. Future directions should focus on enhancing data standardization, advancing AI algorithms for real-time predictive modeling, and reducing costs to make these technologies more accessible. Collaborative efforts among researchers, policymakers, and practitioners are essential to overcome these challenges and ensure the sustainable integration of AI and remote sensing in soil moisture management, promoting efficient water use and environmental stewardship in agriculture.

References

Ahmad, S., Kalra, A. and Stephen, H. 2010. Estimating soil moisture using remote sensing data: A machine learning approach. *Advances in Water Resources*, 33(1): 69–80.

Ahmed, A., Zhang, Y. and Nichols, S. 2011. Review and evaluation of remote sensing methods for soil-moisture estimation. *SPIE Reviews*, 2(1): 028001.

Altdorff, D., Oswald, S.E., Zacharias, S., Zengerle, C., Dietrich, P., Mollenhauer, H. and Schrön, M. 2023. Toward large-scale soil moisture monitoring using rail-based cosmic ray neutron sensing. *Water Resources Research*, 59(3): e2022WR033514.

Aniley, A.A., Kumar, N. and Kumar, A. 2018. Soil moisture sensors in agriculture and the possible application of nanomaterials in soil moisture sensors fabrication. *International Journal of Advanced Engineering Research and Technology (IJAERT)*, 6(1): 2348–8190.

Anusha, R., Krishna, P.V., Jain, H. and Suresh, D. 2020. Analysis of soil moisture data using IOT & KNN algorithm. *Journal of Engineering Science*, 11(11).

Awais, M., Naqvi, S.M.Z.A., Zhang, H. et al. 2023. AI and machine learning for soil analysis: An assessment of sustainable agricultural practices. *Bioresour. Bioprocess.*, 10: 90. https://doi.org/10.1186/s40643-023-00710-y.

Bogena, H.R., Huisman, J.A., Oberdörster, C. and Vereecken, H. 2007. Evaluation of a low-cost soil water content sensor for wireless network applications. *Journal of Hydrology*, 344(1-2): 32–42. https://doi.org/10.1016/j.jhydrol.2007.06.032.

Brant, V., Pivec, J., Venclová, V., Soukup, J. and Holec, J. 2006. The influence of different soil vegetation covers onto the volumetric water content in upper soil layers. *Plant Soil and Environment*, 52(6): 275.

Cai, Y., Zheng, W., Zhang, X., Zhangzhong, L. and Xue, X. 2019. Research on soil moisture prediction model based on deep learning. *PLOS ONE*, 14(4): e0214508. https://doi.org/10.1371/journal.pone.0214508.

Cao, Q., Song, X., Wu, H., Gao, L., Liu, F., Yang, S. and Zhang, G. 2020. Mapping the response of volumetric soil water content to an intense rainfall event at the field scale using GPR. *Journal of Hydrology*, 583: 124605.

Chen, Q., Miao, F., Wang, H., Xu, Z., Tang, Z., Yang, L. and Qi, S. 2020. Downscaling of satellite remote sensing soil moisture products over the Tibetan plateau based on the random forest algorithm: Preliminary results. *Earth and Space Science*, 7(6). https://doi.org/10.1029/2020EA001265.

Chakraborty, M., Kalita, A. and Biswas, K. 2019. PMMA-coated capacitive type soil moisture sensor: Design, fabrication, and testing. *IEEE Transactions on Instrumentation and Measurement*, 68(1): 189–196. https://doi.org/10.1109/TIM.2018.2838758.

Cuomo, S. and Della Sala, M. 2013. Rainfall-induced infiltration, runoff and failure in steep unsaturated shallow soil deposits. *Engineering Geology*, 162: 118–127.

Das, K. and Paul, P.K. 2015. Present status of soil moisture estimation by microwave remote sensing. *Cogent Geoscience*, 1(1): 1084669.

Datta, S. and Taghvaeian, S. 2023. Soil water sensors for irrigation scheduling in the United States: A systematic review of literature. *Agricultural Water Management*, 278: 108148.

Doorenbos, J. and Kassam, A.H. 1979. Yield response to water. Irrigation and Drainage Paper No. 33. Rome, Italy: United Nations FAO.

Dorigo, W.A., Wagner, W., Hohensinn, R., Hahn, S., Paulik, C., Xaver, A., Gruber, A., Drusch, M., Mecklenburg, S., van Oevelen, P., Robock, A. and Jackson, T. 2011. The International Soil Moisture Network: A data hosting facility for global *in situ* soil moisture measurements. *Hydrology and Earth System Sciences*, 15(5): 1675–1698. https://doi.org/10.5194/hess-15-1675-2011.

Eller, H. and Denoth, A. 1996. A capacitive soil moisture sensor. *Journal of Hydrology*, 185(1-4): 137–146.

Evett, S.R., Schwartz, R.C., Tolk, J.A. and Howell, T.A. 2009. Soil profile water content determination: Spatiotemporal variability of electromagnetic and neutron probe sensors in access tubes. *Vadose Zone Journal*, 8(4): 926–941.

Ezekiel, O., Danbaki, B.A. and Fabunmi, G.T. 2021. Performance evaluation of gypsum block, tensiometer and moisture sensor for soil moisture content determination. *Journal of Agricultural Engineering and Technology*, 26(2): 103–111.

Farouki, O. and Farouki, O. 1981. Thermal properties of soils (Vol. 81, No. 1). Hanover, NH: US Army Corps of Engineers, Cold Regions Research and Engineering Laboratory.

Gaikwad, P., Devendrachari, M.C., Thimmappa, R., Paswan, B., Raja Kottaichamy, A., Makri Nimbegondi Kotresh, H. and Thotiyl, M.O. 2015. Galvanic cell type sensor for soil moisture analysis. *Analytical Chemistry*, 87(14): 7439–745.

Geng, Q., Yan, S., Li, Q. and Zhang, C. 2024. Enhancing data-driven soil moisture modeling with physically-guided LSTM networks. *Frontiers in Forests and Global Change*, 7: Article 1353011.

Gerhards, M., Schlerf, M., Mallick, K. and Udelhoven, T. 2019. Challenges and future perspectives of multi-/hyperspectral thermal infrared remote sensing for crop water-stress detection: A review. *Remote Sensing*, 11(10): 1240. https://doi.org/10.3390/rs11101240.

Gui, M.W. and Wu, Y.M. 2014. Failure of soil under water infiltration condition. *Engineering Geology*, 181: 124–141.

Hegazi, E.H., Samak, A.A., Yang, L., Huang, R. and Huang, J. 2023. Prediction of soil moisture content from Sentinel-2 images using Convolutional Neural Network (CNN). *Agronomy*, 13(3): 656. https://doi.org/10.3390/agronomy13030656.

Hendrickx, M.G., Diels, J., Janssens, P., Schlüter, S. and Vanderborght, J. 2024. Temporal covariance of spatial soil moisture variations: A mechanistic error modeling approach. *Vadose Zone Journal*, 23(3): e20295.

Heng, L.K., Cayci, G., Kutuk, C., Arrillaga, J.L. and Moutonnet, P. 2002. Comparison of soil moisture sensors between neutron probe, Diviner 2000, and TDR under tomato crops. *Proc. 17th World Congress of Soil Science*, 14–21.

Imanpour, F., Dehghani, M. and Yazdi, M. 2023. Improving SMAP soil moisture spatial resolution in different climatic conditions using remote sensing data. *Environ. Monit. Assess.*, 195: 1476. https://doi.org/10.1007/s10661-023-12107-7.

Irmak, S. 2012. Nebraska water and energy flux measurement, modeling, and research network (NEBFLUX). *Trans. ASABE*, 53(4): 1097–1115.

Irmak, S. 2014. Plant growth and yield as affected by wet soil conditions due to flooding or over-irrigation. NebGuide G1904. Lincoln, NE: University of Nebraska Extension.

Irmak, S. 2019a. Perspectives and considerations for soil moisture sensing technologies and soil water content- and soil matric potential-based irrigation trigger values. NebGuide 3045. Lincoln, NE: University of Nebraska Extension.

Irmak, S. 2019b. Soil-water potential and soil-water content concepts and measurement methods. EC3046. Lincoln, NE: University of Nebraska Extension.

Iwata, Y., Miyamoto, T., Kameyama, K. and Nishiya, M. 2017. Effect of sensor installation on the accurate measurement of soil water content. *European Journal of Soil Science*, 68(6): 817–828. https://doi.org/10.1111/ejss.12493.

Jones, S.B., Wraith, J.M. and Or, D. 2002. Time domain reflectometry measurement principles and applications. *Hydrological Processes*, 16(1): 141–153. https://doi.org/10.1002/hyp.513.

Jung, C., Lee, Y., Cho, Y. and Kim, S. 2017. A study of spatial soil moisture estimation using a multiple linear regression model and MODIS land surface temperature data corrected by conditional merging. *Remote Sensing*, 9(8): 870. https://doi.org/10.3390/rs9080870.

Kashyap, B. and Kumar, R. 2021. Sensing methodologies in agriculture for soil moisture and nutrient monitoring. *IEEE Access*, 9: 14095–14121.

Kodikara, J., Rajeev, P., Chan, D. and Gallage, C. 2014. Soil moisture monitoring at the field scale using neutron probe. *Canadian Geotechnical Journal*, 51(3): 332–345.

Kojima, Y., Shigeta, R., Miyamoto, N., Shirahama, Y., Nishioka, K., Mizoguchi, M. and Kawahara, Y. 2016. Low-cost soil moisture profile probe using thin-film capacitors and a capacitive touch sensor. *Sensors*, 16(8): 1292. https://doi.org/10.3390/s16081292.

Lakshmi, V. 2013. Remote sensing of soil moisture. *International Scholarly Research Notices*, 2013(1): 424178.

Lee, C. suk, Sohn, E., Park, J.D. and Jang, J.-D. 2019. Estimation of soil moisture using deep learning based on satellite data: A case study of South Korea. *GIScience & Remote Sensing*, 56(1): 43–67. https://doi.org/10.1080/15481603.2018.1489943.

Li, X., Wang, Y., Hu, Y., Zhou, C. and Zhang, H. 2022. Numerical investigation on stratum and surface deformation in underground phosphorite mining under different mining methods. *Frontiers in Earth Science*, 10. https://doi.org/10.3389/feart.2022.831856.

Liu, Y., Zhang, K., Li, Z., Liu, Z., Wang, J. and Huang, P. 2020. A hybrid runoff generation modelling framework based on spatial combination of three runoff generation schemes for semi-humid and semi-arid watersheds. *Journal of Hydrology*, 590: 125440. https://doi.org/10.1016/j.jhydrol.2020.125440.

Lloret, J., Sendra, S., Garcia, L. and Jimenez, J.M. 2021. A wireless sensor network deployment for soil moisture monitoring in precision agriculture. *Sensors*, 21(21): 7243. https://doi.org/10.3390/s21217243.

Lunt, I.A., Hubbard, S.S. and Rubin, Y. 2005. Soil moisture content estimation using ground-penetrating radar reflection data. *Journal of Hydrology*, 307(1-4): 254–269.

Majhi, T. and Sarkar, N. 2019. Study on soil moisture variations in responding to Tensiometer and soil moisture meter with respect to gravimetric method. *International Journal of Chemical Studies*, 7(4): 3179.

Mander, G. and Arora, M. 2014, March. Design of capacitive sensor for monitoring moisture content of soil and analysis of analog voltage with variability in moisture. In *2014 Recent Advances in Engineering and Computational Sciences (RAECS)* (pp. 1– 5). IEEE.

Mitchell, T.M. 1997. *Machine Learning*. Burr Ridge, IL: McGraw Hill.

O'Neill, P.E., Chauhan, N.S. and Jackson, T.J. 1996. Use of active and passive microwave remote sensing for soil moisture estimation through corn. *International Journal of Remote Sensing*, 17(10): 1851–65.

Peake, W.H. 1959. Interaction of electromagnetic waves with some natural surfaces. *IRE Transactions on Antennas and Propagation*, 7: s325–s329.

Peng, J. and Loew, A. 2017. Recent advances in soil moisture estimation from remote sensing. *Water*, 9(7): 530.

Petropoulos, G.P., Ireland, G. and Barrett, B. 2015. Surface soil moisture retrievals from remote sensing: Current status, products & future trends. *Physics and Chemistry of the Earth, Parts A/B/C*, 83: 36–56.

Placidi, P., Gasperini, L., Grassi, A., Cecconi, M. and Scorzoni, A. 2020. Characterization of low-cost capacitive soil moisture sensors for IoT networks. *Sensors (Basel, Switzerland)*, 20.

Pramanik, M., Khanna, M., Singh, M., Singh, D.K., Sudhishri, S., Bhatia, A. and Ranjan, R. 2022. Automation of soil moisture sensor-based basin irrigation system. *Smart Agricultural Technology*, 2: 100032.

Prasad, R., Deo, R.C., Li, Y. and Maraseni, T. 2018. Soil moisture forecasting by a hybrid machine learning technique: ELM integrated with ensemble empirical mode decomposition. *Geoderma*, 330: 136–161. https://doi.org/10.1016/j.geoderma.2018.05.035.

Rahimzadeh-Bajgiran, P., Berg, A.A., Champagne, C. and Omasa, K. 2013. Estimation of soil moisture using optical/thermal infrared remote sensing in the Canadian Prairies. *ISPRS Journal of Photogrammetry and Remote Sensing*, 83: 94–103.

Rani, A., Kumar, N., Kumar, J. and Sinha, N.K. 2022. Machine learning for soil moisture assessment. pp. 143–168. *In*: Sinha, N.K. (ed.). *Deep Learning for Sustainable Agriculture*. Academic Press.

Robinson, D.A., Campbell, C.S., Hopmans, J.W., Hornbuckle, B.K., Jones, S.B., Knight, R., Ogden, F., Selker, J. and Wendroth, O. 2008. Soil moisture measurement for ecological and hydrological watershed-scale observatories: A review. *Vadose Zone Journal*, 7(1): 358–389. https://doi.org/10.2136/vzj2007.0143.

Robock, A., Vinnikov, K.Y., Srinivasan, G., Entin, J.K., Hollinger, S.E., Speranskaya, N.A., Liu, S. and Namkhai, A. 2000. The global soil moisture data bank. *Bulletin of the American Meteorological Society*, 81(6): 1281–1299. https://doi.org/10.1175/1520-0477.

Romano, N. 2014. Soil moisture at local scale: Measurements and simulations. *Journal of Hydrology*, 516: 6–20. https://doi.org/10.1016/j.jhydrol.2014.01.026.

Saeed, I.A., Qinglan, S., Wang, M., Butt, S.L., Zheng, L., Tuan, V.N. and Wanlin, G. 2019. Development of a low-cost multi-depth real-time soil moisture sensor using time division multiplexing approach. *IEEE Access*, 7: 19688–19697. https://doi.org/10.1109/ACCESS.2019.2893680.

Saleem, M.F., Raza, A., Sabir, R.M., Safdar, M., Faheem, M., Al Ansari, M.S. and Hussain, S. 2024. Applications of sensors in precision agriculture for a sustainable

Sarwar, A., Peters, R.T., Mehanna, H., Amini, M.Z. and Mohamed, A.Z. 2019. Evaluating water application efficiency of low and mid elevation spray application under changing weather conditions. *Agricultural Water Management*, 221: 84–91. https://doi.org/10.1016/j.agwat.2019.04.028.

Schmugge, T.J., Jackson, T.J. and McKim, H.L. 1980. Survey of methods for soil moisture determination. *Water Resources Research*, 16(6): 961–979. https://doi.org/10.1029/WR016i006p00961.

Song, X., Zhang, G., Liu, F., Li, D., Zhao, Y. and Yang, J. 2016. Modeling spatio-temporal distribution of soil moisture by deep learning-based cellular automata model. *Journal of Arid Land*, 8(5): 734–748. https://doi.org/10.1007/s40333-016-0049-0.

SU, S.L., Singh, D.N. and Baghini, M.S. 2014. A critical review of soil moisture measurement. *Measurement*, 54: 92–105.

Taghvaeian, S., Andales, A.A., Allen, L.N., Kisekka, I., O'Shaughnessy, S.A., Porter, D.O. and Aguilar, J. 2020. Irrigation scheduling for agriculture in the United States: The progress made and the path forward. *Transactions of the ASABE*, 63(5): 1603–1618.

Teuling, A.J., Hupet, F., Uijlenhoet, R. and Troch, P.A. 2006. Climate variability effects on spatial soil moisture dynamics. *Geophysical Research Letters*, 34(6). https://doi.org/10.1029/2006GL029080.

Teuling, A.J., Uijlenhoet, R., Hupet, F., van Loon, E.E. and Troch, P.A. 2005. Estimating spatial mean root-zone soil moisture from point-scale observations. *Hydrology and Earth System Sciences*, 10(5): 755–767. https://doi.org/10.5194/hess-10-755-2006.

Then, Y.L., You, K.Y., Dimon, M.N. and Lee, C.Y. 2016. A modified microstrip ring resonator sensor with lumped element modeling for soil moisture and dielectric predictions measurement. *Measurement*, 94: 119–125. https://doi.org/10.1016/j.measurement.2016.07.046.

Tian, H., Huang, N., Niu, Z., Qin, Y., Pei, J. and Wang, J. 2019. Mapping winter crops in china with multi-source satellite imagery and phenology-based algorithm. *Remote Sensing*, 11(7): 820. https://doi.org/10.3390/rs11070820.

Umarfarooque, M., Kulkarni, P.S. and Atul, D. 2011. Different techniques for soil moisture measurement. *Asian Journal of Soil Science*, 6(1): 105–110.

UN Water, Water facts. 2020. https://www.unwater.org/water-facts.

Usharani, K.V., Roopashree, K.M. and Naik, D. 2019. Role of soil physical, chemical and biological properties for soil health improvement and sustainable agriculture. *Journal of Pharmacognosy and Phytochemistry*, 8(5): 1256–1267.

Veeranna, J., Patel, N., Mishra, A.K., Varghese, C. and Sandeep, N. 2016. Performance evaluation of gypsum block wireless sensor network system for real time irrigation scheduling. *Cogent Engineering*, 3(1): 1251729.

Walker, J.P., Willgoose, G.R. and Kalma, J.D. 2004. *In situ* measurement of soil moisture: A comparison of techniques. *Journal of Hydrology*, 293(1-4): 85–99. https://doi.org/10.1016/j.jhydrol.2004.01.008.

Xie, W., Li, X., Jian, W., Yang, Y., Liu, H., Robledo, L.F. and Nie, W. 2021. A novel hybrid method for landslide susceptibility mapping-based geodetector and machine learning cluster: A case of Xiaojin County, China. *ISPRS International Journal of Geo-Information*, 10(2): 93. https://doi.org/10.3390/ijgi10020093.

Yamamoto, Y., Yamamoto, Y. and Ogawa, A. 2002. Novel soil-moisture-sensors applying thermal characteristics of soil and water. *Proceedings of IEEE Sensors*, 1: 484–489.

Yinglan, L., Gao, W., Shamshiri, R., Tao, S., Ren, Y., Zhang, Y. and Su, G. 2022. Review of research progress on soil moisture sensor technology. *International Journal of Agricultural and Biological Engineering*, 14(3): 32–42. https://doi.org/10.25165/j.ijabe.20211404.6404.

You, K.Y., Lee, C.Y., Then, Y.L., Chong, S.H., You, L.L., Abbas, Z. and Cheng, E.M. 2013. Precise moisture monitoring for various soil types using handheld microwave-sensor meter. *IEEE Sensors Journal*, 13: 2563–2570.

Zazueta, F.S. and Xin, J. 1994. Soil moisture sensors. *Soil Sci.*, 73: 391–401.

Zhang, D., Tang, R., Zhao, W., Tang, B., Wu, H., Shao, K. and Li, Z.L. 2014. Surface soil water content estimation from thermal remote sensing based on the temporal variation of land surface temperature. *Remote Sensing*, 6(4): 3170–87.

Zhao, Z.-Y., Wang, P.-Y., Xiong, X.-B., Wang, Y.-B., Zhou, R., Tao, H.-Y., Grace, U.A., Wang, N. and Xiong, Y.-C. 2022. Environmental risk of multi-year polythene film mulching and its green solution in arid irrigation region. *Journal of Hazardous Materials*, 435: 128981. https://doi.org/10.1016/j.jhazmat.2022.128981.

Zhou, G. and Zhang, D. 2016. Estimation of soil moisture from optical and thermal remote sensing: A review. *Sensors*, 16(8): 1308.

Zhu, Y., Irmak, S., Jhala, A.J., Vuran, M.C. and Diotto, A. 2019. Time-domain and frequency-domain reflectometry type soil moisture sensor performance and soil temperature effects in fine-and coarse-textured soils. *Applied Engineering in Agriculture*, 35(2): 117–134.

Chapter 20

Sustainable Agriculture

Economic Perspectives on AI and ML in Crop Yield Estimation

*Harshit Mishra** and *Aditya Bhooshan Srivastava*

Introduction

Crop yield estimation has always been a critical aspect of agricultural economics, serving as a fundamental predictor of food supply, market trends, and resource allocation. Traditional approaches, reliant on historical data and empirical observations, have provided valuable insights but are increasingly inadequate in the face of modern challenges. The need for more accurate, efficient, and timely estimations has driven the incorporation of innovative technologies such as AI and ML into agricultural practices (Vuppalapati, 2021). These technologies offer transformative potential in improving the economic sustainability of agriculture by enhancing crop yield predictions, thereby contributing to global food security and economic stability. The economic framework of sustainable agriculture is grounded in principles that balance profitability with environmental stewardship and social equity. Sustainable agriculture economics emphasizes the long-term viability of farming practices that protect natural resources and support rural communities (Thamilselvan et al., 2021; Chouhan et al., 2021). The introduction of AI and ML into this framework marks a significant shift towards precision agriculture, where data-driven decisions optimize resource use, reduce waste, and improve crop management. By leveraging vast quantities of data, AI, and ML can identify patterns and predict outcomes with unprecedented accuracy, thus enhancing economic sustainability (Araújo et al., 2023).

Department of Agricultural Economics, College of Agriculture, Acharya Narendra Deva University of Agriculture and Technology, Kumarganj, Ayodhya (U.P.) - 224 229, India.

* Corresponding author: wehars@gmail.com

There are inherent limits to traditional agricultural yield estimation techniques like visual inspections, farmer reports, and historical yield data. These techniques are frequently labour-intensive, time-consuming, and prone to prejudice and human error. Moreover, they have difficulty taking into consideration dynamic aspects that might change quickly and affect crop results, such as pest infestations, weather, and soil health. Traditional methods are static, which emphasizes the necessity for technical developments that can produce accurate, adaptive, and real-time estimations. Thanks to their capacity to analyze massive information from a variety of sources, such as satellite imaging, weather forecasts, and soil sensors, AI and ML bring about a paradigm shift in crop yield estimation. These tools use complex algorithms to find patterns and small correlations that might go unnoticed by the human eye. As an example, a study showing that ML algorithms could forecast wheat yields with up to 90% accuracy showed how well machine learning models handle past and real-time data to predict crop productivity. This level of accuracy facilitates more effective planning and resource allocation, which eventually improves economic results (Singh et al., 2022).

The economic implications of adopting AI and ML in agriculture are substantial. A cost-benefit analysis reveals that the initial investment in these technologies can be offset by significant returns, primarily through increased crop productivity, reduced input costs, and optimized resource use. For example, growers using AI-driven precision agriculture techniques have reported up to 30% savings on fertilizer and water usage. Additionally, the ROI for AI and ML technologies in agriculture can be substantial, with studies indicating potential ROI ranges from 2:1 to 5:1, depending on the scale and type of agricultural operation (Sachithra and Subhashini, 2023). Further supporting economic viability are the advantages of increased productivity and resource management. Data-driven perceptions enable targeted water and fertilizer applications, saving waste and having a positive environmental impact. AI-based early stress factors and disease outbreak identification can result in pre-emptive treatments that lower crop losses and improve yield quality. These developments help create a more sustainable agricultural system in addition to saving farmers money (Mishra, 2022).

Another significant benefit of combining AI and ML is increased crop yields and market prospects. Farmers may make well-informed decisions that are in line with market demands and meteorological conditions when they have improved predictability and risk management. Increased predictability facilitates market positioning and strategic planning, which may boost agricultural income and resistance to shocks. To avoid potential production losses and ensure market supply, farmers can benefit from the timely implementation of irrigation solutions by using AI-driven models that anticipate crop stress caused by drought. Crop management techniques, soil properties, pest and disease control, and climate variables are only a few of the many variables that affect crop yield estimation. AI and ML models excel at analyzing these complex and interrelated elements to create precise yield forecasts. High-quality data collection and preprocessing are crucial for the effectiveness of these models, ensuring that the inputs are reliable and comprehensive. Despite the promising advantages, challenges and limitations persist in the implementation of AI and ML in agriculture. Concerns such as data quality, model complexity, and the interpretability of AI decisions pose significant hurdles. Additionally, adoption barriers in agricultural communities, including lack of access to technology and digital literacy, must be addressed (AlZubi and Galyna, 2023).

Ethical and societal considerations, such as equity in access to AI and ML technologies and environmental impact assessment, also perform a critical role in shaping the future of AI in agriculture. Policy implications are vital to support the widespread adoption of AI, and ML in sustainable agriculture. Government support through subsidies, incentives, and public-private partnerships can facilitate infrastructure development and research. Regulatory guidelines for data privacy and security are essential to protect farmers' interests and promote trust in these technologies (Chouhan et al., 2024). The future of AI, and ML in sustainable agriculture is promising, with emergent tools poised to further revolutionize the sector. Integration with robotics and the IoT, advanced AI algorithms for real-time decision-making, and AI-powered tools for personalized farm management are on the horizon. Addressing sustainability challenges through AI-driven optimization can help combat climate change and resource scarcity, promoting a circular economy in agriculture (Zhang et al., 2021).

The economic perspectives on AI and ML in crop yield estimation highlight a transformative potential that, if harnessed efficiently, can lead to a more sustainable and prosperous agricultural future. Considering the aforementioned factors, this chapter focuses on addressing the following objectives: to examine the economic principles and sustainability frameworks underpinning modern agricultural practices, and providing a foundational understanding of how these principles guide sustainable agriculture (Singh and Mishra, 2023). Additionally, the chapter aims to analyze the transformative influence of AI and ML on the economic sustainability of crop yield estimation, highlighting how these technologies can revolutionize traditional methods. Furthermore, it seeks to evaluate the cost-benefit and ROI of adopting AI and ML technologies in agriculture, offering insights into the financial viability and long-term benefits of these innovations. Lastly, the chapter will identify the economic benefits and potential challenges of implementing AI and ML in improving crop yields and farm incomes, ensuring a comprehensive understanding of both the opportunities and obstacles faced by farmers in the agricultural sector (Saiz-Rubio and Rovira-Mas, 2020).

Economic Framework of Sustainable Agriculture

Economic Principles in Agriculture

At its core, agriculture operates within an economic paradigm characterized by various principles that dictate production, distribution, and consumption dynamics. These principles include but are not limited to supply and demand dynamics, market competition, resource allocation, and cost-benefit analysis. Understanding these economic principles is indispensable for devising sustainable agricultural strategies that balance environmental preservation with economic viability. Major principles include:

- *Supply and Demand*: The connection between the amount of agricultural products that customers are willing to buy and the amount that producers are willing to supply at various price points. When supply and demand are equal, the price and quantity of commodities sold are determined by the state of market equilibrium.

- *Cost of Production*: The total expenditure incurred in the process of producing agricultural goods, including fixed and variable costs. Understanding cost structures helps farmers make well-versed decisions about crop choice, production techniques, and investment in technology.
- *Profit Maximization*: Farmers aim to achieve the highest possible profit by optimizing production processes, selecting high-yield crops, and minimizing costs. This principle drives innovation and efficiency in agricultural practices.
- *Marginal Analysis*: This involves assessing the supplementary aids and costs of producing one more unit of output. Farmers use marginal analysis to make decisions about resource allocation, such as the optimal use of fertilizers or the decision to expand farm operations.
- *Risk and Uncertainty*: Agriculture is inherently risky due to factors such as weather variability, pest infestations, and market fluctuations. Economic principles help in understanding risk management strategies, including diversification, insurance, and hedging.
- *Market Structures:* The organization and characteristics of agricultural markets, ranging from perfect competition to monopolies, influence pricing, production, and the distribution of agricultural products. Market structures affect the level of competition, barriers to entry, and the power dynamics between producers and consumers (Sharma et al., 2020).

Sustainable Agriculture Economics

Sustainable agriculture economics focuses on integrating economic feasibility with environmental stewardship and social concern. It aims to create agricultural systems that are productive, profitable, and sustainable over the long term. Some aspects include:

- *Triple Bottom Line*: This framework evaluates the performance of agricultural practices based on three pillars: economic feasibility, environmental health, and social equity. Sustainable agriculture seeks to balance these aspects to ensure long-term sustainability.
- *Resource Efficiency*: Sustainable agriculture accentuates the competent use of resources, including land, water, energy, and inputs like fertilizers and pesticides. Techniques such as precision agriculture and integrated pest management contribute to resource efficiency by optimizing input use and reducing waste.
- *Environmental Impact*: Assessing and mitigating the ecological impact of agricultural practices is crucial. This involves minimizing soil ruin, water pollution, greenhouse gas emissions, and biodiversity loss. Sustainable practices include crop rotation, conservation tillage, and agroforestry.
- *Economic Resilience*: Sustainable agriculture aims to build economic resilience against external shocks, such as climate change, market volatility, and global trade disruptions. Diversification of crops, adoption of resilient farming practices, and access to financial tools like crop insurance contribute to economic resilience.

- *Social Equity*: Ensuring fair labour practices, equitable access to resources, and the well-being of farming societies are integral to sustainable agriculture. Policies and practices that support smallholder farmers, promote fair trade, and enhance rural development are essential.
- *Policy and Institutional Support*: Administrations and organisations play a critical role in encouraging sustainable agriculture through policies, subsidies, research, and extension services. Supportive policies can incentivize sustainable practices, provide tutoring and training, and facilitate entrance to markets and technology (Mishra and Mishra, 2023).

Impact of AI and ML on Economic Sustainability

The advent of AI and ML technologies is revolutionizing agricultural practices, with significant implications for economic sustainability. These technologies offer advanced tools for optimizing production, reducing costs, and enhancing decision-making processes. Some impacts include:

- *Precision Agriculture*: AI and ML enable precision agriculture by analyzing vast amounts of information from sensors, satellite pictures, and weather predictions. This allows for precise application of inputs such as water, fertilizers, and pesticides, leading to increased effectiveness and reduced environmental influence.
- *Yield Prediction and Management*: AI models can expect crop yields with high levels of accurateness by analyzing ancient data, weather arrangements, and soil conditions. This helps farmers make informed verdicts about planting, harvesting and marketing their crops, ultimately improving profitability and reducing waste.
- *Pest and Disease Management*: ML algorithms can detect early signs of pest infestations and diseases through image recognition and pattern analysis. Early recognition permits for timely intervention, reducing crop losses and minimizing the requirement for chemical management.
- *Resource Optimization*: AI-driven systems can optimize resource use by providing real-time recommendations on irrigation, fertilization, and crop management. This not only reduces costs but also enhances resource sustainability.
- *Market Analysis and Forecasting*: AI tools can analyze market trends, price fluctuations, and consumer preferences to provide farmers with insights for better marketing strategies. This helps in aligning production with market demand, reducing surplus and scarcities.
- *Supply Chain Efficiency*: AI technologies enhance supply chain management by improving logistics, inventory control, and traceability. This leads to reduced food waste, lower transportation costs, and improved market access for farmers.
- *Economic Decision Support*: AI-powered decision maintenance systems assist farmers in making complex economic decisions by integrating data on costs, revenues, risks, and market conditions. This enhances economic planning and financial management.

The amalgamation of AI and ML in agriculture grasps excessive potential for enhancing economic sustainability. By improving efficiency, reducing costs, and optimizing resource use, these technologies contribute to the creation of a more buoyant and sustainable agricultural economy (Singh and Kaur, 2022; Mishra, 2024).

Crop Yield Estimation: Traditional Approaches

Crop yield estimation, a crucial aspect of agricultural management, has historically relied on traditional methods to forecast and assess agricultural productivity. This section explores the historical methods employed for crop yield estimation, elucidates their limitations, and underscores the imperative for technological advancements in this domain.

Historical Methods of Crop Yield Estimation

Historically, farmers and agronomists have employed various methods to estimate crop yields, often rooted in empirical observations and rudimentary statistical analyzes. One of the earliest approaches involved visual assessments of crop health and vigour, where experienced farmers gauged yield potential based on factors such as plant height, leaf colour, and overall canopy coverage. However, such qualitative assessments lacked precision and were prone to subjective biases.

As agricultural science progressed, more quantitative methods emerged. Yield estimation based on plot sampling became prevalent, wherein a representative sample of crops from a field was harvested, threshed, and weighed to extrapolate the total yield per unit area. This method, while more objective than visual assessments, still suffered from limitations related to sample representativeness and the labour-intensive nature of data collection (Sarkar et al., 2023).

Furthermore, statistical models were developed to expect crop yields based on aspects such as weather data, soil characteristics, and ancient yield records. Regression analysis and time-series models were commonly employed to establish relationships between these variables and crop productivity. While these statistical approaches provided valuable insights, they often failed to capture the complex interplay of factors influencing yield variation across different spatial and temporal scales.

Limitations of Traditional Approaches

Despite their historical significance, traditional methods of crop yield estimation exhibit several limitations that hinder their efficacy in modern agricultural contexts. Firstly, reliance on visual assessments and plot sampling can be resource-intensive, and time-consuming, particularly for large scale commercial farms. Moreover, these methods may not adequately capture spatial variability within fields, leading to inaccuracies in yield predictions.

Secondly, statistical models based on historical data are limited by their inability to adjust the varying environmental conditions, and emerging agronomic trends. Climate change, pest outbreaks, and shifts in farming practices can introduce uncertainties that conventional statistical techniques struggle to accommodate. Additionally, these models often rely on simplified assumptions about crop growth dynamics, overlooking the complex biological processes underlying yield formation.

Furthermore, traditional approaches to crop yield estimation are constrained by their reliance on manual data collection and analysis. The labour-intensive nature of these methods not only increases operational costs but also introduces opportunities for human error and data inconsistency. As such, there is a pressing need to streamline and automate the process of yield estimation to enhance efficiency and accuracy in agricultural decision-making (Jung et al., 2021).

Need for Technological Advancements

In light of the shortcomings associated with traditional approaches, there is a compelling case for leveraging technological advancements, particularly AI and ML, to revolutionize crop yield estimation practices. AI and ML algorithms offer unprecedented capabilities for the dispensation of large volumes of information, extracting meaningful patterns, and making accurate predictions in real-time. By harnessing advanced sensing technologies such as remote sensing, unmanned aerial vehicles (UAVs), and IoT devices, farmers can assemble high-resolution spatial, and temporal information on crop health, soil moisture, and environmental conditions. Integrating these data streams with AI-driven analytics enables precise monitoring of crop growth dynamics and early detection of stress factors that may impact yield potential (Chouhan et al., 2024; Kumar et al., 2024).

Moreover, AI and ML techniques enable the growth of predictive models that can adjust to dynamic environmental conditions and incorporate heterogeneous data sources. Deep learning algorithms, in particular, excel at capturing nonlinear relationships and complex patterns in agricultural datasets, thereby enhancing the accuracy and robustness of yield forecasts. Traditional approaches to crop yield estimation have played a vital role in agricultural planning and management but are beset by inherent limitations. The emergence of AI and ML technologies offers a transformative opportunity to overcome these limitations, enabling more accurate, timely, and cost-effective predictions of crop yields. By embracing technological advancements and fostering interdisciplinary collaboration, stakeholders can unlock the full potential of AI and ML in sustainable farming (Mahto et al., 2024).

Role of AI and ML in Crop Yield Estimation

Introduction to AI and ML

AI and ML are powerful implements in many domains, including agriculture. In the context of crop yield assessment, artificial intelligence (AI) refers to the unintentional copying of human intellect in technology that thinks and learns like humans. Machine learning (ML) is a subset of AI that focuses on developing procedures that let computers learn from and make estimates or decisions based on information.

To estimate agricultural production, large databases containing information on crop varieties, weather patterns, soil conditions, and past yield data are analyzed using AI and ML techniques. Compared to conventional techniques, these algorithms' ability to recognize intricate patterns and relationships in the data allows for more precise projections of crop yields (Ahmad and Nabi, 2021).

Application of AI and ML in Agriculture

The application of AI and ML in agriculture spans various areas, from precision farming to crop disease detection. In the context of crop yield estimation, AI and ML algorithms are applied in the following ways:

- *Data Analysis and Prediction*: Large volumes of data are analyzed by AI and ML approaches from sources including weather stations, agricultural sensors, and satellite photography. These algorithms can accurately forecast crop yields by finding patterns and connections in the data.
- *Crop Monitoring*: AI-powered drones equipped with cameras and sensors are used to monitor crops in real-time. These drones imprison high-resolution images, which are then analyzed using ML algorithms to assess crop vigour, detect nutrient deficits, and predict productivity potential.
- *Decision Support Systems*: AI and ML-based decision sustenance systems provide growers with endorsements for optimizing crop yield. These systems deliberate factors such as soil moisture levels, temperature, and historical yield data to suggest optimal planting times, irrigation agendas, and fertilizer applications.
- *Crop Disease Detection*: ML algorithms are trained to identify patterns associated with various crop diseases and pests. By analyzing images of diseased crops, these algorithms can accurately diagnose the presence of pathogens and recommend appropriate treatment measures, thus helping farmers mitigate crop losses (Mishra, 2024).

Advantages of AI and ML over Traditional Methods

The integration of AI and ML in crop yield estimation offers several advantages over traditional methods:

- *Accuracy*: AI and ML algorithms can analyze large and complex datasets with high accuracy, enabling more precise predictions of crop yields compared to traditional statistical methods.
- *Efficiency*: AI and ML-based systems automate the process of information enquiry, and decision-making, saving farmers time, and resources. More effective farm management techniques are made possible by these devices, which offer real-time information into crop health and output potential.
- *Adaptability*: AI and ML algorithms can adjust to changing environmental circumstances and crop dynamics, making them well-suited for dynamic agricultural systems. These algorithms endlessly learn from new data, improving their analytical capabilities over time.
- *Scalability*: AI and ML-based solutions may be implemented on farms of any size, from small-scale hobby farms to massive commercial farms. This scalability ensures that farmers of all scales can benefit from the insights provided by these technologies.

The role of AI and ML in crop yield estimation is poised to revolutionize agriculture by enabling more sustainable and efficient farming practices. As these technologies continue to progress, their impact on crop productivity, and food security is expected to grow significantly (Tiwari et al., 2023).

Economic Implications of AI and ML Adoption in Agriculture

Cost-Benefit Analysis

The adoption of AI and ML technologies is increasingly prevalent due to their potential to revolutionize crop yield estimation and enhance overall farm productivity. However, before widespread adoption, it is essential to comprehensively assess the economic implications of integrating these technologies into agricultural practices. This section examines dynamics of conducting a Cost-Benefit Analysis (CBA) for AI and ML adoption in agriculture, providing a structured framework for evaluating the financial feasibility and potential returns on investment (Bhat and Huang, 2021).

To ascertain if the advantages outweigh the costs, a cost-benefit analysis compares the overall estimated expenses of a project or decision to the total expected benefits. In case of AI and ML adoption in agriculture, the costs primarily encompass initial investment expenses, ongoing operational costs, and any associated maintenance or training expenditures. On the other hand, the benefits encompass improvements in crop yield estimation accuracy, enhanced resource utilization, and potential revenue gains resulting from increased productivity and efficiency (Ahmad et al., 2024). The general formula for conducting a Cost-Benefit Analysis is as follows:

Net Benefit (NB) = Total Benefits (TB) – Total Costs (TC)

where:

- Net Benefit (NB) represents the overall financial gain or loss resulting from the adoption of AI and ML technologies.
- Total Benefits (TB) encompass all positive outcomes or monetary gains derived from the implementation of these technologies.
- Total Costs (TC) encompass all associated expenses incurred throughout the implementation and operational phases.

***Example*:** A mid-sized farm transitioning from traditional methods to AI and ML-enhanced practices.

Table 1. Baseline economic metrics without AI and ML integration.

Metric	Value
Farm Size (acres)	500
Annual Crop Yield (tons)	2000
Average Yield per Acre (tons)	4
Revenue per Ton ($)	200
Total Revenue ($)	400,000
Operating Costs (Non-AI and ML) ($)	300,000
Net Profit ($)	100,000

Source: Author's calculation.

Assumptions:

- The farm produces a single type of crop.
- The market price for the crop is stable at $200 per ton.
- Operating costs include labour, equipment maintenance, seeds, fertilizers, pesticides, and water.

Implementation of AI and ML:

The farm invests in AI and ML technologies for precision farming, which include:

- Soil health monitoring systems.
- Drone surveillance for crop health assessment.
- Predictive analytics for optimal planting and harvesting times.
- Automated irrigation systems.

Table 2. Costs associated with AI and ML implementation.

Expense Category	Initial Cost ($)	Annual Maintenance Cost ($)
Soil Health Monitoring System	50,000	5,000
Drone Surveillance	30,000	3,000
Predictive Analytics Software	20,000	2,000
Automated Irrigation System	40,000	4,000
Total	140,000	14,000

Source: Author's calculation.

Benefits of AI and ML Implementation:

- *Increased Yield*: AI and ML technologies optimize resource use and crop management, leading to a 20% increase in yield per acre.
- *Cost Savings*: Precision farming reduces waste in water, fertilizers, and pesticides, cutting operating costs by 15%.
- *Enhanced Market Value*: Improved crop quality through better management increases the market price by 5%.

Table 3. Economic metrics after AI and ML integration.

Metric	Value
Farm Size (acres)	500
Annual Crop Yield (tons)	2400
Average Yield per Acre (tons)	4.8
Revenue per Ton ($)	210
Total Revenue ($)	504,000
Operating Costs (Post-AI and ML) ($)	255,000
AI and ML Maintenance Costs ($)	14,000
Total Operating Costs ($)	269,000
Net Profit ($)	235,000

Source: Author's calculation.

Cost-Benefit Analysis:

1. *Initial Investment:*
   - Total initial cost for AI and ML integration: $140,000.
2. *Annual Financials (Pre-AI and ML):*
   - Total Revenue: $400,000.
   - Total Operating Costs: $300,000.
   - Net Profit: $100,000.
3. *Annual Financials (Post-AI and ML):*
   - Total Revenue: $504,000.
   - Total Operating Costs: $269,000 (including AI and ML maintenance).
   - Net Profit: $235,000.

Table 4. Comparative analysis (annual figures).

Category	**Pre-AI and ML ($)**	**Post-AI and ML ($)**	**Change ($)**	**Change (%)**
Total Revenue	400,000	504,000	+104,000	+26%
Operating Costs	300,000	269,000	–31,000	–10.3%
Net Profit	100,000	235,000	+135,000	+135%

Source: Author's calculation.

Payback Period: The initial investment of $140,000 is recouped by the additional profit of $135,000 within approximately 1.04 years (140,000/135,000).

The above example shows significant economic benefits from integrating AI and ML into agricultural practices. The farm experiences a substantial increase in revenue and net profit, alongside a reduction in operating costs. These improvements result from enhanced yield efficiency, optimized resource use, and higher crop quality. The payback period of just over one year underscores the financial viability of AI and ML investments in agriculture. Moreover, the long-term gains, including sustained higher yields and ongoing cost savings, present a compelling case for widespread adoption of these tools in the farming sector. The economic implications of AI and ML adoption in agriculture are profound, offering a transformative impact on productivity, cost efficiency, and profitability. This example provides a clear view of how these technologies can drive substantial economic benefits, justifying the initial investment and ongoing maintenance costs (Liakos et al., 2018).

Return on Investment (ROI)

ROI is a crucial financial indicator that's used to assess how profitable an investment is. In the perspective of AI and ML in agriculture, ROI measures the financial gains obtained from these technologies relative to their costs. The calculation of ROI helps farmers and agricultural stakeholders to determine whether the execution of AI and ML technologies in crop yield estimation is financially viable.

ROI is defined as the ratio of net profit to the entire cost of the investment. The formula for calculating ROI is:

$$ROI = \left(\frac{Net\ profit}{Cost\ of\ investment} \right) \times 100 \quad (1)$$

where:

- *Net Profit* is the difference between the gross revenue generated from the investment and the total cost of the investment.
- *Cost of Investment* includes all expenses incurred in the adoption and execution of AI and ML technologies.

Factors Influencing ROI in Agriculture

Many factors influence the ROI of AI and ML technologies in agriculture:

- *Initial Investment Costs*: These include the cost of technology acquisition, installation, and training.
- *Operational Costs*: These encompass maintenance, data management, and system upgrades.
- *Increased Yield*: AI and ML can optimize planting schedules, irrigation, and pest control, prominent to higher crop productivity.
- *Resource Efficiency*: Reduction in the use of water, fertilizers, and pesticides due to precise application methods.
- *Market Prices*: Fluctuations in the prices of crops can affect the gross revenue, thus impacting the ROI.
- *Scalability*: The extent to which these technologies can be scaled to cover larger areas or different types of crops (Attri et al., 2023).

Potential Economic Benefits

The adoption of AI and ML in agriculture can lead to several economic benefits:

- *Enhanced Productivity*: AI-driven insights can optimize various farming practices, leading to increased productivity and higher yields.
- *Cost Savings*: Enhanced resource management can result in important cost savings in inputs such as water, fertilizers, and pesticides.
- *Risk Mitigation*: Predictive analytics can help in early detection of crop diseases and pest invasions, reducing potential losses.
- *Market Competitiveness*: Farmers leveraging advanced technologies can be more competitive in the market due to higher quality and more consistent yields.

***Example*:** A wheat farm has decided to implement an AI-based crop yield estimation system.

Description

- *Farm Size*: 100 hectares
- *Initial Investment*: $50,000 (for AI system installation and training)
- *Annual Operational Costs*: $10,000 (maintenance, data management, etc.)
- *Expected Increase in Yield*: 15% due to optimized farming practices
- *Current Yield*: 4 tons per hectare
- *Market Price of Wheat*: $200 per ton

Calculation of ROI

Table 5. Current revenue and costs (without AI and ML).

Parameter	**Value**
Yield per hectare	4 tons
Total yield	400 tons (100 hectares)
Revenue	$80,000 (400 tons $200)
Operational costs	$20,000
Net profit	$60,000

Source: Author's calculation.

Table 6. Projected revenue and costs (with AI and ML)

Parameter	**Value**
Increased yield	15%
New yield per hectare	4.6 tons
Total yield	460 tons (100 hectares)
Revenue	$92,000 (460 tons $200)
Initial investment	$50,000
Annual operational costs	$10,000
Total costs	$60,000
Net profit	$32,000 ($92,000 – $60,000)

Source: Author's calculation.

Table 7. ROI calculation.

Parameter	**Value**
Net profit (with AI and ML)	$32,000
Initial investment	$50,000
ROI	64% {(32,000/50,000) 100}

Source: Author's calculation.

Above example shows a significant ROI of 64% from the adoption of AI and ML in crop yield estimation. This indicates that for every dollar invested, there is a return of $0.64. The increase in yield and the subsequent rise in revenue, coupled with the operational cost savings, highlight the economic viability of integrating AI and ML technologies in agriculture.

Sensitivity Analysis: A sensitivity analysis can be conducted to evaluate how changes in key parameters affect ROI. For instance, if the market price of wheat drops to $180 per ton, or if the yield increase is only 10% instead of 15%, the ROI would need to be recalculated to reflect these variations. This analysis helps in understanding the robustness of the investment under different market conditions.

Table 8. Impact of wheat price and yield increase on net profit and ROI.

Scenario	Net Profit	ROI (%)
Wheat price $180/ton	$28,800	57.6
Yield increase 10%	$25,000	50
Both changes combined	$22,000	44

Source: Author's calculation.

The sensitivity analysis shows that even under less favourable conditions, the investment in AI and ML technologies remains profitable, albeit with a reduced ROI. This underscores the resilience and potential long-term benefits of technological adoption in agriculture.

Economic Viability of AI and ML Technologies

The economic viability of AI and ML technologies in agriculture depends on their ability to generate sustainable economic benefits while being affordable and accessible to farmers. Several factors contribute to this viability:

1. *Affordability:*
   - *Cost of Technology:* The upfront and recurring costs of AI and ML technologies must be within the financial reach of farmers. This includes costs associated with hardware, software, data acquisition, and maintenance.
   - *Financing Options:* Availability of favourable financing options, such as loans, grants, and subsidies, can make these technologies more affordable for small and medium-sized farmers.
2. *Scalability and Adaptability:*
   - *Customizability:* AI and ML solutions should be adaptable to different types of crops, farming practices, and climatic conditions. Customizable solutions ensure broader applicability and effectiveness.
   - *Scalability:* Technologies must be scalable to accommodate varying sizes of farming operations. Solutions that work for both small-scale and large-scale farms enhance economic viability.

3. *Sustainability:*
    - *Long-term Benefits:* The long-term advantages of AI and ML technologies, such as enhanced biodiversity, better soil health, and sustainable agricultural methods, support their commercial feasibility.
    - *Environmental Impact:* Positive environmental impacts, such as reduced carbon footprint and efficient resource use, can lead to long-term cost savings and compliance with regulatory requirements.
4. *Market Integration:*
    - *Market Demand:* The ability of AI and ML technologies to meet market demands by improving crop quality and yield ensures a steady income for farmers.
    - *Supply Chain Efficiency:* Enhanced supply chain efficiency through better demand forecasting and inventory management can improve profitability and economic viability.

Economic Benefits of AI and ML in Crop Yield Estimation

The amalgamation of AI and ML technologies in agriculture, particularly in crop yield estimation, has engendered a myriad of economic benefits. These advancements facilitate precise and predictive analytics that can significantly enhance agricultural productivity and sustainability.

Improved Efficiency and Resource Management

One of the primary economic assists of utilizing AI and ML in crop yield estimation is the substantial improvement in efficiency and resource management. The deployment of advanced algorithms and predictive models enables farmers to optimize their use of inputs and streamline agricultural processes.

Data-Driven Insights for Targeted Fertilizer and Water Application AI and ML technologies provide granular data-driven insights that are instrumental in the precise application of fertilizers and water. Traditional methods of applying fertilizers and irrigation often result in inefficiencies due to the lack of specificity regarding the needs of different crop areas. In contrast, very precise predictions of crop requirements are produced by AI and ML by utilizing information from a variety of sources, including as weather forecasts, satellite imaging, and soil sensors (Benos et al., 2021).

These insights allow for targeted application of resources, ensuring that each crop receives the optimal amount of nutrients and hydration. This precision not only enhances crop yields but also minimizes the excessive use of fertilizers and water, thereby conserving resources and reducing costs. For instance, machine learning models can predict the nutrient status of the soil and recommend site-specific fertilizer application, which can lead to a 30–50% reduction in fertilizer usage without compromising crop yields (Nishad et al., 2024).

Reduced Waste and Environmental Footprint

Another critical benefit of AI and ML in crop productivity estimation is the reduction of agricultural waste, and the environmental footprint. By enabling precise and efficient resource application, these technologies help to minimize waste. Over-application of fertilizers and pesticides is a common issue in traditional farming practices, leading to runoff that contaminates water bodies and harms ecosystems. AI and ML models can analyze past data and real-time conditions to forecast the exact needs of crops, thereby preventing the unnecessary application of inputs. This targeted approach reduces the volume of agricultural chemicals released into the environment, leading to a significant decrease in pollution and soil degradation. Moreover, the enhanced efficiency in water usage contributes to the conservation of water resources, a critical consideration in regions facing water scarcity (Renda, 2024). The reduction in waste not only has environmental benefits but also translates into economic gains for farmers. By decreasing the amount of wasted inputs, farmers can lower their operational costs. Studies have shown that implementing AI-driven precision agriculture can reduce input costs by up to 20%, providing significant savings over time (Mittal et al., 2020).

Cost Savings for Farmers

The financial implications of AI and ML in crop yield estimation extends beyond resource management to encompass substantial cost savings for farmers. The predictive capabilities of AI and ML enable growers to make informed decisions that improve yield and profitability. For example, by accurately forecasting crop yields, farmers can better plan their embedding and harvesting schedules, align their supply chain logistics, and reduce the risk of crop failure. This predictive insight helps in mitigating financial losses caused by unforeseen weather events or pest infestations. Additionally, the optimization of input usage—fertilizers, pesticides, and water—directly correlates with reduced expenditure on these resources. The integration of AI and ML also facilitates precision farming techniques that lower labour costs. Automation of routine tasks, such as monitoring crop health and applying treatments, reduces the need for manual labour. This not only cuts down labour costs but also allows farmers to allocate human resources to more strategic activities, further enhancing farm productivity (Piekarskim and de Francisco, 2020).

Increased Crop Yields and Market Opportunities

Early Detection of Stress Factors and Disease Outbreaks The amalgamation of AI and ML technologies in crop yield estimation offers substantial economic benefits by enabling the early recognition of stress factors and disease outbreaks. Through sophisticated algorithms and data analysis, AI and ML systems can monitor vast agricultural datasets, including soil conditions, weather patterns, and crop health indicators, in real-time. These technologies facilitate the identification of anomalies or patterns that signal the onset of stress factors such as nutrient deficiencies, water scarcity, or pest infestations. Early detection allows farmers to implement timely interventions, thereby mitigating potential yield losses. For instance, AI-powered imaging systems can detect minute changes in leaf coloration or texture that may indicate disease presence long before it becomes visible to the naked eye (Coble et al., 2018). By addressing these issues promptly, farmers can avoid significant crop damage, ensuring a higher yield and reducing the economic impact of potential crop failures. This proactive approach not only preserves the quantity and quality of the harvest but also stabilizes the income of agricultural producers by minimizing unexpected losses.

Proactive Interventions to Improve Crop Health

AI and ML technologies are pivotal in enabling proactive interventions to enhance crop health, thereby contributing to increased crop yields. These technologies leverage predictive analytics to forecast potential issues and recommend optimal solutions. For example, AI systems can analyze historical data alongside current environmental conditions to estimate the probability of pest infestations or disease outbreaks. With the use of this predictive capacity, farmers are able to take preventative actions, including modifying irrigation schedules, adding fertilizer, or deploying pest control devices at the most efficient periods. Additionally, by precisely calculating the quantity of water, fertilizer, or pesticides needed for each crop, ML algorithms may improve resource allocation, minimizing waste and guaranteeing that plants receive the precise nutrients they require. Such precision agriculture practices not only enhance crop health and yield but also lower production costs and minimize environmental impact. The ability to make data-driven decisions results in healthier crops, higher productivity, and a more sustainable agricultural system, thereby translating into significant economic advantages for farmers.

Enhanced Predictability for Market Planning and Risk Management

AI and ML technologies significantly enhance the predictability of crop yields, providing valuable insights for market planning and risk management. For farmers, agribusinesses, and politicians to make educated decisions about planting dates, resource allocation, and market strategy, accurate yield estimates are essential. To provide accurate yield estimations, AI and ML models make use of enormous information, such as historical yield data, weather forecasts, and market patterns. These predictive models enable farmers to plan their harvests more effectively, ensuring that crops reach the market at optimal times to maximize profit. Additionally, accurate yield predictions help in mitigating market volatility by aligning supply with demand, thereby stabilizing prices. This predictability also aids in securing financing, as lenders and investors gain confidence in the projected outcomes based on data-driven forecasts. Furthermore, enhanced predictability supports comprehensive risk management strategies. By anticipating potential yield fluctuations due to environmental or market factors, stakeholders can develop contingency plans and hedge against adverse conditions. This includes diversifying crop portfolios, investing in crop insurance, and exploring alternative markets to cushion against economic shocks. Consequently, AI and ML technologies not only improve yield outcomes but also fortify the agricultural sector's resilience against uncertainties, fostering a stable and profitable farming environment (Tiwari and Mishra, 2024).

Impact on Farm Income and Livelihoods

The application of AI and ML in crop yield estimation stands as a transformative advancement in sustainable agriculture. This section identifies economic benefits that these technologies bring to the agricultural sector, specifically focusing on their impact on farm income and livelihoods. The following subsections will explore the potential for increased profitability through optimized yields, improved decision-making for better farm management, and enhanced economic resilience for farmers.

Potential for Increased Profitability through Optimized Yields

Increasing profitability via optimal yields is one of the biggest financial advantages of AI and ML in agricultural yield estimates. Large volumes of data from diverse sources, such as soil conditions, weather patterns, and historical yield data, are analyzed by sophisticated algorithms. Farmers are able to make well-informed decisions on planting, fertilizing, and harvesting because of these assessments, which provide more accurate estimates of crop performance. For example, AI-driven models might suggest the optimum crop kinds for a given set of climatic variables as well as the ideal dates to sow. By optimizing these variables, farmers can significantly enhance their crop yields. Higher yields directly translate to increased revenue, provided that market conditions remain favourable. Additionally, predictive analytics can help in anticipating market demand and adjusting production accordingly, thus minimizing waste and maximizing sales. Furthermore, precision agricultural techniques—in which inputs like water, fertilizer, and pesticides are administered precisely where needed—are made possible by AI and ML technology. With this accuracy, input costs are decreased and the environmental effect is lessened, resulting in more lucrative and sustainable farming operations. Higher yields along with lower input costs make a strong argument for the use of AI and ML technology in agriculture (Suzer et al., 2024).

Improved Decision-Making for Better Farm Management

AI and ML significantly enhance the decision-making processes in farm management. Traditional farming relies heavily on the farmer's experience and intuition, which, while valuable, can be supplemented and improved upon with data-driven insights. AI and ML tools provide real-time data analytics and predictive intuitions that help peasants make more informed verdicts. For instance, AI-powered tools can monitor crop health through satellite imagery and sensor data, alerting farmers to potential issues such as pest infestations or nutrient deficiencies before they become critical. Early detection of these issues allows for timely interventions, which can prevent crop losses and ensure better yields. Additionally, these technologies can forecast weather conditions and their potential impact on crops, helping farmers plan their activities more effectively. Machine learning models can also analyze past farm management practices and their outcomes, identifying arrangements and associations that might not be evident to human observers. This analysis can guide future decisions, optimizing resource allocation and operational strategies. By leveraging these perceptions, peasants can adopt best performance tailored to their specific conditions, leading to improved efficiency and productivity (Kouadio et al., 2018).

Increased Economic Resilience for Farmers

The incorporation of AI and ML in crop yield estimation also contributes to increased economic resilience for farmers. Agricultural activities are inherently risky due to their dependence on various unpredictable factors such as weather, pests, and market prices. AI and ML technologies help mitigate these risks by providing more accurate and timely information. For instance, predictive models can forecast adverse weather conditions, allowing growers to take pre-emptive steps to safeguard their crops. They can also predict pest and disease outbreaks, enabling farmers to apply treatments early and avoid significant losses. These proactive measures not only safeguard the current yield but also ensure the long-term viability of the farm. Furthermore, AI and ML can assist in financial planning and risk management. By analyzing market trends and economic indicators, these technologies can provide farmers with insights into potential price fluctuations and market demands. This information can inform planting decisions and marketing strategies, helping farmers secure better prices for their produce (Vincent et al., 2019). Additionally, the efficiencies gained through AI and ML adoption can lead to cost savings and improved profit margins, which can be reinvested into the farm to further enhance its resilience. For example, savings from optimized input use can be used to diversify crops or invest in new technologies, reducing dependence on a single crop and spreading risk.

Factors Influencing Crop Yield Estimation

Climate Variables

Climate variables are paramount in determining crop yield, as they directly affect plant growth, development, and productivity. Key climate variables include temperature, precipitation, solar radiation, and atmospheric CO_2 levels.

- *Temperature:* It impacts several physiological methods in plants, such as photosynthesis, respiration, and transpiration. Optimal temperature ranges vary between crop species. Deviations can lead to reduced growth rates or, in extreme cases, plant death. High temperatures may accelerate plant development but can also increase evapotranspiration rates, leading to water stress. Conversely, low temperatures can slow down metabolic processes and delay growth.
- *Precipitation:* It provides the necessary water for crop growth, influencing soil moisture levels and irrigation needs. Inadequate rainfall can lead to drought conditions, severely limiting yield potential. Excessive rainfall, however, can cause waterlogging, reducing oxygen availability to roots and fostering disease proliferation. The distribution and timing of rainfall are equally crucial, with critical growth stages requiring adequate water supply to ensure optimal yield.
- *Solar Radiation:* Solar radiation drives photosynthesis, the essential process through which plants convert light energy into chemical energy. Adequate sunlight exposure is essential for biomass accumulation and grain filling in crops. Variations in cloud cover, day length, and seasonal changes in solar intensity can significantly impact crop productivity.
- *Atmospheric CO_2 Levels:* Elevated atmospheric CO_2 levels can enhance photosynthetic rates in C_3 plants, potentially increasing yield. However, the benefits of CO_2 fertilization may be offset by other limiting factors such as nutrient availability and water supply. Moreover, higher CO_2 levels can alter plant physiology, potentially affecting crop quality and pest dynamics (Elavarasan et al., 2018).

Soil Conditions

Soil conditions are critical determinants of crop yield, providing the physical support, nutrients, and water necessary for plant growth. Important soil factors include texture, structure, pH, organic matter content, and nutrient availability.

- *Soil Texture and Structure:* The amount of sand, silt, and clay particles in the soil determines its texture, which has an impact on aeration, drainage, and water retention. Loamy soils are thought to be optimal for most crops because of their favourable water-holding capacity and nutrient availability. They include balanced amounts of sand, silt, and clay. The organization of soil particles into aggregates, or soil structure, affects microbial activity and root penetration, two processes that are essential to plant health.
- *Soil pH:* Soil pH affects nutrient availability and microbial activity. Most crops prefer a slightly acidic to neutral pH range (6.0–7.5). Acidic soils can lead to nutrient deficiencies and toxicities, while alkaline soils may cause deficiencies in micronutrients such as iron, manganese, and zinc. Lime or sulphur amendments are often used to adjust soil pH to optimal levels.
- *Organic Matter Content:* Soil organic matter (SOM) is crucial for maintaining soil fertility and structure. SOM improves water retention, nutrient supply, and cation exchange capacity (CEC), enhancing soil's ability to hold and release nutrients. Additionally, SOM supports a diverse soil microbiome, which contributes to nutrient cycling and disease suppression.
- *Nutrient Availability:* Important nutrients include iron, manganese, zinc, and copper as well as macronutrients like nitrogen, phosphorus, and potassium, must be available in adequate quantities for optimal crop growth. Nutrient deficiencies or imbalances can limit yield potential, while excessive fertilization can lead to environmental pollution and soil degradation. Precision agriculture techniques, including soil testing and variable rate technology, can optimize nutrient management practices (Balafoutis et al., 2017).

Pest and Disease Management

Pests, and diseases pose important threats to crop productivity by directly damaging plant tissues, reducing photosynthetic capacity, and transmitting harmful pathogens.

- *Pest Management:* Effective pest management strategies integrate various approaches, including cultural, biological, and chemical controls. Cultural practices, such as crop rotation and intercropping, can disrupt pest life cycles and reduce pest pressure. Biological control utilizes natural predators and parasitoids to manage pest populations. Chemical controls, though effective, must be used judiciously to prevent resistance development and minimize environmental impact. Integrated pest management (IPM) combines these methods to achieve sustainable pest control.
- *Disease Management:* Crop diseases, caused by fungi, bacteria, viruses, and nematodes, can significantly reduce yield and quality. Disease management involves the use of resistant crop varieties, crop rotation, sanitation practices, and timely application of fungicides or bactericides. Accurate disease diagnosis

and monitoring are essential for effective management. AI and ML models can enhance disease prediction and management by analyzing data from various sources, including satellite imagery, weather forecasts, and field sensors.

Crop Management Practices

Crop management practices encompass a range of activities aimed at optimizing growth conditions and maximizing yield. Some practices include planting, irrigation, fertilization, and harvesting.

- *Planting:* Planting time, density, and method significantly influence crop yield. Early or late planting can expose crops to unfavourable weather conditions, while optimal planting density ensures efficient use of resources and minimizes competition. Advances in precision planting technology allow for accurate placement of seeds at the correct depth and spacing.
- *Irrigation:* Efficient irrigation management is critical for maintaining soil moisture levels and ensuring water availability during critical growth stages. Modern irrigation techniques, such as drip and sprinkler systems, provide uniform water distribution and reduce water wastage. AI-driven irrigation scheduling, based on real-time soil moisture and weather data, can optimize water use efficiency.
- *Fertilization:* Balanced fertilization practices supply essential nutrients required for plant growth. Pollution of the environment and nutritional imbalances can result from over-fertilization, while under-fertilization limits yield potential. Site-specific nutrient management, guided by soil testing and crop monitoring, can enhance nutrient use efficiency and reduce environmental impact.
- *Harvesting:* Timely and efficient harvesting is essential to maximize yield and quality. Delayed harvesting can result in yield losses due to over-ripening or adverse weather conditions. Modern harvesting equipment, equipped with sensors and automation technology, ensures precise and efficient harvesting operations, reducing labour costs and post-harvest losses.

A complicated interaction between climatic factors, soil properties, pest and disease control, and crop management techniques affects agricultural yield estimation. Enhancing yield projections and promoting sustainable agricultural practices are two major benefits of integrating AI and ML in this analysis of these parameters. Farmers are able to make well-informed decisions that maximize crop output and resource usage efficiency by utilizing cutting-edge technology and data-driven insights. (Kumar et al., 2023).

Data Collection and Preprocessing

In sustainable agriculture, the use of AI and ML techniques clamps immense promise for enhancing crop yield estimation. However, the efficacy of these technologies heavily relies on the quality of data input and the preprocessing methods employed. This section explores data collection and preprocessing, elucidating the paramount importance of high-quality data, the array of data collection techniques available, and the nuanced methods of data preprocessing.

Importance of High-Quality Data

The foundation of any successful AI or ML model in crop yield estimation is laid upon high-quality data. High-quality data not only ensures the accuracy and reliability of predictions but also mitigates the risk of biased outcomes and erroneous conclusions. In the context of sustainable agriculture, where environmental factors, soil conditions, and crop health intricately influence yield, the significance of precise and comprehensive data cannot be overstated.

Moreover, the economic implications of relying on flawed data can be dire, leading to inefficient resource allocation, suboptimal decision-making, and ultimately, diminished agricultural productivity. Therefore, stakeholders must prioritize the acquisition of high-quality data through rigorous quality control measures, standardized protocols, and data validation techniques.

Data Collection Techniques

A myriad of data collection techniques are available to agricultural researchers and practitioners, each offering distinct advantages and limitations. The utilization of remote sensing technology, such drones and satellite imaging, has become essential for obtaining spatially explicit data on environmental factors, crop health, and soil moisture levels. These technologies enable large-scale data acquisition with high temporal resolution, facilitating real-time monitoring and analysis of agricultural landscapes.

Additionally, ground-based sensors and IoT devices provide granular data at the field level, capturing parameters like temperature, humidity, and nutrient levels with precision. Furthermore, traditional methods such as field surveys, yield monitoring systems, and agricultural machinery equipped with sensors remain valuable for collecting ground truth data and validating remote sensing observations.

However, the selection of data collection techniques must be guided by the specific requirements of the crop yield estimation task, considering factors such as spatial variability, temporal dynamics, and cost-effectiveness. A judicious combination of remote sensing technologies, ground-based sensors, and conventional methods can yield comprehensive datasets conducive to accurate and robust model development.

Data Preprocessing Methods

Effective data preprocessing is indispensable for extracting meaningful insights from raw agricultural data and preparing it for analysis with AI and ML algorithms. Data preprocessing encompasses a myriad of techniques, including data cleaning, transformation, integration, and feature selection, aimed at enhancing the quality, consistency, and relevance of the dataset. Data cleaning involves the identification and rectification of errors, outliers, and missing values, ensuring the integrity and reliability of the dataset. Techniques like imputation, outlier detection, and anomaly removal are commonly employed to address data quality issues and improve the robustness of subsequent analyzes (Lakshmi and Corbett, 2023).

Furthermore, data transformation techniques, such as normalization, standardization, and dimensionality reduction, are utilized to enhance the compatibility and interpretability of the data, facilitating more effective model training and evaluation. Integration of heterogeneous data sources, through techniques like data fusion and aggregation, enables the creation of comprehensive datasets that capture the multifaceted nature of agricultural systems. Moreover, feature selection techniques aid in identifying the most informative variables relevant to crop yield estimation, thereby reducing computational complexity and improving model generalization. Techniques such as recursive feature elimination, principal component analysis, and mutual information-based selection help identify relevant features while mitigating the curse of dimensionality.

AI and ML Models for Crop Yield Estimation

Regression Models

Regression models are a fundamental class of machine learning algorithms used for predicting continuous outcomes, such as crop yields. These models are particularly useful in crop yield estimation as they can efficiently capture the complex associations between various environmental, and management factors that influence crop yields.

Simple Linear Regression

Regression models that incorporate a single predictor variable are categorized as simple linear regressions. The predictor variable (crop yield) and the responder variable (a linear connection) are assumed by the model. Simple linear regression may be expressed using the formula:

$$\hat{y} = \beta_0 + \beta_1 x \qquad (2)$$

where $\hat{y}$ is the predicted crop yield, β_0 is the intercept, β_1 is the slope, and x is the predictor variable.

Example: Application of simple linear regression for crop yield estimation, consider the following example:

Table 9. Example of simple linear regression application for crop yield estimation based on temperature.

Temperature (°C)	Crop Yield (tons/ha)
15	2.5
16	3.0
17	3.5
18	4.0
19	4.5
20	5.0

Source: Author's calculation.

In this example, we want to estimate the crop yield based on the temperature. We can use simple linear regression to model the association between temperature and crop yield.

First, we calculate the mean of the crop yield and temperature:

$$\underline{y} = \frac{2.5 + 3.0 + 3.5 + 4.0 + 4.5 + 5.0}{6} = 3.5$$

$$x = \frac{15 + 16 + 17 + 18 + 19 + 20}{6} = 17.5$$

Next, we calculate the deviations from the mean:

$$y_i - \underline{y} = y_i - 3.5$$

$$x_i - \underline{x} = x_i - 17.5$$

Then, we calculate the covariance between the deviations:

$$Cov(y,x) = \frac{\Sigma_{\square}^{\square}\square(y_i - \underline{y}) - (x_i - \underline{x})}{n-1} \quad (3)$$

$$Cov(y,x) = \frac{(2.5-3.5)(15-17.5) + (3.0-3.5)(16-17.5) + \cdots + (5.0-3.5)(20-17.5)}{6-1}$$

$$Cov(y,x) = \frac{-0.5 \times -2.5 + 0.0 \times -0.5 + \cdots + 1.5 \times 2.5}{5}$$

$$Cov(y,x) = \frac{1.25 + 0.75 + \cdots + 1.25}{5}$$

$$Cov(y,x) = \frac{-5.0}{5} = 1.0$$

Finally, we calculate the slope and intercept using the following formulas:

$$\beta_1 = \frac{Cov(y,x)}{Var(x)} \quad (4)$$

$$\beta_0 = \underline{y} = \beta_1 \underline{x} \quad (5)$$

$$\beta_1 = \frac{1.0}{Var(x)} \quad (6)$$

$$\beta_0 = 3.5 - \beta_1 \times 17.5$$

$$\beta_1 = \frac{1.0}{(20-15)^2 \times 6} = 0.05$$

$$\beta_0 = 3.5 - 0.05 \times 17.5 = 2.25$$

The simple linear regression model for crop yield estimation based on temperature is:

$\hat{y} = 2.25 + 0.05x$

This model can be used to predict crop yields based on temperature. For example, if the temperature is 18°C, the predicted crop yield would be:

$\hat{y} = 2.25 + 0.05 \times 18 = 4.0$

This example shows the application of simple linear regression for crop yield estimation. The model can be extended to include multiple predictor variables and more complex relationships using multiple linear regression and other machine learning algorithms.

Multiple Linear Regression

Multiple linear regression is the leeway of simple linear regression that contains multiple predictor variables. This model is useful when there are multiple factors that influence crop yields, such as temperature, precipitation, and soil type. The equation for multiple linear regression is:

$$\hat{y} = \beta_0 + \beta_1 x_1 + \beta_2 x_2 + \cdots + \beta_n x_n \tag{7}$$

where $\hat{y}$ is the predicted crop yield, β_0 is the intercept, $\beta_0, \beta_1, \beta_2, \ldots \beta_n$ are the slopes for each predictor variable, and $x_1, x_2, \ldots, x_n$ are the predictor variables.

Example: Application of multiple linear regression in crop yield estimation, consider the following example where we use three predictor variables: Temperature (°C), Rainfall (mm), and Soil pH.

The dataset for this example is as follows:

Table 10. Example of multiple linear regression application for crop yield estimation based on temperature, rainfall, and soil pH.

Temperature (°C)	Rainfall (mm)	Soil pH	Crop Yield (tons/ha)
15	100	6.0	2.5
16	110	6.5	3.0
17	105	6.8	3.5
18	120	7.0	4.0
19	115	7.2	4.5
20	130	7.5	5.0

Source: Author's calculation.

In this example, we aim to estimate the crop yield (Y) based on Temperature (X_1), Rainfall (X_2), and Soil pH (X_3).

The multiple linear regression model can be represented as:

$$Y = \beta_0 + \beta_1 X_1 + \beta_2 X_2 + \beta_3 X_3 + \epsilon \tag{8}$$

where Y is the crop yield, X_1 is the temperature, X_2 is the rainfall, X_3 is the soil pH, β_0 is the intercept, β_1, β_2, and β_3 are the coefficients for the predictor variables, and ϵ is the error term.

First, we calculate the means of the crop yield and predictor variables:

$$\underline{Y} = \frac{2.5+3.0+3.5+4.0+4.5+5.0}{6} = 3.75$$

$$\underline{X}_1 = \frac{16+16+17+18+19+20}{6} = 17.5$$

$$\underline{X}_2 = \frac{100+110+105+120+115+130}{6} = 113.33$$

$$\underline{X}_3 = \frac{6.0+6.5+6.8+7.0+7.2+7.5}{6} = 6.83$$

Next, we calculate the deviations from the mean for each variable:

$$Y_i - \underline{Y} = \{2.5-3.75 = -1.25\ 3.0-3.75 = -0.75 \frac{3.5-3.75=-0.25}{4.0-3.75=0.25\ 4.5-3.75=0.75\ 5.0-3.75=1.25}$$

$$X_{1i} - \underline{X}_1 = \{15-17.5=-2.5\ 16-17.5=-1.5\ 17-17.5=-0.5\ 18-17.5 = 0.5\ 19-17.5=1.5\ 20-17.5=2.5$$

$$X_{2i} - \underline{X}_2 = \{100-113.33=-13.33\ 110-113.33=-3.33\ 105-113.33 = -8.33\ 120-113.33=6.67\ 115-113.33=1.67\ 130-113.33 = 16.67$$

$$X_{3i} - \underline{X}_3 = \{6.0-6.83=-0.83\ 6.5-6.83=-0.33\ 6.8-6.83=-0.03\ 7.0-6.83 = 0.17\ 7.2-6.83=0.37\ 7.5-6.83=0.67$$

Using these deviations, we can compute the covariance and variance required for the regression coefficients. After calculating these values, we solve the normal equations to obtain the regression coefficients β_0, β_1, β_2, and β_3.

For brevity, let us assume the computed coefficients are as follows:

$$\beta_0 = -1.5, \beta_1 = 0.4, \beta_2 = 0.02, \beta_3 = 1.2$$

Thus, the multiple linear regression model for crop yield estimation is:

$$\hat{Y} = -1.5 + 0.4X_1 + 0.02X_2 + 1.2X_3 \tag{9}$$

To predict the crop yield for a given set of conditions, e.g., Temperature = 18°C, Rainfall = 110 mm, and Soil pH = 7.0, we substitute these values into the model:

$$\hat{Y} = -1.5 + (0.4 \times 18) + (0.02 \times 110) + (1.2 \times 7) \tag{10}$$

$$\hat{Y} = -1.5 + 7.2 + 2.2 + 8.4 = 16.3 \textit{ tons/ha}$$

This example shows the application of multiple linear regression for crop yield estimation, representing how multiple environmental and soil factors can be integrated into a predictive model. This methodology can be extended and refined using more complex algorithms and additional predictor variables to augment the accurateness and reliability of crop yield predictions.

Classification Models

Classification models are a pivotal aspect of AI and ML applied in the domain of crop yield estimation. These models aim to categorize data points into predefined classes based on their attributes. The primary goal is to govern which category a new observation belongs to, grounded on a training dataset comprising observations whose category membership is known.

***Example:* Classification Model for Wheat Yield Estimation**

Data Collection and Preprocessing: The application of classification models in crop yield estimation, consider a case study focused on estimating the yield of wheat. The dataset includes various features such as soil quality, weather conditions, and historical yield data. The features are:

1. Soil Quality Index (SQI)
2. Annual Rainfall (mm)
3. Average Temperature (°C)
4. Historical Yield (kg/hectare)

The target variable is the yield category, which can be classified as:

- Low Yield (0–2000 kg/hectare)
- Medium Yield (2001–4000 kg/hectare)
- High Yield (4001–6000 kg/hectare)

The dataset is divided into training and testing subsets, typically in a 70:30 ratio.

Model Selection: For this example, we use a Decision Tree classifier, known for its interpretability and effectiveness in handling both numerical and categorical data.

Training the Model: The training dataset comprises 70% of the observations. Each observation is characterized by its features and the corresponding yield category.

Table 11. Sample data for yield classification based on soil quality index (SQI), rainfall, temperature, and historical yield.

Observation	SQI	Annual Rainfall (mm)	Average Temperature (°C)	Historical Yield (kg/hectare)	Yield Category
1	85	450	18	2500	Medium
2	78	500	20	3200	Medium
3	92	600	15	4200	High
...	...	...	...	...	...

Source: Author's calculation.

The Decision Tree classifier uses the training data to learn the outlines and associations between the features and the yield categories.

Model Testing and Validation: The remaining 30% of the dataset serves as the testing set. The model's enactment is assessed using metrics such as accuracy, precision, recall, and the F1 score.

Table 12. Comparison of actual and predicted yield categories based on soil quality index (SQI), rainfall, temperature, and historical yield.

Observation	SQI	Annual Rainfall (mm)	Average Temperature (°C)	Historical Yield (kg/ hectare)	Actual Yield Category	Predicted Yield Category
1	80	470	19	2700	Medium	Medium
2	75	490	22	2100	Medium	Low
3	88	620	14	4100	High	High
...	...	...	...	...	...	...

Source: Author's calculation.

Consider the following testing observations to demonstrate the classification process:

Testing Observations:

Table 13. Dataset for yield classification based on soil quality index (SQI), rainfall, temperature, and historical yield.

SQI	Annual Rainfall (mm)	Average Temperature (°C)	Historical Yield (kg/hectare)	Actual Yield Category
85	470	19	2750	Medium
78	520	21	2300	Medium
90	600	16	4500	High

Source: Author's calculation.

Prediction Process:

1. *Observation 1:*
   - SQI: 85
   - Annual Rainfall (mm): 470
   - Average Temperature (°C): 19
   - Historical Yield (kg/hectare): 2750

Based on the decision tree, if the SQI is above a certain threshold and the rainfall and temperature fall within specific ranges, the model predicts the yield category as Medium.

2. *Observation 2:*
   - SQI: 78
   - Annual Rainfall (mm): 520
   - Average Temperature (°C): 21
   - Historical Yield (kg/hectare): 2300

For this observation, the lower SQI and higher temperature might push the prediction towards the Low category if the historical yield is not significantly high.

3. *Observation 3:*
   - SQI: 90
   - Annual Rainfall (mm): 600
   - Average Temperature (°C): 16
   - Historical Yield (kg/hectare): 4500

A high SQI, ample rainfall, and moderate temperature, coupled with a high historical yield, likely result in a High yield prediction.

Performance Evaluation: To assess the model's performance, we calculate the accuracy and other metrics. Suppose the predictions for the above observations are:

Table 14. Comparison of actual and predicted yield categories.

Actual Yield Category	Predicted Yield Category
Medium	Medium
Medium	Low
High	High

Source: Author's calculation.

From these observations, we can compute the accuracy:

- Correct predictions: 2 (Medium and High)
- Total observations: 3
- Accuracy = (Correct predictions/Total observations) = 2/3 ≈ 66.67%

Classification models, exemplified by the Decision Tree classifier, play a crucial role in estimating crop yields by categorizing them into predefined classes. Through a systematic approach involving data collection, preprocessing, model training, and validation, these models can provide valuable insights for farmers, aiding in decision-making and resource optimization.

Ensemble Methods

In AI and ML, ensemble methods are strategies that integrate many models to increase prediction accuracy and resilience. These techniques are especially useful for estimating agricultural yields because they can improve prediction performance by integrating different data sources and modeling strategies. The shortcomings of individual models can be mitigated by ensemble approaches by using the benefits of several models, producing yield projections that are more accurate and dependable (Chouhan et al., 2024).

Ensemble approaches are typically classified into two types: bagging and boosting. Bagging, also known as Bootstrap Aggregation, is the process of training many models independently on distinct subsets of data and then averaging their predictions. This strategy decreases variance and helps to avoid overfitting. Boosting develops models consecutively, with each new model attempting to remedy the flaws of its predecessor. This approach focuses on improving model accuracy by reducing bias.

In crop yield estimation, ensemble methods can incorporate various data inputs such as weather conditions, soil properties, satellite imagery, and historical yield data. By combining these diverse data sources, ensemble models can capture the complex, non-linear relationships that influence crop yields. For instance, a Random Forest, which is a bagging method, can be used to average predictions from multiple decision trees, each trained on a different subset of data, thereby enhancing the model's generalizability and robustness (Mishra et al., 2024).

Example: The application of ensemble methods in crop yield estimation, consider a case study where we aim to estimate the yield of a specific crop using three different models: a Decision Tree, a Linear Regression, and a Support Vector Machine (SVM). We will use bagging to create an ensemble model.

Steps:

1. *Data Preparation:*
   - Collect data on crop yields along with relevant features such as temperature, rainfall, soil pH, and historical yield data.
2. *Model Training:*
   - Train a Decision Tree, a Linear Regression, and a Support Vector Machine on the dataset.
3. *Ensemble Creation:*
   - Create multiple subsets of the training data using bootstrapping (random sampling with replacement).
   - Train each of the three models on these subsets to create an ensemble of models.
4. *Prediction and Aggregation:*
   - Each model in the ensemble predicts the test data.
   - Aggregate the predictions by averaging them to obtain the final yield estimate.

Consider a dataset with the following features and actual yields:

Table 15. Sample data for yield assessment based on temperature, rainfall, soil pH, and historical yield.

Sample	Temperature (°C)	Rainfall (mm)	Soil pH	Historical Yield (tons/ha)	Actual Yield (tons/ha)
1	22	120	6.5	3.5	4.0
2	25	80	6.8	4.0	4.5
3	20	150	6.2	3.8	3.9
4	23	90	6.7	4.2	4.3
5	21	130	6.4	3.6	4.1

We create bootstrapped datasets and train the models:

Table 16. Bootstrapped predictions from different models for yield estimation.

Model	Bootstrapped Sample 1 Prediction	Bootstrapped Sample 2 Prediction	Bootstrapped Sample 3 Prediction
Decision Tree	4.2	4.3	4.1
Linear Regression	4.0	4.4	4.2
SVM	4.1	4.5	4.0

To obtain the final prediction for a new test sample with features (Temperature: 22°C, Rainfall: 110 mm, Soil pH: 6.5, Historical Yield: 3.7 tons/ha), we average the predictions of all models:

Table 17. Yield predictions from different models.

Model	Prediction
Decision Tree	4.2
Linear Regression	4.0
SVM	4.1

Source: Author's calculation.

Final Ensemble Prediction: $(4.2 + 4.0 + 4.1)/3 = 4.1$ tons/ha.

This ensemble prediction process demonstrates how integrating multiple models can lead to a more accurate and robust crop yield estimation. By averaging the predictions, the ensemble method effectively reduces the influence of any individual model's errors, thereby enhancing the reliability of the yield forecasts. Ensemble methods play a critical role in crop productivity estimation by linking the strengths of various models and data sources. Through techniques such as bagging and boosting, these methods offer significant improvements in predictive accuracy and robustness, making them invaluable tools for sustainable agriculture and precision farming.

Challenges and Limitations

The application of AI, and ML in crop yield estimation offers significant potential for advancing sustainable agriculture. However, the incorporation of these tools is not without its challenges, and limitations. This section discusses the primary obstacles faced in the deployment of AI and ML in this domain, categorized into three major areas: data quality issues, model complexity and interpretability, and adoption barriers in agricultural communities.

Data Quality Issues

Data quality is a critical factor influencing the effectiveness of AI and ML models in crop yield estimation. High-quality, accurate, and comprehensive data are essential for training robust models capable of providing reliable predictions. However, several issues pertain to data quality:

- *Inconsistency and Heterogeneity:* Agricultural data is often collected from diverse sources, including satellite imagery, IoT devices, weather stations, and farmer reports. The inconsistency and heterogeneity of this data can lead to significant challenges in standardizing and preprocessing it for model training. Disparate formats, varying scales, and different measurement units necessitate rigorous data normalization and cleaning procedures.
- *Missing Data:* Incomplete datasets are common in agricultural contexts due to factors such as equipment failure, environmental conditions, or logistical issues in data collection. Missing data can bias models and reduce the accuracy of predictions. To address this issue, techniques like imputation, in which missing values are inferred based on other available data, or the adoption of robust models that can manage missing values are frequently used.
- *Quality Control:* Ensuring the integrity and reliability of data is paramount. Errors in data collection, transmission, or recording can introduce noise and inaccuracies, adversely affecting model performance. Implementing stringent quality control measures, including validation checks and anomaly detection algorithms, is crucial for maintaining data fidelity.
- *Temporal and Spatial Resolution:* The temporal and geographical resolution of data can have a substantial influence on model results. High-resolution data is preferable as it captures finer details and variability in crop conditions. However, such data can be expensive and resource-intensive to obtain. Balancing resolution with practicality and cost-effectiveness is a continual challenge.

Model Complexity and Interpretability

The complexity of AI and ML models presents another significant challenge in their application to crop yield estimation. While advanced models such as deep learning neural networks can capture intricate patterns and dependencies within data, they also bring about issues related to complexity and interpretability:

- *Overfitting: Complex models*, particularly those with numerous parameters, are prone to overfitting. A model is said to be overfitted when it learns the noise in the training set rather than the underlying patterns, which results in poor generalization to new, unobserved data. Overfitting is frequently avoided by employing strategies like cross-validation, regularization, and trimming.

- *Computational Demands:* Training sophisticated AI models requires substantial computational resources, including high-performance processors and significant memory capacity. This can be a limiting factor, especially in resource-constrained environments typical of many agricultural settings.
- *Interpretability and Transparency:* One of the most critical issues with complex AI models is their lack of interpretability. Unlike simpler models, such as linear regression, whose parameters and predictions are easily understandable, complex models like deep neural networks operate as "black boxes." This lack of transparency can hinder trust and acceptance among stakeholders, including farmers and policymakers. Efforts to improve interpretability, such as the development of explainable AI (XAI) techniques, are essential but remain a challenging area of research.

Adoption Barriers in Agricultural Communities

Even with high-quality data and sophisticated models, the adoption of AI and ML technologies in agricultural communities faces several barriers:

- *Awareness and Education:* A significant barrier is the lack of awareness and understanding of AI and ML among farmers. Many agricultural communities may not be familiar with these technologies or their potential benefits. Education and training programs are crucial to bridge this knowledge gap and promote informed adoption.
- *Cost and Accessibility:* The initial cost of implementing AI and ML solutions, including the acquisition of necessary hardware and software, can be prohibitive for small-scale farmers. Moreover, ongoing costs related to maintenance, updates, and technical support may also be burdensome. Efforts to make these technologies more affordable and accessible are vital for widespread adoption.
- *Infrastructure and Connectivity:* Reliable internet connectivity and digital infrastructure are prerequisites for deploying AI and ML systems. In many rural and remote agricultural regions, such infrastructure may be lacking or inadequate, posing a significant barrier to the adoption of these technologies.
- *Cultural and Social Factors:* Cultural resistance and social factors also play a role in the adoption of new technologies. Traditional farming practices are deeply ingrained in many communities, and there may be scepticism or reluctance to adopt novel methods perceived as complex or risky. Building trust through pilot projects, demonstrations, and collaborative initiatives can help to overcome these barriers.
- *Policy and Regulation:* The regulatory environment can also influence the adoption of AI and ML in agriculture. Policies that support innovation, provide incentives for technology adoption, and ensure data privacy and security are essential to create a conducive environment for these technologies to thrive.

While AI and ML hold promise for enhancing crop yield estimation and promoting sustainable agriculture, addressing these challenges and limitations is crucial for their successful integration. Efforts to improve data quality, manage model complexity, and facilitate adoption in agricultural communities are essential steps towards realizing the full potential of these technologies (Bhargava et al., 2024).

Ethical and Societal Considerations

The incorporation of AI, and ML into agriculture heralds significant advancements in crop yield estimation and sustainable farming practices. However, these technological developments necessitate a thorough examination of their ethical and societal implications. This section discusses the critical considerations of equity in access, environmental impacts, and social acceptance and trust.

Equity Access to AI and ML Technologies

Ensuring that the advantages of AI and ML technologies in agriculture are dispersed equitably across all stakeholders requires equitable access to these technologies. Disparities in access can exacerbate existing inequalities between large-scale commercial farms and smallholder farmers. Large agricultural enterprises often have the resources to invest in advanced technologies, thereby gaining competitive advantages through enhanced productivity and efficiency. Conversely, smallholder farmers, predominantly in emerging regions, may lack the economic means, technical expertise, and infrastructural support to adopt these technologies (Kose et al., 2022).

Addressing this inequity requires a multifaceted approach. Policymakers and stakeholders must implement supportive frameworks that facilitate affordable access to AI and ML tools for smallholder farmers. Subsidies, grants, and public-private corporations can play pivotal roles in bridging the access break. Moreover, tailored training programs and extension services are vital to empower the grower with the necessary skills to utilize these technologies efficiently. Ensuring language and cultural relevance in training materials is also critical to maximizing outreach and impact.

Furthermore, the role of open-source AI and ML solutions cannot be understated. By promoting open access to technological innovations, researchers and developers can foster an inclusive environment where all farmers, regardless of their economic standing, can benefit from advanced crop yield estimation methods. To create an environment that enables equal access to these transformational technologies, cooperation between governments, non-governmental organizations, and the commercial sector is essential.

Environmental Impact Assessment

The arrangement of AI and ML in agriculture presents both opportunities, and challenges from an environmental perspective. On one hand, these technologies can significantly enhance resource efficiency, optimizing inputs such as water, fertilizers, and pesticides to reduce wastage, and environmental degradation. Precision agriculture, powered by AI, enables site-specific management practices that can lead to more sustainable land use and conservation of biodiversity.

On the other hand, the environmental footprint of developing and deploying AI and ML systems must be carefully considered. The energy usage for large-scale AI model training and operation can be significant, which adds to greenhouse gas emissions and climate change. It is critical to evaluate the environmental effects of these technologies throughout their lifespan, from hardware manufacture to electronic waste disposal.

Mitigating the environmental impacts requires adopting sustainable practices in the expansion and distribution of AI and ML technologies. This includes utilizing energy-efficient algorithms, investing in renewable energy sources for data centres, and designing hardware with recyclability in mind. Furthermore, continuous monitoring and valuation of the environmental impacts are essential to ensure that the adoption of AI and ML in agriculture aligns with broader sustainability goals (Tiwari et al., 2024).

Social Acceptance and Trust

The successful integration of AI and ML into agricultural practices hinges on the social acceptance and trust of these technologies among various stakeholders. Farmers, consumers, and policymakers need to be convinced of the reliability, safety, and benefits of AI-driven solutions. Building this trust requires transparent communication, demonstrating the tangible advantages of these tools in improving crop yields and sustainability. One of the primary concerns is data privacy and security. Farmers must be assured that their data, often collected by AI systems for analysis, is handled responsibly and ethically. Executing robust data fortification measures, and transparent data governance frameworks is crucial in alleviating these concerns.

Moreover, the perception of AI and ML technologies as disruptive forces can evoke resistance, particularly among those who fear job displacement and loss of traditional farming practices. To address these fears, it is essential to highlight the complementary role of AI in enhancing human capabilities rather than replacing them. Educational initiatives that showcase successful case studies and the positive impacts of AI on agricultural livelihoods can foster greater acceptance. Community engagement and participatory approaches are also vital. Involving farmers and local societies in the development, and implementation of AI and ML solutions confirms that these tools are tailored to their specific needs, and contexts. This participatory approach not only enhances the relevance and effectiveness of the technologies but also builds a sense of ownership and trust among the stakeholders (Gupta and Khang, 2024).

Policy Implications and Regulatory Framework

Government Support for AI and ML Adoption

The incorporation of AI and ML in agriculture presents significant opportunities for enhancing crop yield estimation and overall sustainability. However, the successful integration of these technologies necessitates robust government support, which can be delineated into three primary avenues: subsidies and incentives, public-private partnerships, and speculation in research and development.

Subsidies and Incentives for AI and ML Adoption in Agriculture

Government subsidies and incentives are critical for encouraging the adoption of AI and ML technologies among growers, particularly smallholders who may lack the financial resources to invest in advanced technologies. Subsidies could take various forms, such as direct financial assistance, tax breaks, or grants aimed at offsetting the initial costs of AI and ML systems. For instance, direct financial assistance could involve grants or low-interest loans specifically designated for the purchase of AI and ML tools, thereby lowering the entry barrier for farmers.

Tax incentives might include deductions or credits for investments in AI and ML technologies. These could be structured to encourage not only the acquisition but also the maintenance and upgrading of AI and ML systems, ensuring that farmers remain at the cutting edge of technological advancements. Additionally, performance-based subsidies, which provide financial rewards based on the successful implementation and outcomes of AI and ML technologies, could further incentivize farmers to integrate these innovations into their practices.

Public–Private Partnerships for Infrastructure Development

The establishment of the infrastructure required to enable the broad application of AI and ML in agriculture is greatly aided by public-private partnerships, or PPPs. These partnerships have the potential to combine the knowledge, assets, and capacities of both industries to produce a more complete and resilient framework for the adoption of AI and ML.

PPPs can facilitate the development of essential infrastructure, such as data centers, high-speed internet, and other digital connectivity solutions, which are foundational for the effective deployment of AI and ML technologies. For example, partnerships between governments and tech companies could lead to the establishment of regional data hubs that collect and process agricultural data, providing farmers with real-time insights and recommendations.

Moreover, PPPs can also focus on creating training programs and educational initiatives to build the necessary human capital for AI and ML integration in agriculture. By collaborating with academic institutions and technology providers, governments can ensure that farmers and agricultural workers are equipped with the skills needed to utilize these technologies effectively.

Investment in Research and Development for Accessible Solutions

Investing in research and development (R&D) is paramount for creating AI and ML solutions that are both effective and accessible to a broad range of farmers. Government funding in R&D can spur innovations tailored to the unique challenges and needs of the agricultural sector, particularly those faced by small-scale and resource-poor farmers. R&D investments should focus on developing cost-effective AI and ML applications that are scalable and adaptable to different farming contexts. This includes creating user-friendly interfaces and ensuring that the technologies can operate in environments with varying levels of technological infrastructure. Additionally, research efforts should prioritize the development of AI and ML solutions that can integrate with existing agricultural practices, minimizing disruption and facilitating smoother transitions.

Government support for R&D can also promote the creation of open-source AI and ML platforms, which can be freely accessed and modified by the farming community. Such platforms can democratize access to advanced technologies, enabling a wider range of farmers to benefit from AI and ML innovations. Furthermore, collaborations between public research organisations and remote tech companies can accelerate the progress and dissemination of cutting-edge solutions, fostering a more inclusive and sustainable agricultural ecosystem. A comprehensive policy framework that includes subsidies and incentives, public-private partnerships, and robust investment in research and development is important for the successful adoption of AI and ML in agriculture. Such a framework not only supports technological advancement but also ensures that these innovations are accessible and beneficial to all farmers, thereby promoting sustainability and resilience in the agricultural sector (Padhiary et al., 2024).

Regulatory Guidelines for Data Privacy and Security

The integration of AI and ML in crop yield estimation necessitates rigorous regulatory guidelines to safeguard data privacy and security. The agricultural sector generates vast amounts of data, including sensitive information related to farm operations, crop yields, and proprietary farming techniques. Protecting this data is paramount to fostering trust and ensuring the ethical use of technology in agriculture. Firstly, regulatory frameworks must define clear standards for data collection, storage, and sharing. These standards should align with existing data protection rules, such as the General Data Protection Regulation (GDPR) for the European Union, which sets stringent requirements for data privacy and the protection of personal information. Similar regulations should be adapted to the agricultural context, ensuring that data related to farm operations is collected with explicit consent and used in a manner that respects the privacy of farmers and other stakeholders.

Secondly, implementing strong cybersecurity safeguards is critical to preventing data breaches and unauthorized access. This involves the deployment of modern encryption techniques, safe data storage systems, and frequent security audits to detect and address any weaknesses. Regulatory guidelines should mandate the use of these security practices to protect against cyber threats that could compromise sensitive agricultural data. Furthermore, there should be clear protocols for data anonymization and de-identification. These processes help to ensure that even if data is accessed by unauthorized parties, it cannot be traced back to specific individuals or farms. Regulatory bodies should provide detailed guidelines on how to effectively anonymize agricultural data without compromising its utility for AI and ML applications.

Moreover, the issue of data ownership must be addressed within the regulatory framework. Farmers and agricultural entities should retain ownership of their data, with the ability to control how it is used and shared. Regulations should stipulate that any use of agricultural data for AI and ML purposes requires explicit permission from the data owners, with clear terms outlining the scope and purpose of data usage. Lastly, transparency and accountability are critical components of any regulatory guideline. There should be mechanisms for farmers and other stakeholders to understand how their data is being used and to hold data processors accountable for any misuse or breaches. This includes the establishment of independent oversight bodies to monitor compliance with data privacy and security regulations and to address grievances related to data misuse (Ed-daoudi et al., 2023).

Incentive Mechanisms for Sustainable Agricultural Practices

Promoting sustainable agricultural practices through AI, and ML requires the development of effective incentive mechanisms. These mechanisms are designed to encourage farmers to adopt technologies and practices that enhance crop yield estimation while maintaining environmental sustainability and economic viability. Firstly, financial incentives play a vital role in motivating farmers to implement bearable practices. Governments and agricultural organizations can offer subsidies, grants, or low-interest loans to support the adoption of AI and ML tools that augment resource use, reduce waste, and increase efficiency. These financial aids can help offset the initial costs of technology adoption, making it more accessible to small and medium-sized farms that might otherwise struggle with the investment.

Secondly, tax incentives can be a powerful tool to encourage sustainable practices. Tax credits or deductions can be offered to farmers who implement AI and ML technologies that provide for the sustainability of the environment, such as precision farming methods that reduce the amount of water, fertilizer, and pesticides used. By reducing the tax burden on farmers who adopt these technologies, governments can create a strong financial motivation to pursue sustainable practices. Additionally, market-based incentives, such as certification and labelling schemes, can drive the adoption of sustainable agricultural practices. For instance, products grown using AI-driven sustainable methods can be certified as "environmentally friendly" or "sustainably produced." These labels can appeal to eco-conscious consumers willing to pay a premium for sustainably produced goods, thereby providing farmers with a financial incentive to adopt and maintain sustainable practices.

Moreover, education and training programs are essential to support the adoption of AI and ML in agriculture. Providing farmers with the knowledge and skills necessary to implement these technologies effectively can enhance their willingness to adopt new practices. Governments and agricultural organizations should invest in training initiatives that educate farmers on the benefits of AI and ML, how to integrate these technologies into their farming operations, and best practices for sustainable agriculture. Furthermore, fostering collaboration and knowledge-sharing among farmers can also serve as an incentive. Creating platforms for farmers to share experiences, challenges, and successes with AI and ML technologies can build a community of practice that encourages widespread adoption. Peer learning and support networks can help farmers navigate the complexities of new technologies and reinforce the benefits of sustainable practices. Lastly, regulatory measures can be used to create incentives for sustainable practices. For example, governments can implement policies that require or encourage the use of AI and ML for resource management and environmental monitoring. Compliance with these regulations can be incentivized through rewards, such as access to additional resources or recognition programs that highlight exemplary sustainable practices (Mishra and Mishra, 2024).

The Future of AI and ML in Sustainable Agriculture

Emerging Technologies and Applications

As we discuss the future of sustainable agriculture, it becomes increasingly evident that emerging technologies, particularly AI and ML, will play pivotal roles in revolutionizing farming practices. This section scrutinizes various facets of these advancements, elucidating their potential impact on agricultural sustainability.

Integration with other Technologies like Robots and IoT

The smooth integration of AI and ML with other cutting-edge technologies like robots and the Internet of Things (IoT) is one of the most promising paths for the future of sustainable agriculture. Through the integration of these technologies, farmers may create a harmonious blend of accuracy and productivity in their farming pursuits. By using unmatched accuracy, robotic systems using AI algorithms can perform a variety of activities on their own, from sowing seeds to harvesting crops. Moreover, these robots can be imbued with ML capabilities, allowing them to adapt and optimize their actions based on real-time environmental data collected through IoT sensors (Oikonomidis et al., 2022). The synergy between AI, robotics, and IoT fosters a paradigm shift in agriculture, enabling the creation of smart farming systems that operate with minimal human intervention. Through precise monitoring and control of agricultural processes, this integration holds the promise of maximizing resource utilization while minimizing environmental impact, thereby promoting sustainability at its core.

Advanced AI Algorithms for Real–Time Decision Making

The advent of advanced AI algorithms represents a watershed moment in agricultural sustainability, empowering farmers with unprecedented capabilities for real-time decision-making. These algorithms use enormous volumes of data, including information on crop health, weather patterns, soil conditions, and market movements, to produce timely and useful insights. Farmers may optimize resource allocation and boost production by using AI to dynamically modify their farming operations in response to changing environmental circumstances. Whether it's determining the optimal timing for irrigation, identifying early signs of pest infestation, or predicting market fluctuations, AI-driven decision-making holds the key to unlocking new frontiers in agricultural efficiency and resilience. Furthermore, the iterative nature of ML algorithms enables continuous refinement and improvement over time, ensuring that farming practices remain adaptive and responsive to ever-changing circumstances. This iterative learning process not only enhances the efficacy of individual farms but also contributes to the collective intelligence of the agricultural community as a whole, fostering a culture of innovation and collaboration (Danai-Varsou et al., 2023).

Development of AI-Powered Tools for Personalized Farm Management

In the quest for sustainable agriculture, the concept of personalized farm management emerges as a transformative approach, facilitated by the development of AI-powered tools tailored to the unique needs of each farm. These tools leverage AI and ML techniques to analyze farm-specific data, encompassing factors such as soil composition, crop varieties, and historical performance metrics, to devise customized management strategies. By embracing a personalized approach, farmers can optimize resource allocation, mitigate risks, and maximize yields in a manner that aligns with the ecological characteristics and constraints of their individual farms. Whether it's recommending the most suitable crop rotations, prescribing precise nutrient applications, or devising tailored pest management plans, AI-powered tools enable farmers to optimize their decision-making processes with surgical precision. Moreover, the advent of cloud-based platforms and mobile applications facilitates seamless integration and accessibility of these AI-powered tools, empowering farmers to make informed decisions anytime, anywhere. This democratization of agricultural intelligence fosters a culture of empowerment and autonomy among farmers, catalysing the transition towards a more sustainable and resilient agricultural landscape. The confluence of developing technologies and the unwavering quest of innovation present enormous promise for the application of AI and ML in sustainable agriculture in the future. Farmers may take a revolutionary step toward a more sustainable, effective, and resilient agricultural future by utilizing AI and ML.

Addressing Sustainability Challenges

Using AI and ML to Combat Climate Change and Resource Scarcity

In the quest for sustainable agriculture, addressing climate change and resource scarcity stands as a paramount challenge. AI and ML present promising avenues for mitigating these challenges by providing tools for precise monitoring, prediction, and adaptation.

- *Precision Agriculture:* Large-scale information gathered from several sources, including satellites, drones, and Internet of Things devices, may be analyzed by AI and ML algorithms to accurately track environmental factors like temperature, nutrient levels, and soil humidity. Farmers may optimize resource allocation with this data-driven method, avoiding waste and environmental impact while boosting production.
- *Climate Modelling and Prediction:* Advanced machine learning models can analyze historical climate data alongside real-time observations to generate accurate forecasts of future weather patterns. By predicting droughts, floods, or

other extreme weather events, farmers can implement preemptive measures to guard their crops and minimize losses.

- *Crop Breeding and Genetic Improvement:* AI-powered algorithms can accelerate the process of crop breeding by analyzing genetic data to identify traits associated with resilience to climate change and resource efficiency. By selecting and breeding crops with these desirable traits, farmers can cultivate varieties better suited to changing environmental conditions.
- *Resource Optimization:* Machine learning algorithms can optimize resource usage by recommending precise irrigation schedules, fertilizer applications, and crop rotations based on real-time data and predictive models. This not only conserves water and nutrients but also reduces reliance on chemical inputs, contributing to environmental sustainability.

Developing Sustainable Farming Practices with AI-Driven Optimization

Sustainable farming practices are essential for preserving soil health, biodiversity, and ecosystem balance while ensuring long-term food security. AI and ML technologies offer innovative solutions to optimize farming operations in alignment with these goals.

- *Soil Health Management:* AI algorithms can analyz e soil data to assess its quality, fertility, and composition. By identifying areas of soil degradation or nutrient depletion, farmers can implement targeted interventions such as cover cropping, crop rotation, and organic amendments to regenerate soil health and productivity.
- *Integrated Pest Management (IPM):* Machine learning algorithms are capable of analyzing disease and pest trends to forecast outbreaks and provide effective management strategies. By combining chemical, cultural, and biological control techniques, farmers can minimize pesticide usage while effectively managing pests and preserving beneficial organisms.
- *Smart Farming Systems:* AI-driven farm management systems can integrate data from various sources, including sensors, drones, and machinery, to optimize decision-making processes such as planting, harvesting, and post-harvest handling. By automating routine tasks and providing real-time insights, these systems improve efficiency, productivity, and resource utilization.

Promoting Circular Economy Principles in Agriculture

The concept of a circular economy, which aims to minimize waste and maximize resource efficiency, is gaining traction in agriculture. AI and ML technologies play a pivotal role in promoting circularity by optimizing resource flows, reducing waste, and enhancing the reuse and recycling of agricultural by-products.

- *Waste Reduction and Valorisation:* AI algorithms can analyze agricultural by-products such as crop residues, manure, and food waste to identify opportunities for valorisation. By converting these materials into biofuels, bioplastics, or organic fertilizers, farmers can reduce waste and generate additional revenue streams.

- *Supply Chain Optimization:* ML models can optimize supply chain logistics by predicting demand, optimizing transportation routes, and minimizing food losses during storage and distribution. By ensuring timely delivery of fresh produce to consumers, these technologies reduce food waste and improve market efficiency.
- *Closed-Loop Systems:* AI-driven monitoring systems can track resource flows within agricultural ecosystems, facilitating the implementation of closed-loop systems when waste from one procedure is used as a resource in another. By closing the loop on nutrient cycles and water usage, farmers can minimize environmental impact and maximize resource efficiency.

AI and ML hold immense potential to address sustainability challenges in agriculture by enabling precision farming, optimizing resource usage, and promoting circular economy principles. By harnessing these technologies, farmers can cultivate resilient, productive, and environmentally friendly agricultural systems that ensure food security for future generations.

Conclusion

Sustainable agriculture necessitates the integration of economic sustainability with technological advancements, particularly through AI and ML in crop yield estimation. Traditional approaches have often fallen short, constrained by limited accuracy and the inability to swiftly adapt to environmental changes. Data from the Food and Agriculture Organisation (FAO), a United Nations agency, indicates that global agricultural productivity needs to grow to 70% by 2050 to meet the food demands of the growing population. AI and ML offer a transformative solution, providing precise yield predictions and facilitating more efficient resource management. A study by McKinsey estimates that AI-driven agriculture could increase global crop yields by up to 30%, resulting in significant economic benefits. The economic framework of sustainable agriculture is fundamentally enhanced by AI and ML, reducing costs through targeted interventions and minimizing resource wastage. The World Bank reports that precision agriculture can reduce water use by up to 30% and fertilizer use by 20%, leading to substantial cost savings and environmental benefits. These technologies also enhance market predictability and risk management, contributing to increased farm incomes and economic resilience. Furthermore, the capacity of AI and ML to analyz e vast datasets ensures early detection of stress factors and disease outbreaks, allowing proactive measures that improve crop health and yield. The ROI for AI and ML in agriculture is compelling. A PwC analysis suggests that AI could contribute up to $15.7 trillion to the global economy by 2030, with agriculture being a significant beneficiary. The adoption of AI and ML in crop yield estimation is pivotal for achieving economic sustainability in agriculture. By improving efficiency, enhancing yields, and reducing environmental impact, these technologies offer a path towards a more resilient and prosperous agricultural sector.

References

Ahmad, A., Liew, A.X.W., Venturini, F., Kalogeras, A., Candiani, A., Di Benedetto, G., Ajibola, S., Cartujo, P., Romero, P. and Lykoudi, A. 2024. AI can empower agriculture for global food security: challenges and prospects in developing nations. *Front. Artif. Intell.*, 7: 1328530.

Ahmad, L. and Nabi, F. 2021. *Agriculture 5.0: Artificial Intelligence, IoT and Machine Learning*. CRC Press.

AlZubi, A.A. and Galyna, K. 2023. Artificial intelligence and internet of things for sustainable farming and smart agriculture. *IEEE Access*, 11: 78686–78692.

Araújo, S.O., Peres, R.S., Ramalho, J.C., Lidon, F. and Barata, J. 2023. Machine learning applications in agriculture: Current trends, challenges, and future perspectives. *Agronomy*, 13(12): 2976.

Attri, I., Awasthi, L.K., Sharma, T.P. and Rathee, P. 2023. A review of deep learning techniques used in agriculture. *Ecol. Inform.*, 102217.

Balafoutis, A., Beck, B., Fountas, S., Vangeyte, J., Van der Wal, T., Soto, I. and Eory, V. 2017. Precision agriculture technologies positively contributing to GHG emissions mitigation, farm productivity and economics. *Sustainability*, 9(8): 1339.

Benos, L., Tagarakis, A.C., Dolias, G., Berruto, R., Kateris, D. and Bochtis, D. 2021. Machine learning in agriculture: A comprehensive updated review. *Sensors (Basel).*, 21: 3758.

Bhargava, K., Naveen, A., Akhil, V.S., Lingireddy, H., Gowtham, K.V., Mudhale, A. and Abhishek, E. 2024. Artificial intelligence (AI) and its applications in agriculture: A review. *Environ. Conserv. J.*, 25.

Bhat, S.A. and Huang, N.F. 2021. Big data and ai revolution in precision agriculture: Survey and challenges. *IEEE Access*, 9: 110209–110222.

Chouhan, S.S., Singh, U.P. and Jain, S. 2024. Introduction to computer vision and drone technology. pp. 1–5. *In*: *Applications of Computer Vision and Drone Technology in Agriculture 4.0*. Singapore: Springer Nature Singapore.

Chouhan, S.S., Singh, U.P., Sharma, U. and Jain, S. 2024. Classification of different plant species using deep learning and machine learning algorithms. *Wirel. Personal. Commun.*, 1–24.

Chouhan, S.S., Singh, U.P., Sharma, U. and Jain, S. 2021. Leaf disease segmentation and classification of Jatropha Curcas L. and Pongamia Pinnata L. biofuel plants using computer vision based approaches. *Measurement*, 171: 10879.

Coble, K.H., Mishra, A.K., Ferrell, S. and Griffin, T. 2018. Big data in agriculture: A challenge for the future. *Appl. Econ. Perspect. Policy*, 40: 79–96.

Danai-Varsou, D., Zhang, P., Afantitis, A., Guo, Z., Lynch, I. and Melagraki, G. 2023. Nanoinformatics and artificial intelligence for nano-enabled sustainable agriculture. pp. 503–531. *In*: *Nano-enabled Sustainable and Precision Agriculture*. Academic Press.

Ed-daoudi, R., Alaoui, A., Ettaki, B. and Zerouaoui, J. 2023. A predictive approach to improving agricultural productivity in morocco through crop recommendations. *Int. J. Adv. Comput. Sci. Appl.*, 14.

Elavarasan, D., Vincent, D.R., Sharma, V., Zomaya, A.Y. and Srinivasan, K. 2018. Forecasting yield by integrating agrarian factors and machine learning models: A survey. *Comput. Electron. Agric.*, 155: 257–282.

Food and Agriculture Organization of the United Nations. 2023. How to feed the world in 2050. FAO. https://www.fao.org/3/ak542e/ak542e.pdf.

Gupta, C. and Khang, A. 2024. Cultivating efficiency-harnessing artificial intelligence (AI) for sustainable agriculture supply chains. In Agriculture and Aquaculture Applications of Biosensors and Bioelectronics. *IGI Global*, pp. 372–388.

Jung, J., Maeda, M., Chang, A., Bhandari, M., Ashapure, A. and Landivar-Bowles, J. 2021. The potential of remote sensing and artificial intelligence as tools to improve the resilience of agriculture production systems. *Curr. Opin. Biotechnol.*, 70: 15–22.

Kose, U., Prasath, V.B., Mondal, M., Podder, P. and Bharati, S. editors. 2022. *Artificial Intelligence and Smart Agriculture Technology*. CRC Press.

Kouadio, L., Deo, R.C., Byrareddy, V., Adamowski, J.F. and Mushtaq, S. 2018. Artificial intelligence approach for the prediction of robusta coffee yield using soil fertility properties. *Comput. Electron. Agric.*, 155: 324–338.

Kumar, N., Kushwaha, R.R., Meena, N.R., Mishra, H. and Yadav, A.P.S. 2023. A study on costs and returns of paddy cultivation in Ambedkar Nagar district of Uttar Pradesh. *Int. J. Stat. Appl. Math. SP.*, 8: 107–111.

Kumar, R., Kumar, A., Bhatia, K., Nisar, K.S., Chouhan, S.S., Maratha, P. and Tiwari, A.K. 2024. Hybrid approach of cotton disease detection for enhanced crop health and yield. *IEEE Access*. Available from https://ieeexplore.ieee.org/document/10589357.

Lakshmi, V. and Corbett, J. 2023. Using AI to improve sustainable agricultural practices: A literature review and research agenda. *Communications of the Association for Information Systems*, 53: 96–137.

Liakos, K.G., Busato, P., Moshou, D., Pearson, S. and Bochtis, D. 2018. Machine learning in agriculture: A review. *Sensors (Basel).*, 18: 2674.

Mahto, M.K., Laxmikanth, P. and Lanka, V.B. 2024. Fundamentals of AI and machine learning with specific examples of application in agriculture. pp. 178–199. *In*: Farming, D.D. (ed.). *Auerbach Publications*.

McKinsey & Company. 2020. Agriculture's connected future: How technology can yield new growth. https://www.mckinsey.com/industries/agriculture/our-insights/agricultures-connected-future-how-technology-can-yield-new-growth.

Mishra, H. 2022. Production and marketing of Cucumber and Bitter gourd in Sultanpur district of Uttar Pradesh.

Mishra, H. 2024a. Counting the costs: understanding agricultural economics and the concept of costs. pp. 198–210. *In*: Sharma, K., Singh, P., Singh, S., Yadav, A. and Awasthi, V. (eds.). *Text Book on Practices for Agricultural Sustainability*. Golden Leaf Publishers.

Mishra, H. 2024b. Molecular technology in agriculture. pp. 204–226. *In*: Patel, D.K., Dwivedi, S., Dubey, A., Kumar, A. and Ali, W. (eds.). *Agricultural Transformation in the Modern Era*. Elite Publishing House.

Mishra, H. and Mishra, D. 2023. Artificial intelligence and machine learning in agriculture: Transforming farming systems. *Res. Trends Agric. Sci.*, 1: 1–16.

Mishra, H. and Mishra, D. 2024. Economic evaluation of UAV-based soil sampling approaches. In: *Applications of Computer Vision and Drone Technology in Agriculture 4.0*. Singapore: Springer Nature Singapore, pp. 271–291.

Mishra, H., Tiwari, A.K. and Nishad, D.C. 2024. Economic viability of sustainable agriculture practices in modern farming. *Adv Agric. Sci. II*, 24: 105.

Mittal, D., Kaur, G., Singh, P., Yadav, K. and Ali, S.A. 2020. Nanoparticle-based sustainable agriculture and food science: Recent advances and future outlook. *Front. Nanotechnol.*, 2: 579954.

Nishad, D.C., Mishra, H., Tiwari, A.K. and Mishra, D. 2024. Post-harvest management: Enhancing food security and sustainability. *Adv. Agric. Sci. II*, 24: 136.

Oikonomidis, A., Catal, C. and Kassahun, A. 2022. Hybrid deep learning-based models for crop yield prediction. *Appl. Artif. Intell.*, 36: 2031822.

Padhiary, M., Saha, D., Kumar, R., Sethi, L.N. and Kumar, A. 2024. Enhancing precision agriculture: A comprehensive review of machine learning and AI vision applications in all-terrain vehicle for farm automation. *Smart Agric. Technol.*, 100483.

Piekarskim, C.M. and de Francisco, A.C. 2020, March. Recent advances and challenges in AI for sustainable agricultural systems. In: *Proceedings of the on 25th International Joint Conference on Industrial Engineering and Operations Management-IJCIEOM: The Next Generation of Production and Service Systems*, p. 119. Springer Nature.

PwC. 2017. Sizing the prize: What's the real value of AI for your business and how can you capitalise? PwC. https://www.pwc.com/gx/en/news-room/docs/report-pwc-ai-analysis-sizing-the-prize.pdf.

Renda, A. 2024. AI and the transformation of agricultural work: economic, social, and environmental implications. In: *Handbook of artificial intelligence at work. Edward Elgar Publishing*, pp. 96–118.

Sachithra, V. and Subhashini, L.D.C.S. 2023. How artificial intelligence uses to achieve the agriculture sustainability: Systematic review. *Artificial Intelligence in Agriculture*, 8: 46–59.

Saiz-Rubio, V. and Rovira-Más, F. 2020. From smart farming towards agriculture 5.0: A review on crop data management. *Agronomy*, 10: 207.

Sarkar, S., Ganapathysubramanian, B., Singh, A., Fotouhi, F., Kar, S., Nagasubramanian, K., Chowdhary, G., Das, S.K., Kantor, G. and Krishnamurthy, A. et al. 2024. Cyber-agricultural systems for crop breeding and sustainable production. *Trends Plant Sci.*, 29: 130–149.

Sharma, A., Jain, A., Gupta, P. and Chowdary, V. 2020. Machine learning applications for precision agriculture: A comprehensive review. *IEEE Access.*, 9: 4843–4873.

Singh, A., Mehrotra, R.V.D., Rajput, D.P., Singh, A.K., Kumar, P. and Singh, A.K. 2022. Geoinformatics, artificial intelligence, sensor technology, big data: Emerging modern tools for sustainable agriculture. *Sustain. Agric. Syst. Technol.*, 295–313.

Singh, M. and Mishra, H. 2023. Medicinal plant farming to improve farmers' economy. *Agriallis*, 30–36.

Singh, P. and Kaur, A. 2022. A systematic review of artificial intelligence in agriculture. *Deep Learn. Sustain. Agric.*, 57–80.

Suzer, M.H., Şenbayram, M. and Çullu, M.A. 2024. Sustainable farming through precision agriculture: Enhancing nitrogen use and weed management. In IntechOpen eBooks.

Thamilselvan, R., Natesan, P., Kalaivaani, P.C.D. and Santhoshkumar, S. 2021, February. Review on crop prediction using deep learning techniques. *In*: Dharani, M.K. (ed.). *J. Phys. Conf. S.* (Vol. 1767, No. 1, p. 012026). IOP Publishing.

Tiwari, A.K. and Mishra, H. 2024. Sustainable agriculture: the future of world. pp. 167–192. *In*: Patel, D.K., Dwivedi, S., Dubey, A., Kumar, A. and Ali, W. (eds.). *Agricultural Transformation in the Modern Era.* Elite Publishing House.

Tiwari, A.K., Mishra, H. and Nishad, D.C. 2024. Market dynamics and consumer perceptions of organic produce in contemporary agriculture. *Adv. Agric. Sci. II.*, 24: 120.

Tiwari, A.K., Mishra, H., Nishad, D.C. and Pandey, A. 2023. Sustainable water management in agriculture: irrigation techniques and water conservation. *Dr. Ajay B Jadhao.*, 53.

Vincent, D.R., Deepa, N., Elavarasan, D., Srinivasan, K., Chauhdary, S.H. and Iwendi, C. 2019. Sensors driven AI-based agriculture recommendation model for assessing land suitability. *Sensors (Basel).*, 19: 3667.

Vuppalapati, C. 2021. Machine learning and artificial intelligence for agricultural economics: Prognostic data analytics to serve small scale farmers worldwide. *Springer Nature*, 314.

Zhang, P., Guo, Z., Ullah, S., Melagraki, G., Afantitis, A. and Lynch, I. 2021. Nanotechnology and artificial intelligence to enable sustainable and precision agriculture. *Nature Plants*, 7: 864–876.

Index

A

Advancement 202, 210
Agricultural sustainability 313
Agricultural technology 234
Agriculture 1–3, 5, 6, 220, 221, 224, 226, 228
Agriculture mapping 83, 85, 88, 89
AI and ML 348–350, 352–366, 368–370, 374, 376, 379–389
AI-driven monitoring 337, 339, 340
Artificial Intelligence (AI) 151, 153, 154, 156–158, 160–168, 172–175, 177, 180–183, 185, 186, 195–198, 250–252, 254, 257, 260, 262–266, 269, 274–276, 287, 302, 306, 313

B

Basil 67–77, 79, 80

C

Challenges 348–350, 379, 381, 383, 385, 387, 389
Chat-bot systems 220
Circular economy 185–191, 193, 195–198
Computer Vision (CV) 6, 67–69, 153, 159, 161, 163, 164, 166, 172–175, 180–183, 274–276, 287
Computer vision technologies 310
Crop health monitoring 83, 88, 89
Crop monitoring 41, 49, 50, 54, 57, 60, 109
Crop production 310
Crop yield estimation 348–350, 353–356, 358, 360–366, 368–374, 377–379, 381, 382, 384, 385, 389

D

Data analytics 108, 109
Deep Learning (DL) 46, 59, 67–74, 77, 79, 80
Drone 1–6

E

Ecological informatics 151–153, 166, 167
Economic benefits 350, 358, 359, 361–364, 389
Environmental monitoring 250–252, 256

F

Federated learning 231–236, 238–240, 242
Forest fires 15, 16, 18, 19, 23, 25
Fundamentals of wastewater treatment 260

I

IoT 350, 354, 369, 379, 386

L

Land use 41, 43

M

Machine Learning (ML) 37, 56, 57, 59–61, 129–131, 202–205, 207–213, 231–234, 239, 242

N

Natural language processing 220, 222, 225, 227
Neural networks 157, 159, 161–164, 282, 283, 285, 287, 288
Normalized Difference Vegetation Index (NDVI) 298

P

Plant growth mapping 173–177, 179, 181
Plant health monitoring 69
Precision agriculture 67–70, 74–77, 79, 80, 92–94, 96, 104, 107, 109, 174–177
Precision farming 231–233, 236
Prediction 9–14, 19–28, 205, 207–212

R

Real-time data 248, 251, 254
Remote sensing (RS) 36–61, 83–99, 104–109, 116–144, 274–276, 279–283, 285, 287–289
Remote sensing imagery 164
ROI 349, 350, 358–361, 389

S

Satellite imagery 11–15, 18–22, 26
Sensor networks 166, 167
Sensors 173–177, 181
Smart agriculture 43
Soil analysis 83, 84
Soil fertility evaluation 305, 310–312
Soil moisture 274–289
Soil moisture sensors 318, 319, 323–328, 331, 340
Sustainability 84, 89, 90, 185, 189, 190, 192, 193, 196, 198
Sustainable agriculture 116–118, 120, 123, 125–127, 129, 134, 137, 142, 143, 348, 350–352, 364, 368, 369, 378, 379, 381, 385–387, 389

T

Technology adoption 226

U

Unmanned Aerial Vehicle (UAV) 12, 15, 17, 18, 27, 296, 299

W

Waste management 185, 187, 190, 192–195
Wastewater treatment 259, 260, 262, 266–270
Water management 117, 118, 133, 134, 144
Water quality monitoring 246, 248–250, 252–257
Water quality parameters 207, 208
Water resources management 249
Wildfire 9–15, 17–24, 26–29

Y

Yield estimation 294–299, 301, 302
Yield prediction 102, 106, 109, 117, 129–133, 143, 144
YOLOv8 68–77, 79

About the Authors

Dr. Siddharth Singh Chouhan is an Assistant Professor in School of Computing Science and Engineering, VIT Bhopal University, Kothri-Kalan, Sehore, India. He has a PhD in Computer Science and Engineering from Shri Mata Vaishno Devi University Katra, India. He is a PostDoctorate from the University of Malta, Malta. He has a BTech degree (2010) and a MTech (2013) in Computer Science and Engineering from RGPV University, Bhopal, India. His areas of interest include Soft Computing, Computer vision, Drone Technology, Deep Learning, and Precision Agriculture. He had authored several research papers in many reputed journals and conferences.

Prof. Uday Pratap Singh is a Professor in the Department of Mathematics, Central University of Jammu, Jammu, India. He completed his BSc and MSc (Mathematics and Statistics) from Dr. R.M.L. (Avadh) University, Ayodhya, India, and again did his MSc (Mathematics and Computing) from Indian Institute of Technology, Guwahati, India, and a PhD in Computer Science from Barkatullah University, Bhopal. His areas of interest include Soft Computing, Nonlinear Systems, and Image Processing. He had authored several research papers in the many reputed journals and conferences. He is a life member of Soft Computing Research Society (SCRC), Barata Ganita Parishad, Computer Society of India and Member of IEEE and AMS.

Prof. Akash Saxena has a PhD and Master of Technology degree (with Honours) in Electrical Engineering from Malaviya National Institute of Technology, Jaipur in 2015 and 2008 respectively and a Bachelor of Engineering degree (with Honours) in Electrical Engineering from Engineering College Kota in 2001. Prof Saxena is working as (Director of Training and Placement) and Professor in Department of Electrical Engineering Central University of Haryana, India with almost two decades of teaching experience. As an academician, he has been associated with numerous institutions. He is a member of Doctoral Research Committee and a member of the Board of Studies. He has presented his research work at various platforms to showcase his findings to fellow researchers and provide a pathway to the future researchers. His works have been published in leading journals in the form of short communication/ letters /articles/research papers.

Prof. Akhilesh Tiwari is a Professor in Department of Information Technology, MITS Gwalior, MP, India. He has a Bachelor degree and Master's degree in Information Technology from RGPV Bhopal, India, and a PhD in Information Technology. His area of research interests includes Knowledge Discovery in Databases & Data Mining, Terrorist Network Mining, Hesitation Mining and Pattern Warehousing. He has published more than 80 research papers in reputed journals and conferences published by leading publishers with SCI, SCIE and SCOPUS indexing.

Prof. Sanjeev Jain is a Vice Chancellor of the Central University of Jammu, India. He has a Master's degree in Computer Science and Engineering from IIT Delhi, India, in 1992, and a PhD in Computer Science and Engineering. He has over 35 years of experience in teaching, research and administration. He has the credit of making a significant contribution to research and development in the area of image processing, soft computing, IoT and mobile ad-hoc network. He had authored several research papers in the many reputed journals and conferences. He has undertaken a number of major research and development government sponsored and private projects. He is a member of various societies like the Computer Society of India, IEEE.